Advances in Wastewater Research

Series Editor

Maulin P. Shah, Lovely Professional University, Phagwara, Punjab, India

The Advances in Wastewater Research highlights the prospective future development of wastewater treatment techniques, research and field applications. The series also focuses on the limitations and challenges associated with some generally accepted wastewater strategies and evaluates the possible applications of these corrective strategies to eliminate toxic pollutants from the environment through integrated technologies in wastewater treatment. The series covers the research areas including, but not limited to, toxic pollutants removal from wastewater, methods and procedures, materials, management of wastewater, valorisation possibilities of waste resulting from the removal of toxic pollutants from wastewater, wastewater recycling, environmental impact, and wastewater policies. The series contribute to remove, reduce, or neutralize water contaminants that threaten human health, and ecosystem productivity and integrity. The series includes peer-reviewed monographs, edited volumes, textbooks, and conference proceedings.

Ashootosh Mandpe · Nadeem Khan ·
Maulin P. Shah
Editors

Membrane Filtration and Membrane Bioreactors for Wastewater Treatment

Springer

Editors
Ashootosh Mandpe
Department of Civil Engineering
Indian Institute of Technology Indore
Indore, Madhya Pradesh, India

Nadeem Khan
King Fahd University of Petroleum and Minerals
Dhahran, Saudi Arabia

Maulin P. Shah
Department of Research Impact and Outcome Research and Development Cell
Lovely Professional University
Phagwara, Punjab, India

ISSN 3004-9725 ISSN 3004-9733 (electronic)
Advances in Wastewater Research
ISBN 978-981-95-6705-8 ISBN 978-981-95-6706-5 (eBook)
https://doi.org/10.1007/978-981-95-6706-5

This Springer imprint is published by the registered company Springer Nature Singapore Pte Ltd.
The registered company address is: 152 Beach Road, #21-01/04 Gateway East, Singapore 189721, Singapore

Preface

Growing concerns over water scarcity, pollution, and sustainable resource management have increased the demand for advanced water treatment technologies. Among these, membrane technology has emerged as a highly effective and versatile approach for water purification, wastewater treatment, desalination, and resource recovery. Its ability to produce high-quality effluent with compact system design and operational flexibility makes it an important component of modern environmental engineering solutions.

This book presents a comprehensive overview of membrane-based technologies and their diverse applications in water and environmental management. The chapters cover fundamental principles of membrane systems, the role of microorganisms in wastewater treatment, membrane characterization techniques, drinking water purification, seawater desalination, and industrial applications. In addition, emerging concepts such as circular economy approaches and challenges like membrane fouling are discussed.

The book aims to provide researchers, students, engineers, and policymakers with an integrated understanding of membrane technologies and their role in addressing global water and environmental sustainability challenges.

Indore, India
Dhahran, Saudi Arabia
Ankleshwar, India

Ashootosh Mandpe
Nadeem Khan
Maulin P. Shah

About This Book

Membrane technology has become a key solution for addressing global challenges related to water scarcity, wastewater treatment, and sustainable resource management. This book presents a comprehensive overview of membrane-based technologies and their applications in modern environmental and industrial systems. It explores the principles of membrane processes, the role of microorganisms in wastewater treatment, membrane characterization techniques, drinking water purification, seawater desalination, and diverse industrial applications. The book also discusses emerging perspectives such as circular economy approaches and challenges like membrane fouling and mitigation strategies. Designed for researchers, engineers, and graduate students, this volume provides valuable insights into the advancements and future potential of membrane technologies in sustainable water management.

Contents

Chapter 1
Membrane Bioreactors for Wastewater Treatment

Ashootosh Mandpe, Devaanshi Jagwani, and Lalitesh Sinha

Abstract Membrane Bioreactors (MBRs) are a new and effective technology for wastewater treatment, combining membrane filtration with biological treatment to produce high-quality effluent. In this chapter, a thorough overview of MBR technology is presented, including its basic principles, system configuration, operational mechanism, and most significant advantages over traditional wastewater treatment processes. It delves into critical factors such as membrane fouling, energy utilization, and expense thinking, in addition to recent breakthroughs in hybrid MBRs, nanomaterial-modified membranes, and automated process management. MBR usage in municipal and industrial sewage water treatment is described, including case histories and performance assessments from actual implementations. In addition, problems and future directions such as sustainability, resource capture, and developing contaminants are discussed in order to influence further study and technological breakthroughs. This chapter is an important resource for researchers, engineers, and policymakers who are interested in knowing and maximizing the use of MBR systems in sustainable water management.

Keywords Membrane bioreactors · Wastewater treatment · Sustainable water management

1.1 Introduction

A Membrane Bioreactor (MBR) is a type of innovative wastewater treatment technology which combines a suspended growth bioreactor and a microfiltration (MF) or ultrafiltration (UF) membrane system. It ensures the better separation of solid and liquid through enhancing efficiency, allowing for high-quality effluent production

A. Mandpe
Indian Institute of Technology, Indore, India

D. Jagwani (✉) · L. Sinha
Institute of Engineering and Science, IPS Academy, Indore, India
e-mail: devaanshijagwani@ipsacademy.org

A. Mandpe et al. (eds.), *Membrane Filtration and Membrane Bioreactors for Wastewater Treatment*, Advances in Wastewater Research,
https://doi.org/10.1007/978-981-95-6706-5_1

to reuse or discharge in an environmentally friendly way. Based on its performance superior to the rest, the technology of MBR is popular nowadays among both industrial and municipal wastewater treatment plants (WWTPs) (Yuvaraju and Deena 2024). The demand for Membrane Bioreactor technology is spurred by environmental sustainability, regulatory requirements, water scarcity, and efficiency in operations. Its capacity to generate high-quality effluent, minimize sludge generation, facilitate water reuse, and fit also function efficiently into small spaces makes MBRs a perfect solution for contemporary wastewater treatment challenges (Rahman et al. 2023). MBRs are necessitated by numerous environmental, operating, and regulatory issues that have been noted with the conventional wastewater treatment technologies.

(1) Strict Water Quality Regulations: Governments and pollution control authorities all over the globe impose stringent requirements on discharge limitations for contaminants like suspended solids (TSS), biological oxygen demand (BOD), chemical oxygen demand (COD), nitrogen, and phosphorus. MBRs generate high-quality effluent of quality that equals or surpasses these regulatory standards, thus suiting very sensitive environments like rivers, lakes, and coastlines. In a few parts of the world, MBR-treated water is usable for irrigation purposes, industrial usage, or even drinking water uses (Galinha et al. 2018).
(2) Rising Water Scarcity and Reuse Requirement: Freshwater supply is diminishing with climate change, urbanization, and industrialization. MBR technology facilitates reuse and recycling of wastewater, assisting industries and cities in minimizing freshwater use. Treated water from MBRs is low in turbidity and bacterial contamination and is ideal for uses such as agricultural irrigation, groundwater recharge, and cooling towers (Nguyen et al. 2022).
(3) Enhanced Treatment Performance Compared to Conventional Systems: Conventional activated sludge processes (ASP) involve secondary clarifiers and tertiary filtration, while MBR combines these processes in one system, enhancing efficiency. Improved organic matter and nutrient removal (BOD, COD, nitrogen, and phosphorus) over traditional processes. Less sludge production because of the extended sludge retention time (SRT), which reduces sludge disposal and treatment requirements (Pervez et al. 2020).
(4) Space-Saving and Compact Design: MBR systems have reduced space requirements in relation to traditional wastewater treatment plants. They do away with large sedimentation tanks, thus suitable for densely populated urban zones, decentralized treatment facilities, and industries with limited space. Beneficial for retrofitting and rehabilitation of existing wastewater treatment plants to increase capacity without increasing the size of the plant (Yuvaraju and Deena 2024).
(5) Efficient Elimination of Pathogens and Micropollutants: MBRs employ ultrafiltration (UF) or microfiltration (MF) membranes, which are capable of removing bacteria, viruses, and protozoa without the need for additional chemical disinfection. Enhanced removal of emerging pollutants like pharmaceuticals residues, endocrine-disrupting chemicals (EDCs), and microplastics, rendering them suitable for advanced wastewater treatment (Rahman et al. 2023).

(6) Low Membrane Fouling and Enhanced Lifespan: The development of membrane technology has given rise to low-fouling membranes, backwashable systems, and enhanced aeration technologies, lowering the cost of maintenance. The service life of MBR membranes has been significantly enhanced, thus making them more economical in the long term (Nguyen et al. 2022).
(7) Compliant with Industrial and Municipal Wastewater Treatment: MBRs find extensive applications in food and beverage, pharmaceuticals, petrochemicals, textile, and pulp and paper industries where high-strength wastewater needs to be treated at a high level. MBRs are employed in municipalities for the treatment of sewage, particularly in metropolitan cities where space and water quality are of utmost importance (Hai et al. 2018).

The Membrane Bioreactor (MBR) process was initially developed in 1969 by Dorr-Oliver Inc. Nevertheless, its initial application was restricted because of the high energy requirements and cost of membrane materials. With the passage of time, improvements in membrane materials, system design, and operational efficiency have greatly improved the viability of MBR technology for large-scale applications. The extensive commercial use of MBR technology picked up pace with its successful application in industrial and municipal wastewater treatment.

Presently, the worldwide market for MBR is about USD 3.35 billion and is expected to reach USD 8.78 billion at a compound annual growth rate (CAGR) of 7.6%. Submerged MBR configurations in the 1990s and the innovation of high-performance membranes resulted in the inauguration of several medium- to large-scale wastewater treatment plants. Now, more than 5000 wastewater treatment plants globally utilize MBR technology, with China witnessing the most growth (Rahman et al. 2023).

As beneficial as it is proven to be in terms of efficiency and sales, more must be done for MBR operations to become less energy-intensive and more cost-friendly, increasing MBR's role as a candidate for a greater variety of wastewater treatment processes.

1.2 Insight into Membrane Bioreactor

1.2.1 Bioreactor Component

A bioreactor in a wastewater treatment plant is a specially constructed chamber that encourages the growth of microorganisms, like bacteria and protozoa, which decompose and eat pollutants in raw wastewater (Visvanathan et al. 2000). The biological process may differ depending on oxygen levels, giving rise to three main forms of bioreactors:

[1] Aerobic Bioreactors: Use oxygen to degrade organic material and oxidize ammonia (NH_3) into nitrate (NO_3^-) through nitrification.

[2] Anoxic Bioreactors: Function under oxygen-free conditions to enable denitrification, where nitrates are reduced to nitrogen gas (N_2) and discharged into the atmosphere.
[3] Anaerobic Bioreactors: Function in the absence of oxygen and are employed to break down organic materials, commonly yielding methane (CH_4) as a product in applications such as anaerobic digestion.

Membranes are usually added after aerobic or anaerobic bioreactors to remove suspended solids, increasing the efficiency of the treatment. The MBR process commonly utilizes an aerobic bioreactor together with a membrane system.

Types of Bioreactors Utilized in MBR Technology.

Bioreactors are categorized into three general types (Visvanathan et al. 2000) depending on the manner in which microbial biomass is sustained:

- Suspended Growth Bioreactors: The biomass (microorganisms) is suspended freely within the liquid phase, accumulating as flocs. This is the most prevalent type employed in MBR operations.
- Attached Growth (Biofilm) Bioreactors: Microorganisms develop on solid surfaces or carrier materials within the reactor. Applied in certain applications but not very prevalent for MBRs.
- Hybrid Bioreactors: They incorporate both suspended and attached growth mechanisms to enhance treatment performance. If properly designed, they can be employed in MBR systems (Fig. 1.1).

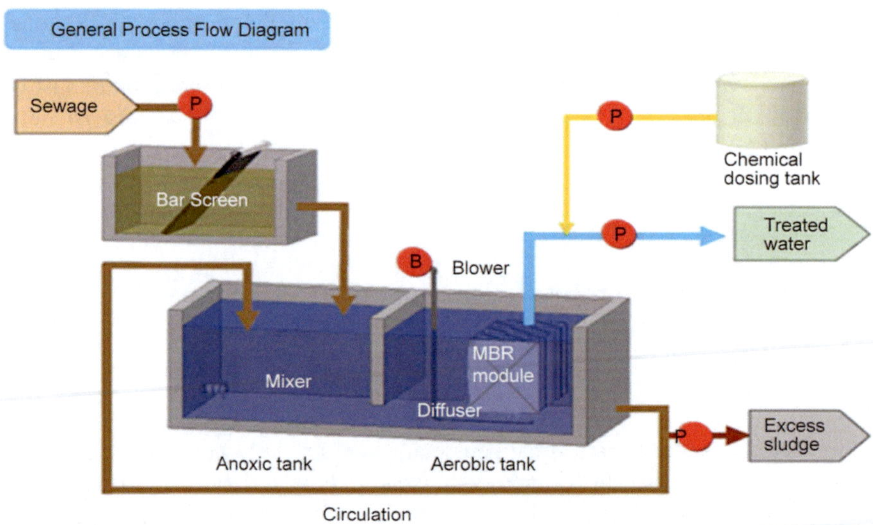

Fig. 1.1 General process flow diagram of wastewater treatment with MBR. *Source* https://b2bpurchase.com/membrane-bioreactor-delivers-superior-water-quality-and-process-control/

1.2.2 Membrane Component

Membranes in an MBR system are a solid–liquid separation unit that replaces traditional clarifiers used in conventional activated sludge processes (CAS). They trap the microbial biomass inside the bioreactor while permitting treated water to permeate through (Galinha et al. 2018). This leads to improved removal of pollutants and minimizes the treatment plant footprint (Fig. 1.2).

Types of Membranes Used in MBR Systems.

MBR applications often incorporate microfiltration (MF) or ultrafiltration (UF) membranes, and UF membranes are generally favored because:

i. A smaller pore size that provides additional removal of suspended solids, bacteria, and certain viruses.
ii. A lower fouling tendency, leading to lower maintenance and increased operation efficiency.

Membrane Geometries in MBR Systems.

MBR membranes exist in diverse geometries depending on the nature of the treatment requirements (Galinha et al. 2018):

(a) Hollow Fiber (HF) Membranes: Look like slender, straw-shaped fibers with minuscule pores. Possess a high surface-area-to-volume ratio, making them dense and efficient. Can be operated in vacuum-driven or pressure-driven modes.
(b) Flat Sheet (FS) Membranes: Are composed of flat polymeric sheets with uniform pore structures. Usual installation in submerged modes within the bioreactor. Easy to clean and maintain but may have lower packing density than HF membranes.
(c) Tubular (Multi-tubular, MT) Membranes: Have a tubular structure with an enlarged inner diameter. Applied in high-solids-concentration applications or industrial wastewater treatment.

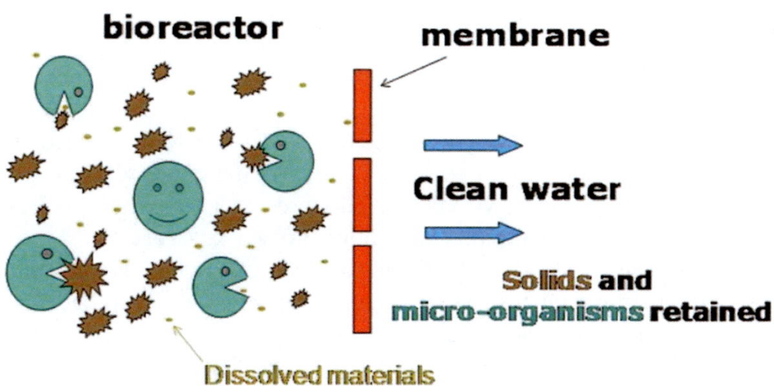

Fig. 1.2 MBR schematic. *Source* https://polarpedia.eu/en/membrane-bioreactor/

Table 1.1 Types of MBR

Type	Description
Submerged MBR (sMBR)	Membranes are immersed in the bioreactor; suction is applied to draw permeate through membranes
Side-stream MBR (sMBR)	Membranes are placed outside the bioreactor, and wastewater is pumped through them at high pressure

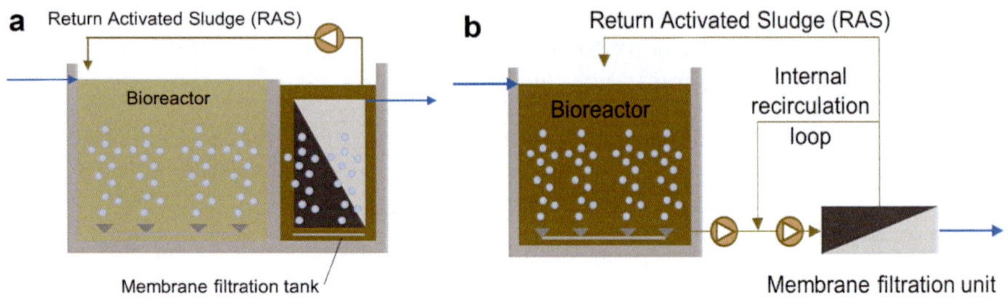

Fig. 1.3 **a** Submerged MBR. **b** Side-stream (or external) MBR. *Source* https://www.pcimembranes.com/articles/membrane-bioreactors-mbr-for-wastewater-treatment/

(d) Other forms of membranes, like spiral-wound (SW) membranes, are not MBR-compatible as they are vulnerable to high concentrations of solids.

Types of MBR.

Usually, pressure-driven membranes are used for smaller installations and/or tough-to-treat industrial wastewaters, while submerged membranes are used for medium and large installations (Fig. 1.3).

1.3 Components of an MBR System

1.3.1 Working of MBR

The treatment process starts with pretreatment (Fig. 1.4), in which coarse and fine screens (1–3 mm) are used to remove large debris, rags, and grit to avoid clogging of the membrane and prolong system life. This process is important as it shields downstream equipment and minimizes operational disturbances (Fig. 1.4).

After pretreatment, wastewater is fed into the bioreactor, where organic materials are degraded by microorganisms and pollutants are eliminated. The process is under aerobic, anoxic, and anaerobic conditions to support biochemical processes such as nitrification (ammonia to nitrate conversion), denitrification (nitrate to nitrogen gas conversion), and removal of phosphorus. In contrast to traditional activated sludge systems, MBRs maintain a greater biomass concentration (8000–12,000 mg/L MLSS), increasing the efficiency of treatment and minimizing the footprint overall.

Table 1.2 Key Units and Functions of a Membrane Bioreactor (MBR) System

Unit	Function	Details
Biological treatment unit (bioreactor)	Removes organic matter, nutrients, and pathogens	Uses microorganisms to break down organic pollutants via aerobic, anoxic, or anaerobic processes. **Typical MLSS Concentration**: **8000–12,000 mg/L** (higher than conventional processes)
Membrane filtration unit	Separates treated water from solids, bacteria, and pathogens	**Types of Membranes**: • Microfiltration (MF) (0.1–0.4 μm) • Ultrafiltration (UF) (0.01–0.1 μm) **Configuration**: • **Submerged MBR (sMBR)**—Membranes are immersed in the bioreactor • **Side-Stream MBR (sMBR)**—Membranes are installed externally, requiring high-pressure pumps
Aeration system	Provides oxygen for microbial activity and prevents membrane fouling	**Types of Aeration**: • **Fine bubble diffusers**—Improve oxygen transfer • **Coarse bubble aeration**—Reduces membrane clogging
Pumping and control system	Regulates flow, pressure, and membrane cleaning cycles	**Automation**: Reduces operator intervention

Subsequent to biological treatment, the wastewater is subjected to membrane separation in which treated water is pulled through membranes by vacuum or low-pressure pumping. Pore-sized membranes of 0.01–0.4 μm can efficiently capture suspended solids, bacteria, and viruses, yielding transparent, high-quality effluent. MBR uses microfiltration (MF) and ultrafiltration (UF) membranes that do not require secondary clarifiers and enable direct reuse of water. The membranes are set up as either submerged (sMBR), in which they are contained within the bioreactor, or side-stream (sMBR), in which there is external high-pressure pumping.

The last step is disinfection and effluent reuse or discharge. Because MBR-treated water is already free of most pathogens, very little disinfection is needed, and this is normally done through chlorination or ultraviolet (UV) treatment. The effluent of high quality can be discharged into natural bodies of water or reused for purposes like irrigation, industrial processes, and flushing toilets. Furthermore, with additional treatment through reverse osmosis (RO), the water can be made potable.

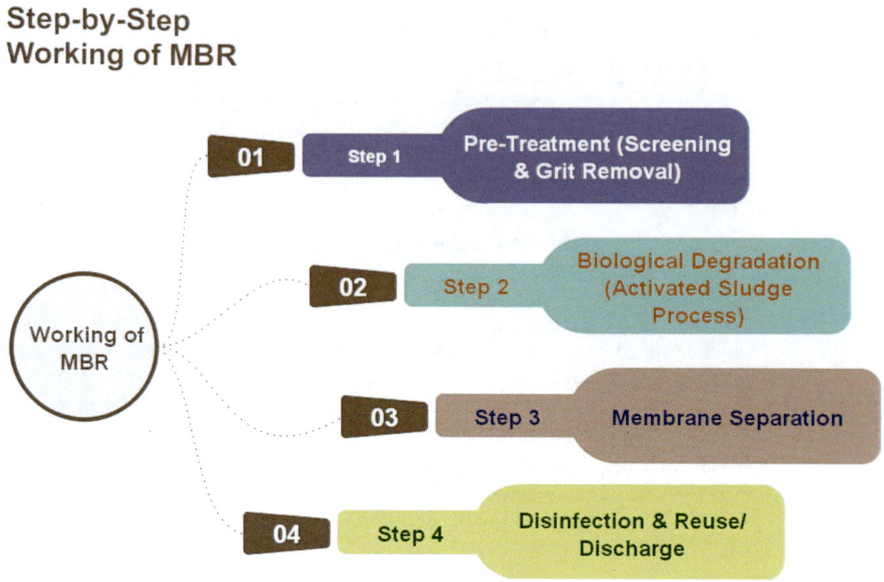

Fig. 1.4 Working of MBR technology

In general, the MBR process provides a small footprint, better effluent quality, and greater automation potential, making it a perfect solution for contemporary wastewater treatment. Its capability to effectively remove contaminants while allowing water reuse makes it a sustainable option for municipal and industrial applications.

1.4 Benefits of MBR Technology

Membrane Bioreactors (MBRs) are distinguished from the traditional Conventional Activated Sludge (CAS) process in some significant advantages they have (Rahman et al. 2023). These include varying from CAS by varying the Solid Retention Time (SRT) and Hydraulic Retention Time (HRT). MBR systems run on an immensely high SRT, having more time available for microbial function and destruction of organic components. This longer SRT facilitates greater fragmentation of high molecular weight organic contaminants, resulting in more stable and effective biological treatment. In contrast, the HRT of MBR is lower than that of CAS, implying that wastewater treatment can be carried out within less time, which enhances the process efficiency overall (Rahman et al. 2023).

Another significant strength of MBR technology is that it has greater sludge separation efficiency. Sludge separation in the CAS process is by sedimentation in secondary clarifiers, which may cause operational problems like sludge bulking and solid carryover in the effluent. MBRs avoid this issue by incorporating microfiltration or ultrafiltration membranes, which are a physical barrier for solid–liquid separation. This leads to highly clarified effluent with very low BOD, suspended solids, and turbidity concentrations as compared to CAS. The enhanced effluent quality makes the water treated with MBR more appropriate for reclamation and reuse, such as in industrial processes, irrigation, and even indirect potable reuse. MBR systems also consume less space than CAS plants because they do not require huge clarifiers, thus being especially advantageous for urban centers with land limitations.

Apart from its superiority over CAS, MBR technology can also be used in anaerobic wastewater treatment, taking over from traditional anaerobic digestion processes. As alternatives to conventional anaerobic digesters, MBRs can be coupled with Up-Flow Anaerobic Sludge Blanket (UASB) reactors, Expanded Granular Sludge Bed (EGSB) reactors, or Anaerobic Baffled Tank (ABT) reactors. The integration refines the anaerobic digestion process through better solid–liquid separation, maintaining high biomass concentration, and reducing wastage of sludge. In Anaerobic Membrane Bioreactors (AnMBRs), the membrane system facilitates controlled biomass concentration, resulting in less Chemical Oxygen Demand (COD) in the treated effluent than traditional anaerobic processes. AnMBRs are especially beneficial for treating high-strength industrial wastewater, minimizing sludge production, and producing biogas as a byproduct of value.

Less Footprint and More Hydraulic Throughput

One of the greatest benefits of MBR systems is that they can eliminate large secondary clarifiers in favor of compact membrane cassettes, minimizing the overall plant footprint. In contrast with conventional systems, where secondary clarifiers are sized on the basis of hydraulic and solids loading rates, MBRs operate with high mixed liquor suspended solids (MLSS) levels in bioreactors (typically 8–12 g/L) versus CAS systems (3–5 g/L). This enables the same mass of biomass to be housed in a much smaller tank, saving land by as much as 50% (Judd 2017).

For new WWTPs, MBR retrofitting provides enhanced hydraulic capacity without the need for expansion. Experiments have proven that MBR retrofitting can raise the treatment capacity of CAS systems by 30–50% with still better effluent quality (Meng et al. 2017).

High-Quality Effluent Free of Bacteria and Pathogens

The microfiltration or ultrafiltration membranes of MBRs create a robust physical barrier that eliminates suspended solids, bacteria, and most viruses from the effluent to be treated. MBR effluent generally fulfills strict discharge criteria with minimal post-treatment compared to CAS systems, which involve sedimentation and additional disinfection processes (Krzeminski et al. 2017).

Table 1.3 Key Features and Advantages of Membrane Bioreactor (MBR) Technology over Conventional Treatment Plants

Feature	Advantage
Compact design	Reduces plant footprint by up to **50%**
Superior effluent quality	Suitable for **reuse** or sensitive discharge areas
Higher MLSS concentration	Improves treatment efficiency (8000–12,000 mg/L vs. 3000–5000 mg/L in conventional plants)
Eliminates secondary clarifiers	Reduces land and infrastructure requirements
Fully automated operation	Minimizes operator intervention

Its superior quality makes MBR-treated water appropriate for reuse purposes including:

- Urban irrigation
- Industrial utility water
- Toilet flushing
- Reverse osmosis (RO) feed water for drinking water reuse.

With growing stringent regulations on water reuse globally, MBRs provide a sustainable option that complies with rigorous effluent discharge standards, satisfying environmental requirements (Le-Clech 2010).

Increased Automation and Decentralized Installation

MBR processes are highly automated, minimizing manual handling and labor expenses. Major automation aspects include:

- Real-time monitoring of membrane integrity and transmembrane pressure (TMP)
- Automated backwashing and cleaning cycles
- Remote operation and control via SCADA (Supervisory Control and Data Acquisition) systems

Such automation technologies facilitate decentralized wastewater treatment, so MBRs are well-suited for remote communities, industrial plants, and small community treatment facilities (Drews 2010). Moreover, sophisticated control techniques maximize membrane performance, energy efficiency, and chemical dosing, further maximizing operating efficiency (Le-Clech 2010).

1.5 Operational Parameters and Challenges in MBR-Based Wastewater Treatment

The effectiveness of Membrane Bioreactor (MBR) technology in wastewater treatment is dependent on various operational parameters (Bhattacharyya et al. 2022). These are membrane material, pretreatment processes, food-to-microorganism (F/

M) ratio, permeate flux, temperature, aeration, solid retention time (SRT), hydraulic retention time (HRT), and cleaning procedures. These parameters need to be optimized for effective treatment with cost-effectiveness (Jijingi et al. 2024).

1.5.1 Challenges in MBR Systems

Despite its merits, MBR technology also encounters several important challenges to overcome to further boost its competitiveness over traditional wastewater treatment techniques:

a. **Membrane Fouling**: Perhaps the greatest weakness of MBR is membrane fouling caused by solids building up on the surface of the membrane, causing lowered permeability and raised energy requirement. Fouling depends on factors such as membrane material, influent characteristics, and process conditions. Implementing effective fouling control strategies, such as improved membrane materials, better pretreatment, and optimized cleaning protocols, can mitigate this issue.

   Membrane fouling in MBR processes is determined by three primary factors: membrane characteristics (pore size, material, hydrophobicity), solution characteristics (particle concentration, component interactions), and operating conditions (pH, temperature, flow rates). Hydrophobic materials or smaller membrane pores will enhance fouling, and changes such as hydrophilic coatings may be required (Fig. 1.5).

   Ceramic membranes, especially in food and dairy applications, provide greater stability and resistance than polymeric membranes (Asif and Zhang 2021). The properties of the influent solution, such as solid content, pH, and ionic strength, also influence fouling by adding to pore blocking or cake deposition. Increased temperature and higher crossflow velocity contribute to fouling reduction by enhancing permeate flux and reducing foulant deposition.

   In order to manage fouling, several methods have been tried:

   - Antifouling membranes (e.g., PVDF, PAN, PES) provide resistance but at a high cost (Erkan et al. 2020).
   - Surface coatings alter membranes to decrease fouling (Erkan et al. 2020).
   - Mechanical cleaning techniques, like biofilm carriers, vibrating membranes, and rotating modules, avert foulant buildup (Bagheri and Mirbagheri 2018).
   - Chemical cleaning by means of acids, bases, oxidants, and surfactants is employed for extensive foulant buildup. Mild fouling is best treated in situ, and extensive foulant recovery employs ex situ cleaning (Bagheri and Mirbagheri 2018).

   Studies are ongoing to refine membrane materials and operating strategies for reducing fouling and enhancing the efficiency of MBR systems.

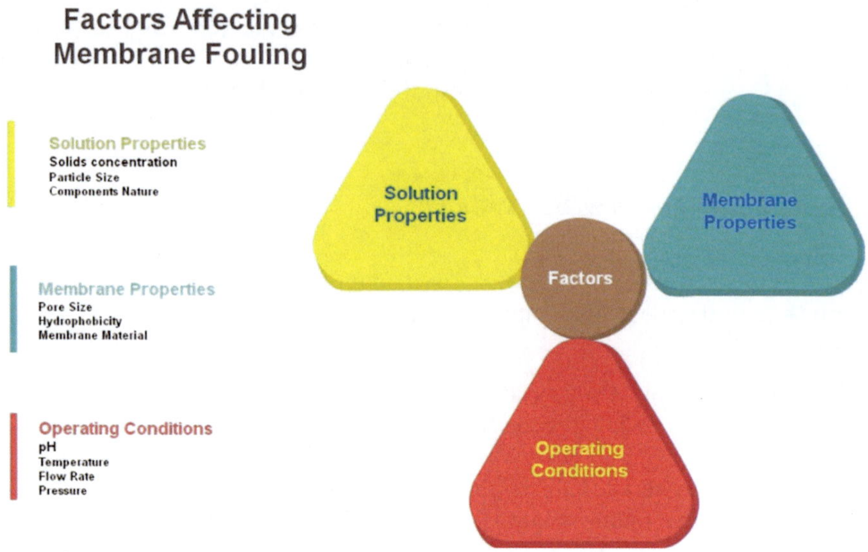

Fig. 1.5 Factors affecting membrane fouling

Fouling Mitigation Strategies in Membrane Bioreactors (MBRs)

Membrane fouling is among the most significant issues in Membrane Bioreactor (MBR) operation, as it results in increased energy consumption, maintenance costs, and decreased treatment efficiency. Fouling is caused by the deposition of suspended solids, microbial biofilms, extracellular polymeric substances (EPS), and organic and inorganic deposits on the membrane surface. To mitigate such problems, different fouling control measures have been created, such as the application of coagulants, adsorbents, quorum quenching (QQ), and membrane reciprocation (Ryu et al. 2024; Kim et al. 2023). Coagulants and adsorbents facilitate the removal of suspended solids and organic materials, whereas QQ inhibits microbial signaling to impede biofilm development. Membrane reciprocation also enhances hydrodynamic conditions, minimizing foulant deposition and increasing membrane life. Through integrating these innovative techniques, MBRs are capable of enhanced operating efficiency, lower maintenance expenses, and better long-term performance, thus constituting a more competitive and sustainable option for wastewater treatment.

Coagulants and Adsorbents: Chemical Approaches for Fouling Prevention

The use of coagulants and adsorbents is a common practice to minimize fouling by altering wastewater properties and avoiding membrane blockage.

Coagulants

Coagulants like polyaluminum chloride (PAC), ferric chloride ($FeCl_3$), and alum ($Al_2(SO_4)_3$) improve flocculation, lowering the level of fine suspended matter and dissolved organic materials that cause fouling. These chemicals allow the development of larger, more settleable flocs, thus reducing blockages through pores and enhancing filtration efficiency.

Adsorbents

Adsorbents like powdered activated carbon (PAC), biochar, and zeolites assist in the removal of organic pollutants and dissolved impurities prior to reaching the membrane surface. PAC-supported MBR systems have been observed to enhance membrane permeability and minimize soluble microbial products (SMPs) accumulation, thereby postponing biofouling and organic fouling.

Through the application of coagulants and adsorbents, MBRs are able to sustain higher flux rates and extended filtration cycles, ultimately leading to enhanced treatment efficiency and operational cost reduction.

Quorum Quenching (QQ): A Biological Strategy for Biofouling Control

Quorum Quenching (QQ) is a sophisticated biofouling control method that inhibits microbial communication, thus inhibiting biofilm development on the membrane surface. The technique inhibits quorum sensing (QS), a cell-to-cell communication process by which bacteria control biofilm formation and EPS production (Kim et al. 2023).

Mechanism of QQ

Enzymatic Inhibition: QQ enzymes like acylase and lactonase degrade signaling molecules (acyl-homoserine lactones, AHLs) that facilitate bacterial coordination during biofilm development.

Bacterial Interference: QQ bacteria (e.g., Rhodococcus sp., Pseudomonas sp.) are added to the MBR system to break down quorum sensing signals, minimizing biofilm formation on membrane surfaces.

Benefits of QQ Technology:

- Avoids biofilm-induced membrane fouling without impairing beneficial microbial activity.
- Decreases cleaning frequency and chemical consumption, resulting in reduced maintenance costs.
- Increases membrane life by reducing irreversible fouling due to microbial growth.

Membrane Reciprocation: Increasing Hydrodynamic Shear for Fouling Prevention

Membrane reciprocation is a mechanical approach involving the oscillatory motion of the membrane module to minimize particle deposition and biofilm growth. In contrast to the more traditional aeration-based cleaning processes, membrane reciprocation enhances membrane surface shear forces directly, inhibiting the deposition of foulants (Ryu et al. 2024).

Table 1.4 Challenges and Corresponding Solutions in Membrane Bioreactor (MBR) Technology

Challenges	Solutions
Membrane fouling	Regular aeration, chemical cleaning, and antifouling coatings
Energy consumption	Energy-efficient membranes and optimized aeration reduce costs
High initial cost	Long-term savings from reduced land use, labor, and maintenance

Process of Membrane Reciprocation:

The membrane module is subjected to reciprocating motion (back and forth movement), which generates turbulent flow and shear forces that remove contaminants. This method improves mass transfer, avoiding pore clogging and biofilm adhesion.

Advantages of Membrane Reciprocation:

- Decreases aeration requirement, resulting in reduced energy usage relative to traditional air scouring processes.
- Reduces biofouling and organic fouling, increasing membrane operating life.
- Enhances overall MBR performance by sustaining a higher flux-to-energy ratio.

b. **High Energy Consumption**: MBR systems require a substantial amount of energy, primarily for membrane filtration and aeration, which makes them more energy-intensive than conventional activated sludge (CAS) systems. Nonetheless, research suggests that CAS in conjunction with tertiary treatment uses comparable amounts of energy to MBR, which renders MBR as a viable, energy-efficient alternative.

c. **Cost of Membrane Materials**: Membrane initial investment and replacement cost is a major concern. Enhancements in high-flux membranes, reduced fouling, and longer membrane life can mitigate capital and operational costs.

1.6 Techno-Economic Assessment

1.6.1 Large-Scale MBR Plants

A research assessing the economic efficiency and viability of 35 large Membrane Bioreactor (MBR) plants with a minimum treatment capacity of 10,000 m^3/day was done by Gao et al. (2021). Applying cost–benefit analysis (CBA) and shadow pricing, the research estimates profitability, cost efficiency (CE), and energy efficiency (EE). The results indicate that MBR plants yield an average net profit of about 35 CNY/m^3 (~4.9 USD/m^3), which verifies their economic viability. Shadow pricing based on environmental benefits also adds further financial strength to the merits of MBR technology. The research also points out that MBRs based on more stringent local effluent standards are more profitable than those using the general national standard. Geographical differences in net profit indicate that eastern region MBR plants are more profitable than middle and western region plants because of higher water treatment needs, superior economic conditions, and more advanced industrial infrastructure. Underground MBRs are also marginally more profitable than overground MBRs, even discounting land-saving advantages, possibly because of the higher price of land in urban areas.

To examine the cost and energy efficiency of MBR plants, the research uses a Non-Radial Directional Distance Function (NDDF)-based Data Envelopment Analysis (DEA). On average, MBR plants' cost efficiency (CE) is 0.77, or 77% efficient compared with an ideal model. MBRs under stricter effluent guidelines have lower CE, primarily resulting from higher operation costs of additional treatment processes. The average energy efficiency (EE) of MBR plants is 0.66, implying that it saves just 66% of the available energy savings. Underground MBRs might have inferior EE due to increased aeration and pumping energy demands. Yet, a significant improvement in EE post-2015 implies that technology innovations and improved operational practices have resulted in superior energy performance.

Panel model analysis does single out several drivers of CE and EE, both external in origin. These are more rapid economic development through access to adequate infrastructure, advanced training of manpower, and better finances for the day-to-day functioning. High density of people is essential as greater density brings huge, intensive sewage treatment complexes necessary to utilize a lot but manage the lot better. Beyond that, it alters MBR's performance to varying degrees of impact on costs as well as power consumption.

The research highlights the need for an adjustment of MBR technology according to local conditions for maximum application. MBR plants ought to be engineered to suit economic status, population density, and industrial wastewater patterns. While stricter effluent requirements make profitability higher, they also bring with them increased treatment costs, requiring sophisticated operation methods to maximize cost and energy effectiveness. Government policies like the provision of subsidies for energy-efficient wastewater treatment further enable the enhancement of CE and EE. Further, innovations in low-energy membranes, process optimization, and control systems powered by artificial intelligence can enable MBR efficiency to be enhanced.

In the future, developed nations must focus on maximizing MBR performance in order to close the gap between actual and ideal CE and EE. Sophisticated control methods, such as real-time monitoring, automation, and machine learning-based process improvement, will lower operating expenses and energy use by a considerable margin. Policy initiatives and economic incentives will also be essential to speed up the introduction of high-efficiency MBR technologies, guaranteeing future-proof wastewater treatment solutions.

1.6.2 Comparative Economic and Sustainability Analysis

A number of studies have compared the economic viability and sustainability of MBR with traditional wastewater treatment:

Economic Analysis in Tikrit, Iraq (Samin et al. 2014). A comparative economic analysis of three wastewater treatment plants—CAS, CAS with pre-denitrification, and MBR—showed that MBR had a greater present worth cost for the same influent treatment capacity. Yet MBR had a better quality of effluent and consumed less land area, which made it a more desirable choice for densely populated urban areas.

Sustainability Assessment in China (Chen et al. 2018). A 60,000 m^3/day MBR plant was compared with a local conventional wastewater treatment plant (WWTP). Although MBR had greater technical performance, it also had greater energy consumption and environmental loading. These issues can be addressed in the long term by developments in membrane technology, effective aeration, better fouling control, and sludge treatment methods.

1.7 Investigating the Use of Membrane Bioreactors in Wastewater Treatment Processes

1.7.1 Efficiency and Environmental Impact

Membrane Bioreactors (MBRs) are a state-of-the-art technology in wastewater treatment that brings tremendous benefits over traditional treatment technologies by integrating biological breakdown with membrane filtration (Namdeti et al. 2025). The hybrid process efficiently separates organic material, nutrients, and pathogens, generating high-quality effluent that is in compliance with rigorous water quality requirements and in many cases suitable for direct reuse in industrial, agricultural, and even drinking purposes. In contrast to conventional activated sludge systems, MBRs substitute secondary clarifiers and tertiary filtration units with microfiltration or ultrafiltration membranes, resulting in reduced treatment footprint and increased process efficiency (Banti et al. 2020).

One of the main strengths of MBRs is their capability to treat at higher mixed liquor suspended solids (MLSS) concentrations, usually between 8000 and 12,000 mg/L, as opposed to 2500 and 4000 mg/L for conventional systems. This enables better pollutant degradation and a smaller reactor size. The membranes serve as a physical barrier, keeping suspended solids and microorganisms from being carried away with the treated effluent, thus improving effluent quality and minimizing sludge generation. In addition, MBRs offer superior pathogen removal, minimizing the requirement for lengthy post-treatment disinfection (Bera et al. 2022).

Even though they have several advantages, MBR systems have drawbacks like increased capital and operating expenses, mainly because of membrane fouling and energy usage. Membrane fouling, brought about by suspended solids, microbial biofilm, and dissolved organic matter, can result in more frequent maintenance needs and decreased permeability over time. To reduce this problem, regular membrane cleaning, aeration, and utilization of advanced antifouling membrane materials are necessary (Judd 2004). Moreover, energy usage in MBRs is more compared to that in conventional systems because of the requirement for aeration of membranes and suction pressure for filtration. Nevertheless, advancements in membrane technologies, more efficient energy-saving aeration practices, and the implementation of energy recovery technologies are all contributing to improving the cost-effectiveness and sustainability of MBR applications continuously (Osman et al. 2024).

Because of their better effluent quality and lower environmental footprint, MBRs are gaining acceptance in municipal wastewater treatment plants, industrial effluent treatment, and decentralized wastewater management (Arias et al. 2019). Their capability to treat fluctuating loads, simplicity of automation, and nutrient recovery potential make them a first choice for areas with water scarcity and stringent environmental regulations. With research focusing on improving membrane durability, decreasing energy consumption, and optimizing process performance, MBRs are poised to play a crucial role in future sustainable wastewater treatment and water resource management (Rahman et al. 2023).

1.7.2 Efficiency of Membrane Bioreactors

1.7.2.1 High Effluent Quality and Stability

MBRs can guarantee very high-quality effluent that makes them applicable for water reuse even in industrial, municipal, and decentralized wastewater treatment applications. MBRs provide much higher detoxification levels when combined with biological processes along with membrane filtration. This will ensure that suspended solids, organic matters, and microorganisms are removed and will yield a stable, continuous quality of the effluent, meeting or exceeding water reuse standards, including those adopted for the World Health Organization (WHO) and U.S. Environmental Protection Agency (EPA) (Jalilnejad et al. 2020).

One of the key advantages of MBRs is that they produce pathogen-free, low-turbidity effluents that require limited post-treatment disinfections. Membrane filtration incorporating microfiltration (MF) and ultrafiltration (UF) membranes constructs a physical barrier for bacteria, viruses, and protozoa removal. Such a high level of treatment qualifies effluent from MBRs for various non-potable reuse activities, including irrigation, toilet flushing, and industrial cooling, as well as possible indirect potable reuse after additional treatment. Moreover, it increases quality from optimized biological processes, especially nitrogen and phosphorus removal, through MBR technology.

Among the most advanced configurations of the MBR technology is the Anaerobic Membrane Bioreactor (AnMBR), which has provided standout performance meteorological data during wastewater treatment trials, particularly with reference to treatment capabilities in very high-strength industrial applications. Using anaerobic microorganisms for organic matter degradation while generating biogas as a byproduct, AnMBRs operate differently from normal aerobic MBRs and are processes run under anaerobic conditions. Studies have shown: AnMBRs remove between 85 and 95% of chemical oxygen demand (COD) while producing a completely clarified effluent without suspended solids (Stuckey 2024; Zielinska and Ojo 2023). It reduces sludge production and also augments energy recovery, making AnMBRs attractive for the sustainable treatment of wastewater.

The increased water scarcity and rigorous regulatory requirements spell doom for water reuse and hence add to the popularity of MBR technology. With innovative developments in membrane materials, fouling control strategies, and efficient energy operations, MBRs in the near future would provide a crucial contribution toward sustainable water management goals. Due to high-rank and reliable effluent made possible with the MBR systems, they offer the best approach for combating water scarcity and procuring circular water reuse practices across different sections.

1.7.2.2 Compact Design and Operational Flexibility

Bioreactors with membranes (MBRs) enjoy a significant advantage in the area of compact design, thus substantially reducing the land area needed for wastewater treatment in comparison with conventional activated sludge (CAS) systems. Large secondary clarifiers and tertiary treatment units are typically required for conventional plants to produce high-quality effluent. The MBR integrates biological treatment and membrane filtration into one system that does not require any separate clarification and filtration stages. Therefore, MBRs are especially suited to urban centers, industrial facilities, and decentralized wastewater treatment systems in which land available for such systems is limited (Jalilnejad et al. 2020). Studies indicate that the footprint of MBRs could be 30–50% smaller than in conventional systems, making them attractive for retrofit applications to existing treatment plants or new treatments in land-constrained areas.

In addition to their compactness, MBRs have exceptional operational flexibility and can therefore treat the wastewater effectively under varying influent conditions. These systems can accommodate high organic loads and high hydraulic shocks without significant disturbances to the operation, predominantly due to the ability to maintain a high concentration of microorganisms within the bioreactor (Stuckey 2024; Zielinska and Ojo 2023). Unlike conventional activated sludge systems, working typically at mixed liquor suspended solids (MLSS) concentrations of 2500–4000 mg/L, MBRs may achieve up to and over an MLSS of 8000–12,000 mg/L. Enhanced microbial density will stimulate aerobic degradation of organic pollutants and also instill stability to the process when sudden fluctuations in influent composition are encountered.

Further, MBRs can also be configured and be flexible in various treatment scenarios. For example, they can be submerged (immersed) systems, where membranes are immersed directly into the bioreactor for filtration, or side-stream systems, where membranes are placed in external modules and mixed liquor is pumped to them. This flexibility allows operators to design MBRs that best suit particular wastewater characteristics, energy constraints, and effluent quality requirements. Moreover, advanced control and automation systems allowing real-time monitoring and optimization of the process further enhance the flexibility to accommodate changing environmental and operational conditions.

MBRs are considered more and more due to their compact footprint, high resilience, and adaptability in municipal wastewater treatment plants, industrial effluent treatment, decentralized sanitation systems, and water reuse applications. The ability to tackle variable loads, space constraints, and stringent effluent quality prerequisites makes them the option of choice for sustainable and efficient wastewater treatment in both the developed and developing world.

1.7.2.3 Energy Efficiency and Resource Recovery

Recently, Membrane Bioreactors (MBRs) have attracted increased recognition as devices to improve energy efficiency and resource recovery in wastewater treatment. Anaerobic Membrane Bioreactors are an emerging way to make MBRs energy-neutral or even energy-positive in operation by integrating anaerobic biological processes with membrane filtration, which will cover and eliminate the disadvantages of energy-intensive membrane aeration and filtration in conventional MBRs, according to which they need high input energy. Such systems utilize anaerobic digestion, which is the process of micro-organisms breaking down organic matter in the absence of oxygen and producing methane-rich biogas as a byproduct. The methane can then be captured and used on-site for energy production, either as standalone electricity generation or direct heating applications, thus greatly reducing the imports of the plant from the outside world (Stuckey 2024; Li et al. 2022). Properly optimized AnMBRs can sometimes yield more power than what is needed, hence become energy independent or even net energy producers.

Another merit of this fermentation system is that it permits shorter hydraulic retention times (HRTs) and longer solid retention times (SRT) than those of the corresponding aerobic MBRs. Hence, generating less sludge minimizes energy demand for sludge handling/disposal. Also, AnMBRs do not require oxygen aeration for the treatment of wastewater, one of the major energy cost-consuming operations in conventional wastewater treatment. Therefore, minimizing aeration energy requirements as well as maximizing biogas recovery constitutes an extremely efficient and sustainable approach, especially concerning industrial wastewater treatment and decentralized specific site wastewater management.

Beyond energy recovery, MBR can also be useful in obtaining nutrients in a manner with structural approaches toward a circular economy in wastewater management. Nutrient recovery technologies, using advanced systems with MBR, will enable the extraction of nitrogen (N) and phosphorus (P) as valuable ingredients for agricultural and industrial fertilizers (Zielinska and Ojo 2023). There are ammonia recovery techniques such as air stripping, ion exchange, and electrochemical processes, and there is also precipitation of phosphorus as struvite, which is magnesium ammonium phosphate, a slow-release fertilizer applicable in agriculture.

Resource recovery strategies integrated with MBR systems prove the global thrust toward sustainable wastewater treatment and resource circularity as regards MBR configuration. Such biogas would be used in energy production, with essential nutrients recovered through MBR technology to mitigate environmental impacts but, in addition, to create economic opportunities through the generation of valued byproducts from what could otherwise be waste streams. With continuous improvements in research and technology in membrane materials, fouling control, and energy-efficient operation, the scope of application for MBRs in the field of sustainable wastewater treatment and resource recovery would expand further toward more robust resilience and sustainability.

1.8 Environmental Impact of Membrane Bioreactors

1.8.1 Reduction of Greenhouse Gas Emissions

Membrane Bioreactors (MBRs), especially Anaerobic Membrane Bioreactors (AnMBRs), are crucial in mitigating greenhouse gas (GHG) emissions from wastewater treatment. Conventional wastewater treatment plants, especially aerobic biological treatment-based plants, are major contributors to GHG emissions, mainly carbon dioxide (CO_2) and nitrous oxide (N_2O), released as a result of aeration and microbial respiration. Conversely, AnMBRs operate anaerobically, utilizing anaerobic microbial activity to degrade organic matter and produce methane-rich biogas as a byproduct (Stuckey 2024; Li et al. 2022). This recovered methane can be utilized as a renewable fuel source, replacing fossil fuel consumption and radically diminishing the carbon footprint of treatment.

In contrast to traditional aerobic treatment systems, which involve extensive energy input to aerate water, AnMBRs do not necessitate oxygen supplementation, and hence there is a considerable decrease in energy-related CO_2 emissions. Furthermore, when the biogas generated in AnMBRs is efficiently utilized for on-site power generation or as heat, it avoids the uncontrolled release of methane—a greenhouse gas with 25 times the global warming potential of CO_2. Through the use of gas-tight reactors and effective biogas collection systems, AnMBRs ensure that methane is utilized for energy generation instead of being emitted into the atmosphere, thus making wastewater treatment plants more energy-efficient and sustainable.

In addition to methane recovery, MBRs also help in GHG mitigation through low sludge production. Traditional wastewater treatment processes generate large volumes of excess sludge, which require further treatment and disposal, often through landfilling or incineration—both of which contribute to carbon emissions. In contrast, MBRs operate at higher solid retention times (SRTs), resulting in significantly lower sludge yield (Jalilnejad et al. 2020). The smaller volume of sludge reduces the energy required for dewatering, transportation, and disposal, in addition to reducing the environmental footprint of the treatment process. The sludge in anaerobic systems also gets more stabilized, limiting methane emissions that are generally associated with conventional sludge digestion and disposal practices.

Through the combination of biogas recovery, energy efficiency optimization, and sludge volume reduction techniques, MBRs, particularly AnMBRs, provide a climate-neutral wastewater treatment option. With wastewater treatment plants across the globe working toward carbon neutrality and minimizing their environmental impact, the implementation of MBR technology offers a sustainable route toward decreased emissions, energy autonomy, and resource recovery (Alevizos et al. 2023). Advances in membrane technology, process innovation, and biogas upgrading in the future will further elevate the GHG abatement potential of MBRs and make them an essential element in global efforts toward mitigating climate change.

MBRs reduce emissions in several ways:

a. Less Discharge of Wastewater = Reduced Energy Demand: As MBRs facilitate reuse of water, less discharge of treated water is required to rivers or to be transported over long distances. This lowers pumping, distribution, and discharge energy requirements, which reduces related CO_2 emissions.
b. Treatment with Reduced Carbon Footprint: In comparison to traditional wastewater treatment processes, MBRs have higher biomass concentrations, which enable quicker and more effective degradation of organic matter. This minimizes the energy needed for aeration, resulting in lower operational emissions.
c. Anaerobic Membrane Bioreactors (AnMBRs) for Methane Recovery: AnMBRs recover methane generated through anaerobic digestion, which can be upgraded to biogas for energy production. This minimizes the use of fossil fuels, making the wastewater treatment process energy-neutral or even energy-positive.
d. Reduced Sludge Formation and Handling Emissions: MBRs generate less surplus sludge than traditional activated sludge systems, minimizing the requirement for transportation, treatment, and landfilling of sludge, all of which involve GHG emissions.

1.8.2 Efficient Removal of Emerging Pollutants

Membrane Bioreactors (MBRs) have been attracting considerable interest as they can effectively remove emerging contaminants such as pharmaceuticals, personal care products (PPCPs), pesticides, industrial chemicals, and endocrine-disrupting compounds (EDCs), which are usually not removed by conventional wastewater treatment techniques. These pollutants, which are abundantly present in municipal and industrial wastewater, create severe environmental and human health impacts owing to their persistence, bioaccumulation, and possible toxicity (Lopez-Herguedas et al. 2023; Pathak et al. 2020). The traditional WWTPs are dependent mainly on sedimentation and biodegradation, which are inadequate to eliminate such micropollutants effectively, resulting in their repeated efflux into aquatic environments.

Conversely, MBRs integrate biological treatment with membrane filtration, providing a sophisticated and effective solution for the enhanced removal of such contaminants.

The efficiency of MBRs in removing emerging contaminants is mainly due to three mechanisms (Asif et al. 2019):

1. **Biodegradation**: The intensified microbial concentration and extended solid retention times (SRTs) in MBRs enable strengthened biodegradation of intricate organic pollutants, such as pharmaceuticals and pesticides. The extended interaction of wastewater with specialized microbial groups enables enzymatic degradation of micropollutants that are otherwise resistant to conventional treatment plants.
2. **Membrane Filtration**: Microfiltration (MF) or ultrafiltration (UF) membranes in MBRs provide physical barriers, blocking the transfer of particulate-bound pollutants and bigger organic compounds. This diminishes the load of contaminants in treated effluent and produces higher-quality treated water than in conventional activated sludge processes.
3. **Adsorption and Retention**: Hydrophobic pollutants, like some pharmaceuticals and EDCs, have the tendency to adsorb on the sludge particles in the bioreactor and thus get accumulated and degraded and not released to the effluent (Table 1.5).

Table 1.5 Comparison of key MBR types, efficiency, and environmental impact

MBR type	Efficiency highlights	Environmental impact
Conventional MBR	High effluent quality, stable operation, and low sludge production Qi (2024), Jalilnejad et al. (2020)	Reduces greenhouse gas emissions and minimizes sludge disposal issues Qi (2024), Jalilnejad et al. (2020)
Anaerobic MBR (AnMBR)	Achieves 85–95% COD removal and produces methane for energy recovery Stuckey (2024), Zielinska and Ojo (2023)	Captures methane, reducing carbon footprint and operational costs Stuckey (2024), Zielinska and Ojo (2023)
Hybrid MBR Systems	Enhanced removal of emerging contaminants and improved fouling resistance Khan et al. (2023), Millanar-Marfa et al. (2020)	Facilitates water reuse and reduces environmental discharge of harmful contaminants Khan et al. (2023), Millanar-Marfa et al. (2020)
Electrochemical MBR	Improves treatment efficiency and reduces fouling through electrochemical processes Castrogiovanni (2024), Mao et al. (2024)	Promotes sustainable treatment with minimal chemical use and lower energy consumption Castrogiovanni (2024), Mao et al. (2024)

1.8.3 *Hybrid and Advanced Membrane Bioreactor Systems*

Membrane Bioreactors (MBRs) are highly efficient in wastewater treatment, but their efficacy can be maximized by coupling them with other innovative treatment processes. This hybridization overcomes operational drawbacks, including membrane fouling, energy requirements, and removal of resistant pollutants, while it also promotes resource recovery and sustainability. Different MBR-based hybrid systems have been proposed, including membrane distillation (MD), advanced oxidation processes (AOPs), and electrochemical treatments, each of which has enhanced performance in certain applications.

Combining MBRs with complementary treatment technologies offers a multifunctional strategy for addressing wastewater treatment issues, including energy usage, fouling, and micropollutant removal. MBR-MD systems enhance the recovery of water while harnessing waste heat, MBR-AOP hybrids improve the degradation of persistent pollutants, and MBR-electrochemical systems allow for effective contaminant removal while potentially recovering energy. As research and technological advancements continue, these hybrid MBR systems will play a critical role in improving wastewater treatment efficiency, facilitating water reuse, and advancing circular economy principles in contemporary water management (Capodaglio 2021).

1.8.3.1 MBR-Membrane Distillation (MD) Hybrid Systems

MBR-MD integration is a potential technique, especially in applications with access to waste heat. Membrane distillation is a thermal-coupled separation process employing a hydrophobic membrane for water vapor transport with the exclusion of contaminants. MBR-MD integration presents various advantages (Pathak et al. 2020) such as:

- High-Quality Water Recovery: MD can separate salts, heavy metals, and other non-volatile impurities while letting clean water vapor pass through, resulting in high-purity distillate for reuse (Glover et al. 2022).
- Use of Low-Grade Waste Heat: MD is lower-temperature operated (40–80 °C) than conventional distillation, enabling it to utilize waste heat from industrial processes or renewable energy, for example, solar thermal energy. This improves energy efficiency and cuts carbon emissions.
- Decrease of Environmental Footprint: Because MD does not involve high-pressure pumps (as opposed to reverse osmosis), it possesses lower operating energy requirements, further enhancing the sustainability of wastewater treatment.

1.8.3.2 MBR-Advanced Oxidation Process (AOP) Hybrid Systems

Advanced Oxidation Processes (AOPs) constitute the production of strongly oxidizing hydroxyl radicals (•OH), which have the capability of degrading resistant organic compounds, including pharmaceuticals, pesticides, and industrial chemicals, which are hard to be degraded using typical biological treatment. The coupling of MBRs with AOPs has presented a number of benefits (Alevizos et al. 2023):

- Improved Degradation of Micropollutants: AOPs efficiently break down pharmaceutical residues, endocrine-disrupting chemicals (EDCs), and other toxic organics, stopping them from entering the environment (Millanar-Marfa et al. 2020).
- Less Membrane Fouling: AOPs break down organic foulants prior to reaching the membrane, thus extending membrane life and lowering maintenance costs.
- Enhanced Pathogen Inactivation: The intense oxidative character of AOPs provides complete disinfection, killing viruses, bacteria, and antibiotic-resistant genes.

Typical AOPs combined with MBRs are:

i. Ozone-Based Systems (O_3/MBR): Efficient in degrading stable organic pollutants (Khan et al. 2023).
ii. UV/H_2O_2-MBR: Utilizes ultraviolet light and hydrogen peroxide to form hydroxyl radicals for degradation of contaminants (Augsburger et al. 2021).
iii. Fenton and Photo-Fenton Reactions: Uses Fe^{2+} and H_2O_2 to catalytically degrade complex organic molecules (Kösem et al. 2024).

1.8.3.3 MBR-Electrochemical Hybrid Systems

Electrochemical treatment technologies, including electrocoagulation (EC), electrooxidation (EO), and microbial electrochemical systems (MESs), are being combined with MBRs to enhance efficiency in wastewater treatment (Chen et al. 2023). These technologies provide:

a. Efficient Removal of Heavy Metals and Dissolved Organics: Electrocoagulation facilitates the flocculation of contaminants, enhancing turbidity, colloidal particle, and dissolved metal removal (Castrogiovanni 2024).
b. Minimized Membrane Fouling: Electrochemical oxidation can degrade organic foulants, inhibiting biofouling and inorganic scaling on the membrane surface (Ensano et al. 2016).
c. Energy Recovery Potential: Microbial Electrochemical Systems (MESs) like Microbial Fuel Cells (MFCs) and Microbial Electrolysis Cells (MECs) enable direct bioelectricity production or hydrogen production from organic wastewater, rendering the treatment process more sustainable (Arun et al. 2024).

In addition to improving removal of recalcitrant and low-biodegradability pollutants, MBRs can also be integrated with advanced treatment technologies, such as:

a. Nanofiltration (NF) and Reverse Osmosis (RO): Both are membrane treatment processes that give a secondary stage of treatment to remove even finer organic molecules, heavy metals, and salts for an even purer water (Khan et al. 2023).

b. Advanced Oxidation Processes (AOPs): The integration of ozonation, UV irradiation, Fenton reactions, or photocatalysis with MBRs accelerates the degradation of recalcitrant pollutants through the generation of highly reactive hydroxyl radicals (•OH), which disintegrate complex pollutants into harmless byproducts (Millanar-Marfa et al. 2020).
c. Activated Carbon Adsorption: The incorporation of powdered activated carbon (PAC) or granular activated carbon (GAC) into MBR systems improves the adsorptive capacity for hydrophobic pollutants, diminishing their concentration in the effluent.

1.8.4 Advancements in Membrane Bioreactor (MBR) Technology: New Configurations and Materials

The ongoing development of Membrane Bioreactor (MBR) technology has contributed to the advancement of new membrane materials and novel reactor designs that improve treatment efficacy, extend membrane life, and minimize operating expenses. New materials for membranes, including mixed matrix membranes (MMMs), and advanced system architectures, such as the electro-encapsulated self-forming dynamic membrane bioreactor (e-ESFDMBR), have indicated great promise in addressing issues related to membrane fouling, stability, and energy requirements.

Recent developments in membrane materials and system designs are revolutionizing the realm of MBR technology. Mixed Matrix Membranes (MMMs) demonstrate remarkably enhanced permeability, fouling resistance, and durability, placing them among the most efficient options for difficult wastewater treatment applications. In parallel, future-generation MBR designs such as the e-ESFDMBR bring self-forming dynamic membranes with electrochemical augmentation, resulting in improved treatment efficacy, reduced operational costs, and enhanced sustainability. As studies go on, these new MBR technologies will be a key to propelling wastewater treatment to greater efficiency, resource recovery, and sustainability.

1.8.4.1 Mixed Matrix Membranes (MMMs): Improving Membrane Performance

Mixed Matrix Membranes (MMMs) are a significant innovation in membrane technology, combining polymeric materials with inorganic fillers to enhance membrane hydrophilicity, fouling resistance, and structural stability. The membranes incorporate nanomaterials like metal–organic frameworks (MOFs), graphene oxide (GO), carbon nanotubes (CNTs), and silica nanoparticles, which are responsible for enhanced performance in MBR systems (Abdullah et al. 2023).

Main Advantages of MMMs:

- **Enhanced Hydrophilicity and Permeability**: The addition of hydrophilic inorganic fillers increases water permeability, lowering membrane resistance and energy expenditure.
- **Increased Fouling Resistance**: Nanomaterial-supported MMMs inhibit organic pollutant, microbe, and biofilm adhesion, thus reducing membrane fouling and membrane lifespan.
- **Increased Mechanical and Chemical Stability**: MMMs display greater chemical breakdown resistance, thermal stability, and operational stress, which are beneficial in long-duration wastewater treatment operations.

Examples of advanced MMMs are

Graphene Oxide (GO)-Based MMMs: Increase antimicrobial activity, lower biofouling, and enhance mechanical strength (Quan et al. 2017).

Zeolite-Loaded MMMs: Have excellent heavy metal and organic pollutant adsorption capacity, enhancing effluent quality (Zambare and Nemade 2021).

Carbon Nanotube (CNT)-Doped MMMs: Provide higher mechanical stability and antifouling performance, rendering them applicable to industrial wastewater treatment (Pacheco et al. 2021).

1.8.4.2 Electro-Encapsulated Self-forming Dynamic Membrane Bioreactor (e-ESFDMBR): A Next-Generation MBR Configuration

The electro-encapsulated self-forming dynamic membrane bioreactor (e-ESFDMBR) is a novel MBR design that has proven to have improved treatment performance and efficient fouling management (Millanar-Marfa et al. 2022). The system combines self-forming dynamic membranes (SFDMs) and electrochemical improvement in solving some of the biggest shortcomings of traditional MBRs.

How e-ESFDMBR Works:

[1] Self-Forming Dynamic Membrane (SFDM) Layer: A biological or particulate-based filtration layer naturally forms on a support medium as an alternative to using a fixed synthetic membrane. This decreases dependence on costly membranes and enables low-cost operation with self-repair mechanisms.

[2] Electrochemical Enhancement: The system uses an electric field to manage biofilm growth, increase pollutant degradation, and decrease membrane fouling. The technology also allows for electrochemical oxidation of pollutants for better water quality.

Major Advantages of e-ESFDMBR:

1. Enhanced Fouling Control: The self-forming membrane constantly regenerates itself, avoiding irreversible fouling and minimizing cleaning frequency.
2. Increased Pollutant Removal Efficiency: Electrochemical promotion assists in the degradation of recalcitrant organic matter and pathogens, making it ideal for the treatment of industrial wastewater.
3. Reduced Operating Costs: The self-forming membrane saves money on membrane replacements, whereas electrochemical treatment avoids the frequent use of chemical cleaning chemicals.
4. Energy Efficiency: Electrochemical process allows for direct oxidation and energy recovery, resulting in a more energy-efficient wastewater treatment process.

1.8.5 Ensuring Safer Effluent for Environmental Discharge and Reuse

By effectively eliminating pharmaceuticals, pesticides, industrial chemicals, and other micropollutants, MBR-treated effluent is of high-enough quality to meet strict water quality standards for direct environmental discharge or reuse in irrigation, industrial cooling, and groundwater recharge. With water scarcity and pollution issues on the rise, the contribution of MBRs to the production of safe, high-quality reclaimed water is becoming more significant. Future research and technology advancements in membrane materials, hybrid treatment processes, and microbial engineering will further enhance the efficacy of MBRs in the removal of emerging contaminants to provide sustainable wastewater treatment options that are environmentally friendly and beneficial to human health.

1.8.6 Water Reuse

Membrane bioreactors (MBRs) are central to water reuse and resource recovery, offering a cutting-edge treatment technology that can facilitate the safe and effective recycling of wastewater for numerous purposes. By generating high-quality effluent that is equal to or better than regulatory requirements, MBR-treated water can be reused in agricultural irrigation, industrial applications, landscape irrigation, groundwater recharge, and even potable reuse (Glover et al. 2022; Jalilnejad et al. 2020). This ability is especially crucial in water-scarce areas, where freshwater resources are becoming more stressed due to climate change, population increase, and industrialization.

Uses of MBR-Treated Reclaimed Water

a. **Agricultural Irrigation**: MBR effluent is high in nutrients like nitrogen and phosphorus, which can favorably impact agricultural development by lowering the requirement for commercial fertilizers. Because MBRs efficiently eliminate pathogens, suspended solids, and organic chemicals, the treated water can be put to beneficial use in food crop irrigation with negligible health hazards for consumers as well as farm workers (Racar et al. 2020; Biswas et al. 2021).
b. **Industrial Applications**: Industrial applications include textiles, pharmaceuticals, electronics, and power plants that need significant amounts of high-quality water. Treated wastewater using MBRs can be safely used in cooling systems, boiler feedwater, and process water, minimizing industrial dependence on fresh water resources. The closed-loop water recycling by MBRs assists industries in reducing operational costs and meeting environmental discharge standards (Jijingi et al. 2024).
c. **Urban and Landscape Irrigation**: Cities and urban planners increasingly employ MBR-treated water for park irrigation, golf courses, highway landscaping, and green roofs. This alleviates the pressure on municipal water supplies while preserving urban greenery, enhancing air quality, and making cities more resilient to climate change (Asif and Hai 2019).
d. **Groundwater Recharge and Indirect Potable Reuse**: Well-treated MBR effluent can be safely discharged into aquifers for recharging groundwater reserves. This avoids over-extraction, minimizes land subsidence, and provides protection against seawater intrusion in coastal regions. In indirect potable reuse facilities, water from MBR is further treated (e.g., advanced oxidation, activated carbon treatment, or reverse osmosis) and then mixed with natural water sources and finally supplied for consumption as drinking water (Firozjaee et al. 2024).
e. **Direct Potable Reuse (DPR)**: Some advanced MBR systems, when combined with multi-barrier purification technologies, can produce water clean enough for direct human consumption. Successful projects in Singapore, Namibia, and parts of the United States demonstrate the feasibility of using MBR-treated wastewater for safe, high-quality drinking water production (Capodaglio 2021).

1.9 Future Directions and Challenges for Membrane Bioreactors (MBRs)

Membrane Bioreactors (MBRs) are a technology that has the potential to revolutionize wastewater treatment, providing high-quality effluent, space reduction, and higher contaminant removal. Despite that, there are some challenges involved in scaling up, lowering operation costs, prevention of membrane fouling, minimization of energy consumption, and increasing the removal of emerging pollutants. Overcoming these challenges using technological innovation and research breakthroughs will be important in making the large-scale implementation of MBRs in municipal and industrial wastewater treatment possible.

1.9.1 Scaling up and Commercialization

Although MBR technology has numerous advantages, its high cost of capital and operational expenses continue to be major hindrances to large-scale adoption. The high initial costs of MBRs are mainly linked to the production, installation, and energy-driven aeration systems of membrane modules (Jalilnejad et al. 2020; Li et al. 2022). Moreover, the replacement and maintenance of membranes contribute to long-term costs.

Important Strategies for Scaling Up MBRs:

1. **Cost-Effective Membrane Production**: Research should aim at creating less expensive, long-lasting, and high-performance membrane materials with increased service life and better fouling resistance. Mixed matrix membranes (MMMs) and nanocomposite membranes have the potential for improving membrane longevity at lower costs.
2. **Process Optimization for Large-Scale Systems**: The use of hybrid treatment systems integrating MBRs with traditional CAS or AOPs has the potential to increase cost-effectiveness and treatment efficiency. Decentralized MBR systems are capable of delivering affordable solutions to remote and rural communities, saving infrastructure investment and wastewater transportation.
3. **Industrial and Municipal Adoption**: Governments, industries, and municipalities need economic incentives, subsidies, and policy assistance to promote MBR adoption. Cost-sharing schemes and the incorporation of MBRs in large-scale wastewater treatment can be fostered through partnerships among governments, research institutions, and private organizations.

1.9.2 Addressing Fouling and Energy Consumption

Fouling of membranes is still the biggest operational concern in MBRs, requiring increased energy inputs, more frequent cleaning, and shorter membrane lifetime (Ryu et al. 2024; Kim et al. 2023). Energy-requiring aeration of membranes for scours and microbial processes also lends to high operation costs in MBRs.

Fouling Control Strategies

Next-Generation Membranes: The formation of antifouling membranes, such as hydrophilic coatings, materials resistant to biofilm formation, and dynamic self-cleaning membranes, can make a significant decrease in fouling. Electrically improved membranes and nanomaterials exhibit self-cleaning behavior to enhance operating stability.

Highly Developed Cleaning and Fouling Avoidance Techniques

Emerging techniques in fouling control, such as quorum quenching (QQ), membrane reciprocation, and chemical-free backwashing, may be used to increase membrane efficiency and life. The membrane bioreactor hybridization technique combined with granular media filtration or electrochemical treatment may lower membrane clogging.

1.9.3 Techniques for Energy Savings

Low-Energy Aeration Methods

- Integrating intermittent aeration, bubble-free aeration, or membrane diffusers may optimize oxygen transfer efficiency and lower energy consumption without compromising high biological treatment efficiency.
- Applying computational fluid dynamics (CFD) modeling may assist in designing aeration for minimal energy consumption.

Integration of Renewable Energy Sources

- Solar-powered MBR systems, anaerobic digestion to produce biogas, and microbial fuel cells (MFCs) may enhance energy efficiency.
- Anaerobic Membrane Bioreactors (AnMBRs) are especially promising as they generate biogas (methane) that can be utilized as an energy source, and hence they can be considered a potential energy-neutral or energy-positive technology.

1.9.4 Improving Removal of Emerging Contaminants

Although MBRs are quite efficient in eliminating suspended solids, organic matter, and pathogens, their effectiveness against pharmaceuticals, endocrine-disrupting chemicals (EDCs), pesticides, and microplastics is an area of particular concern (Khan et al. 2023; Pathak et al. 2020). Such emerging pollutants can be harmful to human health as well as aquatic life, hence their removal from wastewater is very important for sustainable wastewater treatment.

Strategies for Enhanced Contaminant Removal:

- **Advanced Membrane Materials**: The integration of nanomaterials, graphene oxide, and molecularly imprinted polymers (MIPs) into MBR membranes enhances selectivity and micropollutant removal. Membrane distillation (MD) and forward osmosis (FO) technologies can supplement MBRs in enhancing low-molecular-weight contaminant removal efficiency.
- **Hybrid and Integrated Treatment Systems**: The integration of MBRs with activated carbon adsorption, ozonation, and electrochemical oxidation improves the destruction and adsorption of recalcitrant contaminants. MBR-AOP hybrid systems (such as MBR with UV/H_2O_2, Fenton's reagent, or photocatalysis)

have shown increased rates of degradation for pharmaceuticals and other micropollutants.
- **Monitoring and Process Optimization**: Artificial intelligence (AI) and machine learning (ML)-based real-time monitoring systems have the potential to optimize MBR processes for efficient contaminant removal. AI-based predictive modeling has the potential to enhance effluent quality control, decrease chemical dosing, and save energy (Table 1.6).

While significant research is being carried out worldwide, India is also advancing in wastewater treatment using MBR technology as seen by case studies in Table 1.7.

Key Findings from Applications of MBR in India:

- Municipal sewage treatment by MBRs is extensively used across metropolitan cities with high organic and nutrient removal efficiency.
- Treatment of textile wastewater in areas such as Tiruppur (a textile city) aims at color removal and high COD destruction.
- Treatment of pharmaceutical and hospital wastewater is a priority in cities such as Hyderabad owing to excess amounts of micropollutants.
- Landfill leachate treatment with MBR-RO hybrid systems is efficient for the removal of heavy metals and organic pollutants.
- Wastewater from the dairy and food industries is treated efficiently with HF-MBRs, with high degradation of organic matter.
- Treatment of petrochemical wastewater with ceramic membranes is efficient for the removal of hydrocarbons and persistent organic pollutants.

1.10 Life Cycle Assessment of Membrane Bioreactor in Wastewater Treatment Plant

Membrane Bioreactors (MBRs) integrate membrane filtration with biological treatment, greatly enhancing effluent quality and making water reuse possible. MBR technology, though, has greater energy demand and material consumption than traditional wastewater treatment practices. Membrane Bioreactor (MBR) technology is a new wastewater treatment method that offers improved operating flexibility, a smaller environmental impact, and high-quality effluent. Membrane Bioreactors (MBRs) are modern wastewater treatment systems that combine membrane filtration and biological treatment. There are two broad categories of MBRs: Anaerobic MBR (AnMBR) and Conventional MBR (CMBR), i.e., operates under aerobic conditions. The LCA of both categories is outlined below.

Table 1.6 Summary of recent research work on membrane bioreactors

Reference	Insights	Methods used
Jalilnejad et al. (2020)	Membrane Bioreactor (MBR) technology enhances wastewater treatment efficiency by producing high-quality effluent, reducing sludge output, and promoting stability. Its environmental impact is positive, supporting resource reuse and low energy consumption, though challenges like membrane fouling and temperature sensitivity exist	Membrane Bioreactor (MBR) technology Anaerobic Membrane Bioreactor (AnMBR) technology Membrane Aeration Bioreactor (MABR) technology Anaerobic ammonium nitrogen membrane bioreactor technology
Glover et al. (2022)	The study demonstrates that membrane bioreactor-membrane distillation (MBR-MD) systems significantly reduce environmental impacts for indirect potable reuse, achieving a 53.7% reduction when utilizing waste heat, while conventional systems show higher impacts, particularly from sludge disposal and energy use	ReCiPe midpoint assessment method for life cycle impact assessment 18 environmental impact categories studied for LCIA
Ryu et al. (2024)	Membrane bioreactors (MBRs) enhance wastewater treatment efficiency by reducing fouling and energy consumption. Utilizing non-aerated membrane modules can decrease energy demand significantly, promoting environmentally friendly technology development and improving the economic feasibility of MBR processes in wastewater management	Analyzing factors influencing MBR process and fouling mechanisms Comparing energy consumption of flat-sheet and hollow-fiber MBRs
Lopez-Herguedas et al. (2023)	Membrane bioreactors (MBRs) showed enhanced removal of only 16 micropollutants, with similar performance to conventional activated sludge treatment. Despite their potential, MBRs may not be efficient enough for widespread use, necessitating further research for more effective alternatives	Liquid chromatography tandem high-resolution mass spectrometry (LC-HRMS) Gas chromatography-mass spectrometry (GC–MS)
Khan et al. (2023)	Membrane Bioreactor (MBR) technology enhances wastewater treatment efficiency, achieving 40–79% removal of high organic loads and up to 90% for emerging contaminants. However, traditional MBRs may not effectively remove hazardous micropollutants, necessitating complementary treatment strategies for environmental protection	High-retention membranes instead of porous membranes Hybrid MBR-nanofiltration systems and advanced oxidation processes

(continued)

Table 1.6 (continued)

Reference	Insights	Methods used
Jalilnejad et al. (2020)	Membrane bioreactors (MBRs) offer higher efficiency in solid–liquid separation compared to conventional activated sludge processes, enhancing pollutant degradation. However, membrane fouling remains a significant challenge, impacting performance and costs, necessitating further research on coagulants and adsorbents for mitigation	Membrane fouling reduction through addition of coagulants and adsorbents Identification of future perspectives for MBR applications and fouling mitigation
Pathak et al. (2020)	Membrane bioreactors (MBRs) effectively remove organic micropollutants (OMPs) from wastewater, demonstrating high efficiency and selectivity. Their implementation offers an environmentally friendly solution, addressing stringent legislative requirements while minimizing potential environmental hazards associated with persistent pollutants in municipal and industrial wastewater	Membrane separation and membrane bioreactors (MBRs) are used for OMP removal The paper discusses the performance of various membrane processes and operational parameters
Millanar-Marfa et al. (2020)	Membrane bioreactors (MBRs) effectively remove organic micropollutants (OMPs) from wastewater, producing high-quality effluent with low sludge production. However, challenges like OMP recalcitrance and membrane fouling can limit efficiency, which can be mitigated by integrating MBRs with other processes	Integration of MBR with advanced oxidation, ozonation, adsorption, reverse osmosis, and electrochemical processes Quorum sensing inhibition through quorum quenching for membrane fouling mitigation
Yogarathinam et al. (2024)	Membrane bioreactors (MBR) enhance wastewater treatment efficiency by recovering water from sewage effluent, addressing water scarcity. However, membrane fouling poses significant operational challenges, necessitating innovative solutions to mitigate its impact on performance and environmental sustainability in sewage treatment applications	Types and novel configurations of MBR outlined Fouling mechanisms and mitigation strategies discussed
Millanar-Marfa et al. (2020)	Membrane bioreactors (MBRs) enhance wastewater treatment efficiency by integrating membrane filtration with biological processes, achieving high contaminant removal rates. However, membrane fouling increases operational costs, necessitating innovative solutions like the electro moving bed membrane bioreactor (eMB-MBR) for improved performance	Application of electrochemical processes Utilization of moving bed biofilm reactor (MBBR) technology

(continued)

Table 1.6 (continued)

Reference	Insights	Methods used
Stuckey (2024)	Anaerobic membrane bioreactors (AnMBRs) achieve 85–95% COD removal, produce low sludge yields, and tolerate shock loads. They can be net energy positive, but require further research to optimize performance, reduce energy use, and minimize environmental impact through life cycle analysis	Managed gas sparging, mechanical shaking, addition of flocculants/ adsorbents Quorum quenching to ameliorate fouling in AnMBRs
Li et al. (2022)	Anaerobic membrane bioreactors (AnMBRs) enhance wastewater treatment efficiency by reducing energy consumption and sludge production. However, challenges like membrane fouling and dissolved methane recovery impact their environmental benefits, necessitating further research for large-scale applications and improved performance	The paper provides an overview of AnMBR technology and mechanism The paper discusses improvements in the performance of membrane materials and modules
Zielinska and Ojo (2023)	Anaerobic membrane bioreactors (AnMBRs) efficiently treat wastewater by retaining biomass, producing energy, and recovering nutrients. However, challenges like methane loss and fouling impact their environmental benefits, necessitating further research to optimize performance and minimize greenhouse gas emissions	Rotating membrane and particle sparging for fouling control Recovery of methane, hydrogen, and ethanol for energy production
Millanar-Marfa et al. (2022)	Membrane bioreactors (MBRs) efficiently remove organic matter and nutrients from wastewater, reducing environmental pollutants like nitrogen and phosphorus. The e-ESFDMBR shows lower fouling rates and improved treatment performance, enhancing overall efficiency and minimizing negative environmental impacts	Electrocoagulation processes Self-forming dynamic biomembrane intercalated between polyester fabrics
Kim et al. (2023)	The study highlights that membrane bioreactors (MBRs) effectively treat wastewater while facing fouling challenges. By combining membrane reciprocation and quorum quenching, MBRs achieved significant energy savings and extended service time, enhancing efficiency without negatively impacting microbial characteristics or treatment outcomes	Membrane reciprocation at 30 rpm Quorum quenching with 100 or 200 mg/L BH4

(continued)

Table 1.6 (continued)

Reference	Insights	Methods used
Abdullah et al. (2023)	Membrane bioreactors (MBRs) are highly efficient in wastewater treatment due to their combination of membrane filtration and biological activity, resulting in high effluent quality and reduced space requirements, thus positively impacting environmental sustainability by addressing clean water scarcity	Evaluation of polymers, ceramics, and mixed matrix membranes Discussion of various fillers for membrane applications
Razman et al. (2022)	Membrane bioreactors (MBRs) offer excellent effluent quality and reduced membrane costs, making them efficient for wastewater treatment. Their adoption addresses environmental concerns related to wastewater disposal, enhancing sustainability and compliance with stringent discharge regulations globally	Modification of conventional MBR to hybrid MBR Focus on enhancing performance and reducing technology footprints
Castrogiovanni (2024)	Membrane bioreactors (MBRs) enhance wastewater treatment efficiency by reducing fouling and improving purification performance. Their application, particularly in electrobioreactors (eMBRs), avoids chemical additives, thus minimizing environmental impact while promoting sustainable wastewater reuse and management simplicity	Application of electrochemical processes to MBRs Review of membrane electrobioreactors (eMBRs) for wastewater treatment
Mao et al. (2024)	The study highlights that the electrochemical membrane bioreactor (EMBR) improves wastewater treatment efficiency by reducing membrane fouling and enhancing effluent quality, with increased removal rates of pollutants, thus presenting a sustainable and environmentally friendly option for sewage treatment	Exploration of operation parameters in EMBR system Introduction of a weak electric field for fouling control
Stuckey (2024)	Membrane Bioreactors (MBRs) enhance wastewater treatment efficiency, achieving 98.59% effluent quality and 98.70% nutrient removal. However, challenges like membrane fouling exist. This study presents an integrated control strategy to optimize MBR performance, minimizing environmental impact and computational load	Artificial bee colony optimization (ABC) Recurrent neural network (RNN)

(continued)

Table 1.6 (continued)

Reference	Insights	Methods used
Kim et al. (2024)	Membrane bioreactors (MBRs) enhance wastewater treatment by decoupling hydraulic and solid retention times, improving effluent quality. However, challenges like fouling and energy consumption persist, necessitating integration with desalination technologies to optimize efficiency and minimize environmental impact	Membrane bioreactor (MBR) technology for wastewater treatment Integration of nanofiltration (NF) and reverse osmosis (RO) Use of forward osmosis (FO) and membrane distillation (MD)
Andersson et al. (2024)	Membrane bioreactors (MBRs) enhance wastewater treatment efficiency by meeting strict water quality standards. Innovative cleaning strategies can reduce chemical use by up to 75%, leading to cost reductions of 70% and environmental impact reductions of 95% for certain indicators	Innovative membrane-cleaning strategies explored in pilot trials Alternative cleaning strategies reduce chemical use significantly
Issa et al. (2022)	The study demonstrates that the novel anaerobic fluidized membrane bioreactor (AnFMBR) significantly improves fouling control and energy efficiency in wastewater treatment, achieving 52–94% lower energy requirements and extended operating periods, thus enhancing overall environmental sustainability	Novel anaerobic fluidized membrane bioreactor system tested Granular activated carbon used for fouling control
Kimura (2022)	Membrane bioreactors (MBRs) enhance wastewater treatment efficiency but face challenges from membrane fouling. Control of physically reversible fouling through aeration design is crucial for optimizing performance and minimizing environmental impact, as current methods inadequately address the complexities of multiphase flow	Physically reversible fouling control through intensive aeration Application of advanced research methods for multiphase flow
Unal et al. (2021)	The paper highlights that membrane bioreactors (MBRs) effectively treat wastewater but face challenges like membrane fouling. The electro-cleaning method in the study shows promise in enhancing MBR efficiency and reducing environmental impact by controlling membrane fouling	Electromembrane bioreactor (E-MBR) with platinum-plated titanium anode and activated carbon cloth membrane cathode Electro-cleaning with applied voltage and direct current exposure time

(continued)

Table 1.6 (continued)

Reference	Insights	Methods used
Jolis (2015)	Membrane bioreactors (MBRs), particularly submerged anaerobic MBRs (AnMBRs), enhance wastewater treatment efficiency by reducing energy demand and sludge production while improving effluent quality. AnMBRs demonstrate lower environmental impacts compared to aerobic systems, especially in global warming potential and ecotoxicity	Plant-wide energy model for energy demand assessment Life cycle cost analysis for optimal AnMBR configurations Life Cycle Assessment of AnMBR technology at various temperatures Sensitivity and uncertainty analysis for sustainability evaluation
Nguyen et al. (2020)	Anaerobic Membrane Bioreactors (AnMBR) effectively treat industrial wastewater, addressing conventional treatment limitations while producing biogas for renewable energy. Their application enhances efficiency and mitigates environmental impacts, ensuring compliance with stringent water quality standards to protect human health and ecosystems	Anaerobic Membrane Bioreactor (AnMBR) for wastewater treatment Biogas production from industrial wastewater
Zhao et al. (2021)	Membrane bioreactors (MBRs) exhibit high removal efficiency for organic compounds and pathogens, contributing to reduced environmental impact. They offer advantages like low energy consumption and less chemical use, although challenges like fouling and scaling can limit their broader application	Membrane bioreactor (MBR) technology for wastewater treatment Integrated MBR-membrane systems for secondary purification
Banti et al. (2020)	Membrane bioreactors (MBRs) demonstrate superior environmental performance compared to conventional activated sludge systems, with lower global warming potential (0.002 kg CO_2-eqFU-1) and reduced excess sludge removal impact (0.50 kg CO_2-eqFU-1), enhancing efficiency in municipal wastewater treatment	Life Cycle Assessment (LCA) methodology Comparison of MBR and CAS units for wastewater treatment
Baresel (2017)	Membrane Bioreactor (MBR) technology demonstrates efficient nutrient removal, effective elimination of microplastics and bacteria, and reduced greenhouse gas emissions compared to traditional methods. Additionally, when combined with complementary treatments, MBRs can effectively remove pharmaceutical residues and other micropollutants	Pilot-scale experiments for MBR process evaluation Combined treatment for micropollutant removal efficiency

(continued)

Table 1.6 (continued)

Reference	Insights	Methods used
Ashraf et al. (2024)	Membrane bioreactors (MBRs) enhance wastewater treatment efficiency by integrating biological processes with membrane separation, achieving precise contaminant removal and superior effluent quality. Their compact design contributes to environmental sustainability, making MBRs indispensable in modern municipal wastewater treatment plants	Membrane bioreactors (MBRs) for efficient contaminant removal Electrocoagulation, advanced oxidation processes, constructed wetlands, anaerobic digestion
Frontistis et al. (2023)	Membrane bioreactors (MBRs) demonstrate superior efficiency in wastewater treatment by combining biological and physical processes. Their complexity poses operational challenges, but integrating artificial neural networks can optimize performance and enhance reliability, ultimately improving environmental outcomes in advanced water treatment	Monitoring, simulating, optimizing, and controlling membrane bioreactors Integrating ANNs with IoT devices and new-generation sensors
Hu et al. (2020)	The AnMBR offers a viable alternative for municipal wastewater treatment, focusing on pollutant removal and biogas production. However, challenges like membrane fouling and low bioenergy recovery efficiency must be addressed to enhance its environmental benefits and operational feasibility	Review of AnMBR process and configurations Analysis of membrane fouling mechanisms and control measures
Ni et al. (2024)	Membrane bioreactors (MBRs) are efficient for wastewater treatment, effectively degrading pollutants and inactivating bacteria. However, membrane fouling poses significant challenges, necessitating advanced oxidation processes (AOPs) to enhance performance and mitigate environmental impacts associated with traditional methods	Electrochemical oxidation for membrane foulant removal Photocatalysis, Fenton, and ozonation for mixed-liquor improvement
Banti et al. (2021)	The paper highlights that adjusting operating parameters in membrane bioreactors (MBRs) can effectively control membrane fouling, improving efficiency and reducing transmembrane pressure, thus enhancing overall performance and potentially minimizing environmental impacts in wastewater treatment processes	Adjustment of HRT, F/M loading, and DO concentration Control of filamentous bacteria population for membrane fouling reduction

(continued)

Table 1.6 (continued)

Reference	Insights	Methods used
Kim et al. (2023)	Anaerobic membrane bioreactors (AnMBRs) enhance wastewater treatment efficiency by achieving higher organic loading rates and reducing energy requirements. They also facilitate resource recovery through biogas production, contributing positively to environmental sustainability while addressing challenges like membrane fouling and operational costs	Review of AnMBR principles and advantages Analysis of recent advancements and challenges
Kitanou et al. (2019)	Membrane bioreactors (MBRs) effectively remove organic contaminants and nutrients from wastewater, achieving up to 100% TSS removal and 90% COD reduction. Their compact design and high removal efficiency contribute positively to environmental sustainability in wastewater treatment processes	Ultrafiltration membrane bioreactor (UF-MBR) system Analysis of physical and chemical parameters of wastewater samples
El-Ghorab et al. (2025)	Membrane bioreactors effectively remove emerging contaminants from wastewater, enhancing treatment efficiency. They address various pollutants while minimizing environmental impact, though challenges remain in optimizing operational parameters like solids retention time and transmembrane pressure for improved performance and sustainability	Advanced oxidation processes for contaminant removal Membrane separation, adsorption, and photocatalytic processes discussed

1.10.1 Life Cycle Assessment of AnMBR

The Life Cycle Assessment (LCA) of anaerobic membrane bioreactors (AnMBR) in wastewater treatment emphasizes energy consumption as a crucial factor in enhancing environmental performance (Corominas et al. 2013). The LCA of AnMBRs examines their environmental impacts across four primary stages: raw material extraction and manufacturing, construction, operation, and end-of-life disposal or recycling (Jiménez-Benítez et al. 2023).

1. Raw Material Extraction and Manufacturing

   The manufacture of MBR components, including membranes, pumps, and reactors, involves high energy and material consumption. The environmental effects at this stage are mainly related to resource depletion and greenhouse gas (GHG) emissions. For example, the production of polymeric membranes leads to fossil resource scarcity and global warming potential (Zilling and Lestander 2019).

Table 1.7 Types of wastewater successfully treated by Membrane Bioreactors (MBRs) in India

Type of wastewater	Location (India)	Membrane type	Treatment efficiency (COD/BOD/nutrient removal)	Reference
Municipal wastewater	Delhi, Mumbai, Bengaluru	Hollow fiber, UF and MF	90–98% COD/BOD removal, 85–90% TN removal	Sharma et al. (2021)
Industrial wastewater	Gujarat, Maharashtra	Flat-sheet PVDF membranes	85–95% COD/BOD reduction	Patel and Desai (2020)
Textile wastewater	Tiruppur, Surat	HF-MBR and UF membranes	90% color removal, 80% COD reduction	Ramesh et al. (2019)
Pharmaceutical wastewater	Hyderabad, Pune	Hybrid MBR-NF system	95% COD reduction, > 90% micropollutant removal	Rao et al. (2018)
Food and beverage wastewater	Punjab, Tamil Nadu	HF-MBR and Ceramic membranes	95% BOD reduction, 80% COD removal	Mehta and Verma (2022)
Landfill leachate	Ghazipur (Delhi), Pune	HF and RO Hybrid MBR	85–90% COD reduction, heavy metal removal	Gupta et al. (2020)
Hospital wastewater	AIIMS Delhi, Chennai	PVDF-based UF MBR	> 90% removal of pharmaceuticals and pathogens	Srivastava et al. (2021)
Dairy wastewater	Anand (Gujarat), Punjab	HF-MBR	98% BOD removal, 80% nitrogen removal	Singh et al. (2023)
Pulp and paper wastewater	Uttarakhand, West Bengal	Submerged UF-MBR	90% COD removal, 75% color reduction	Banerjee et al. (2019)
Petrochemical wastewater	Jamnagar, Mumbai	Ceramic UF-MBR	92% COD reduction, 85% hydrocarbon removal	Kumar and Reddy (2022)

2. Construction Phase

   MBR facility construction entails civil works, including the construction of reactors and the establishment of infrastructure. This process may have profound environmental effects, specifically in mineral resource scarcity and water use. The construction process can contribute 51–71% of mineral resource scarcity effects and 18–27% of fossil resource scarcity effects, as indicated by studies (Jiménez-Benítez et al. 2023).

3. Operation Phase

   The operation phase is the most important in terms of energy use and environmental performance. MBRs need constant energy for operations like aeration, pumping, and cleaning of the membrane. Yet, certain configurations, including anaerobic MBRs (AnMBRs), recover energy in the form of biogas production, which renders them net energy producers (Jiménez-Benítez et al. 2023). The capture of dissolved methane and nutrients can also further improve the environmental performance of MBRs (Jiménez-Benítez et al. 2023).
4. End-of-Life Phase

   The end-of-life phase involves the disposal or recycling of MBR components. While recycling can reduce resource depletion impacts, improper disposal of materials like membranes and chemicals can lead to environmental contamination. Currently, there is limited research on the end-of-life management of MBRs, highlighting the need for further studies (Tian et al. 2023).

The assessment covers a number of effect types, including eutrophication potential, energy, resource, and global warming potential (GWP), and it complies with ISO 14040/14044 requirements as outlined below.

1. Global Warming Potential (GWP)

   The GWP of MBRs is influenced by energy consumption and biogas production. Anaerobic MBRs can act as carbon sinks due to their ability to recover methane, resulting in emissions credits of up to 0.223 $kgCO_2eq$ per kg COD removed (Jiménez-Benítez et al. 2023). However, aerobic MBRs may have higher GWP due to energy-intensive aeration processes (Sari et al. 2023).
2. Eutrophication

   Eutrophication impacts are primarily associated with nutrient discharge in the effluent. MBRs generally have high nutrient removal efficiency, but anaerobic configurations may produce nutrient-rich effluent, which can be beneficial for reuse in agriculture (Razman et al. 2022). The inclusion of nutrient recovery systems can further mitigate eutrophication impacts (Jiménez-Benítez et al. 2023).
3. Acidification

   Acidification impacts are linked to energy consumption and the release of nitrogen oxides during operation. Studies have shown that MBRs can have lower acidification impacts compared to conventional activated sludge systems, particularly when renewable energy sources are used (Banti et al. 2020; Zilling and Lestander 2019).
4. Resource Depletion

   Resource depletion impacts are significant during the manufacturing and construction phases. The use of materials like steel and plastics in MBR components contributes to fossil and mineral resource scarcity. However, the recovery of resources such as water, energy, and nutrients can offset these impacts (Jiménez-Benítez et al. 2023).

1.10.2 Life Cycle Assessment of CMBR

The Life Cycle Assessment (LCA) of conventional membrane bioreactors (MBRs) in wastewater treatment highlights their environmental advantages over traditional activated sludge systems (CAS). MBRs demonstrate superior performance in various impact categories, including global warming potential (GWP), eutrophication, and acidification. The assessment reveals that MBRs not only provide high-quality effluent but also reduce overall environmental impacts throughout their life cycle.

1. Environmental Impact Comparison

   - Global Warming Potential: MBRs show a GWP of 0.002 kg CO_2-eq per functional unit (FU), significantly lower than CAS at 0.14 kg CO_2-eq/FU without land area contributions (Banti et al. 2020).
   - Eutrophication and Acidification: MBRs contribute less to eutrophication and acidification due to more efficient nutrient removal processes (Viotti et al. 2024).

2. Energy Consumption

   MBRs are energy-intensive, but their design allows for higher mixed liquor suspended solids (MLSS), leading to reduced hydraulic retention time (HRT) and smaller footprint (Le-Clech 2010). The integration of renewable energy sources, such as biogas recovery, can further mitigate energy demands (Jiménez-Benítez et al. 2023).
3. Operational Efficiency

   MBRs require advanced operational controls and maintenance strategies to manage membrane fouling, which can impact their overall efficiency (Song et al. 2020). The compact design of MBRs eliminates the need for sedimentation tanks, enhancing space efficiency (Le-Clech 2010).

The comparison of AnMBR and CMBR for Life Cycle Assessment in wastewater treatment is shown in Table 1.8.

Table 1.7 provides a comparative analysis of Anaerobic Membrane Bioreactors (AnMBR) and Conventional Membrane Bioreactors (CMBR) across different life cycle stages based on environmental impact, energy consumption, and operational efficiency. AnMBR is more sustainable as a result of energy recovery (biogas generation), less GWP, acidification, and resource depletion, and thus is an eco-friendly option. CMBR, although energy-consuming, is superior in nutrient removal efficiency and hence is a viable option where high-quality effluent is desired. The selection between AnMBR and CMBR would be based on energy availability, treatment objectives, and disposal aspects.

Table 1.8 Comparison of LCA of AnMBR and CMBR in wastewater treatment plant

Category	AnMBR	CMBR	References
Raw material extraction and manufacturing	High energy and material consumption due to membrane, pump, and reactor production	Similar high material use, but with polymeric membranes and steel structures	Zilling and Lestander (2019)
Construction phase	Significant mineral and fossil resource depletion (51–71% mineral scarcity, 18–27% fossil scarcity)	Similar environmental impact but varies with scale and material sourcing	Jiménez-Benítez et al. (2023)
Operation phase	Net energy producer due to biogas recovery; dissolved methane and nutrients can be captured to enhance environmental performance	High energy demand due to aeration, requiring optimization to reduce environmental burden	Jiménez-Benítez et al. (2023)
End-of-life phase	Potential for material recycling, but limited research on disposal impacts	Recycling potential exists but also faces disposal challenges for membranes	Tian et al. (2023)
Global warming potential (GWP)	Lower GWP due to methane recovery (-0.223 kgCO_2eq per kg COD removed)	Higher GWP (0.002 kg CO_2eq per functional unit) but still lower than conventional activated sludge	Banti et al. (2020)
Eutrophication	Nutrient-rich effluent can be beneficial for agricultural reuse	More efficient in nutrient removal, reducing eutrophication impact	Viotti et al. (2024)
Acidification	Lower acidification potential due to energy recovery and less NOx emissions	Also lower than conventional activated sludge, but energy-intensive aeration remains a factor	Viotti et al. (2024)
Resource depletion	Manufacturing and construction contribute significantly to resource scarcity, but energy recovery offsets this impact	High material demand but benefits from efficient water recovery and smaller footprint	Jiménez-Benítez et al. (2023)

(continued)

Table 1.8 (continued)

Category	AnMBR	CMBR	References
Energy consumption	More energy-efficient due to biogas recovery	Energy-intensive, though MLSS allows for reduced hydraulic retention time (HRT)	Le-Clech (2010)
Operational efficiency	Requires careful handling of membrane fouling and gas management	Requires advanced fouling control and maintenance but benefits from compact design	Song et al. (2020)

1.10.3 LCIA of MBR and CAS

The Life Cycle Inventory Analysis (LCIA) by Banti et al. (2020) is a comparison of the environmental effects of Membrane Bioreactor (MBR) and Conventional Activated Sludge (CAS) wastewater treatment processes (Fig. 1.6).

The comparison is made using five environmental impact indicators: Acidification Potential (AP), Eutrophication Potential (EP), Global Warming Potential (GWP), Ozone Depletion Potential (ODP), and Photochemical Ozone Creation Potential (POCP).

1. Acidification Potential (AP): Acidification is a process where emissions like sulfur dioxide (SO_2) and nitrogen oxides (NO_x) react with atmospheric constituents to give rise to acidic compounds, causing acid rain. The AP values

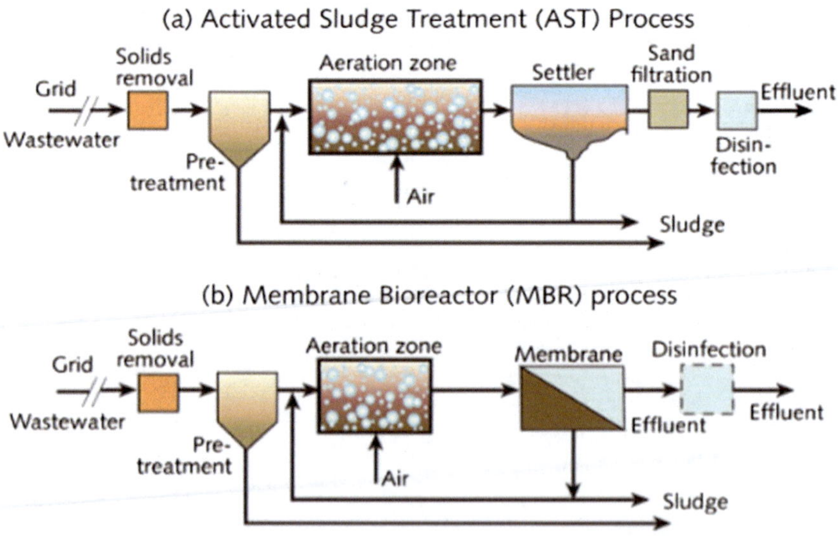

Fig. 1.6 **a** Activated sludge process, **b** membrane bioreactor process. *Source* Pandey and Singh (2014)

of MBR and CAS are 1.98×10^{-3} kg SO_2-eq/FU and 1.15×10^{-2} kg SO_2-eq/FU, respectively. This means that the MBR system has much lower acidification potential compared to CAS, implying lower emissions of acidifying pollutants and less impact on ecosystems and infrastructure.

2. Eutrophication Potential (EP): Eutrophication takes place when a surplus of nutrients like phosphorus and nitrogen is deposited in water bodies, resulting in algal blooms and a reduction in oxygen levels. The EP values of MBR and CAS are 8.60×10^{-4} kg PO_4-eq/FU and 4.77×10^{-3} kg PO_4-eq/FU, respectively. The lower EP of MBR indicates its higher nutrient removal efficiency, hence less risk of water body contamination and disturbance of the ecosystem.
3. Global Warming Potential (GWP): GWP quantifies the greenhouse gas emissions' influence in CO_2 equivalents. GWP for MBR and CAS is 4.96×10^{-1} kg CO_2-eq/FU and 2.68 kg CO_2-eq/FU, respectively. This indicates that MBR is more likely to cause global warming because of higher energy consumption, mainly for aeration, membrane filtration, and cleaning operations. The increased energy requirement results in higher carbon emissions unless renewable resources are used.
4. Ozone Depletion Potential (ODP): ODP measures the contribution of emissions that lead to depletion of the stratospheric ozone layer. ODP for MBR and CAS are 3.13×10^{-8} kg CFC-11-eq/FU and 1.70×10^{-7} kg CFC-11-eq/FU, respectively. The very low ODP value for MBR reflects its low contribution toward the depletion of the ozone layer, which can be attributed to the low emission rates of ozone-depleting substances.
5. Photochemical Ozone Creation Potential (POCP): POCP is the potential of emissions to form ground-level ozone (smog). POCP of MBR and CAS are 1.50×10^{-4} kg C_2H_4-eq/FU and 8.40×10^{-4} kg C_2H_4-eq/FU, respectively. The lower POCP of MBR signifies lower VOCs and nitrogen oxides emissions, resulting in lower smog formation and improved air quality.

The Life Cycle Assessment (LCA) outcomes show that the Membrane Bioreactor (MBR) system possesses numerous environmental benefits over the Conventional Activated Sludge (CAS) system. The MBR possesses lower acidification potential, eutrophication potential, ozone depletion potential, and photochemical ozone creation potential and is an environmentally superior option in the majority of impact categories. Nevertheless, the MBR system has higher global warming potential because of its greater energy consumption, resulting in higher carbon emissions.

For enhanced sustainability of MBR systems, optimization of energy consumption, renewable energy integration, and optimization of membrane efficiency to minimize operating effects can be undertaken. Through optimization of its energy requirements, MBR technology can be more sustainable and eco-friendly in wastewater treatment.

1.11 Future Prospects and Optimization Strategies

To further improve the sustainability of MBR technology and obtain better water quality than traditional practices, additional research and technological development are required. The areas that need to be improved are:

- Creation of next-generation membrane materials with greater permeability, lower fouling rate, and longer lifespan.
- Improved aeration methods to save energy while ensuring process efficiency.
- Process optimization techniques, such as real-time monitoring, automation, and control systems based on artificial intelligence (AI), to enhance cost-effectiveness and process stability.
- Policy incentives and subsidies to encourage the use of MBR technology, especially in areas with high water quality standards.

Through overcoming these challenges, MBR technology has the potential to be a more sustainable, cost-saving, and more popular solution to wastewater treatment and water reclamation.

1.12 Conclusion

Membrane Bioreactors (MBRs) are a very efficient technology for wastewater treatment, combining membrane filtration with biological treatment. Comparative Life Cycle Assessment (LCA) and performance analyses of various MBR configurations—Anaerobic MBR (AnMBR) and Conventional MBR (CMBR)—emphasize their respective strengths and weaknesses regarding energy efficiency, environmental footprint, and treatment performance.

Salient Findings:

Higher Effluent Quality: MBRs show improved contaminant removal compared to traditional CAS systems, generating effluent of good quality for reuse. CMBRs show exceptional nutrient removal capability, mitigating eutrophication effects, whereas AnMBRs produce nutrient-rich effluent that is good for agricultural reuse.

Energy and Resource Efficiency: AnMBR is energy efficient and has the potential to become a net energy producer based on biogas recovery through anaerobic digestion. CMBR is high in energy consumption for aeration, which alone contributes greatly toward its environmental footprint.

Environmental Impact and Sustainability: AnMBR has lower Global Warming Potential (GWP) due to methane recovery, minimizing the greenhouse gas emissions. Both the systems have less acidification and ozone depletion potential than CAS, but CMBR is still more energy-intensive. End-of-life disposal is challenging, especially in terms of membrane waste management and recycling of material.

Operational Considerations: Membrane fouling is a persistent problem in both AnMBR and CMBR and needs measures for fouling control as well as maintenance. Small footprint and low hydraulic retention time (HRT) render MBRs ideal for space-limited applications.

Life Cycle Assessment (LCA) Insights: AnMBR reduces resource depletion from biogas production but has significant environmental impacts during construction and raw material extraction. CMBR requires high material consumption but has efficient water recovery and footprint minimization.

1.13 Summary

- AnMBR is preferred where energy recovery and sustainability are high priorities, and thus it is best suited for decentralized wastewater treatment and where biogas use is possible.
- CMBR is more suitable for high-quality effluent applications with stringent discharge requirements, even though it has higher operational energy requirements.
- Advances in membrane technology, fouling prevention, and energy efficiency in the future will establish whether MBRs are going to be long-term mainstream wastewater treatment.
- Thus, MBRs represent a promising alternative to conventional treatment systems, with AnMBR leading in sustainability and CMBR excelling in effluent quality. The choice between the two depends on site-specific conditions, regulatory requirements, and operational cost considerations.

References

Abdullah R, Astira D, Widiyanto AR, Dharma HNC, Hidayat A, Santoso L, Sulistiono DO, Rahmawati Z, Gunawan T, Jaafar J, Kusumawati Y, Othman MHD, Fansuri H (2023) Recent development of mixed matrix membrane as a membrane bioreactor for wastewater treatment: a review. Case Stud Chem Environ Eng. https://doi.org/10.1016/j.cscee.2023.100485

Alevizos V, Georgousis I, Kapodistria A (2023) Toward climate neutrality: a comprehensive overview of sustainable operations management, optimization, and wastewater treatment methods. Pollutants 3(4):521–543. https://doi.org/10.3390/pollutants3040033

Andersson SL, Baresel C, Andersson S, Westling K, Eriksson M, Carranza Munoz A, Persson G, Narongin-Fujikawa M, Johansson K, Rydberg T (2024) Chemical-saving potential for membrane bioreactor (MBR) processes based on long-term pilot trials. Membranes 14(6):126. https://doi.org/10.3390/membranes14060126

Arias A, Vallina I, Lorenzo Y, Komesli OT, Katsou E, Feijoo G, Moreira MT (2019) Water footprint of a decentralized wastewater treatment strategy based on membrane technology. In: Environmental water footprints: agricultural and consumer products. Elsevier, pp 85–119. https://doi.org/10.1016/B978-0-12-819852-0.00007-5

Arun J, SundarRajan P, Pavithra KG, Priyadharsini P, Shyam S, Goutham R, Le QH, Pugazhendhi A (2024) New insights into microbial electrolysis cells (MEC) and microbial fuel cells (MFC) for simultaneous wastewater treatment and green fuel (hydrogen) generation. Fuel 355:129530. https://doi.org/10.1016/j.fuel.2023.129530

Ashraf MV, Rather SA, Khan MAH, Ahmad S (2024) Innovative treatment technologies for municipal wastewater, pp 95–128. https://doi.org/10.1201/9781003454199-5

Asif MB, Ansari AJ, Chen SS, Nghiem LD, Price WE, Hai FI (2019) Understanding the mechanisms of trace organic contaminant removal by high retention membrane bioreactors: a critical review. Environ Sci Pollut Res 26:34085–34100. https://doi.org/10.1007/s11356-019-06619-5

Asif MB, Hai FI (2019) Wastewater reuse applications and MBR. Membr Biol React 183–205. https://doi.org/10.1016/B978-0-12-816809-7.00007-6

Augsburger N, Zaouri N, Cheng H, Hong PY (2021) The use of UV/H_2O_2 to facilitate removal of emerging contaminants in anaerobic membrane bioreactor effluents. Environ Res 198:110479. https://doi.org/10.1016/j.envres.2021.110479

Bagheri M, Mirbagheri SA (2018) Critical review of fouling mitigation strategies in membrane bioreactors treating water and wastewater. Biores Technol 258:318–334. https://doi.org/10.1016/j.biortech.2018.03.040

Banerjee S, Das T, Mukherjee A (2019) Treatment of pulp and paper mill effluent using submerged UF-MBR: a case study from West Bengal. Chem Eng J 368:456–468

Banti DC, Tsangas M, Samaras P, Zorpas AA (2020) LCA of a membrane bioreactor compared to activated sludge system for municipal wastewater treatment. Membranes 10(12):421. https://doi.org/10.3390/MEMBRANES10120421

Banti DC, Mitrakas M, Samaras P (2021) Membrane fouling controlled by adjustment of biological treatment parameters in step-aerating MBR. Membranes 11(8):553. https://doi.org/10.3390/MEMBRANES11080553

Baresel C (2017) Membrane bioreactor processes to meet today's and future municipal sewage treatment requirements. J Environ Sci 3(2). https://doi.org/10.16966/2381-5299.140

Bera SP, Godhaniya M, Kothari C (2022) Emerging and advanced membrane technology for wastewater treatment: a review. J Basic Microbiol 62(3–4):245–259. https://doi.org/10.1002/jobm.202100610

Bhattacharyya A, Liu L, Lee K, Miao J (2022) Review of biological processes in a membrane bioreactor (MBR): effects of wastewater characteristics and operational parameters on biodegradation efficiency when treating industrial oily wastewater. J Mar Sci Eng 10(9):1229. https://doi.org/10.3390/jmse10091229

Biswas A, Mailapalli DR, Raghuwanshi NS (2021) Treated municipal wastewater to fulfill crop water footprints and irrigation demand—a review. Water Supply 21(4):1398–1409. https://doi.org/10.2166/ws.2021.003

Capodaglio AG (2021) Fit-for-purpose urban wastewater reuse: analysis of issues and available technologies for sustainable multiple barrier approaches. Crit Rev Environ Sci Technol 51(15):1619–1666. https://doi.org/10.1080/10643389.2020.1813272

Castrogiovanni F (2024) Application of electrochemical processes to membrane bioreactors for sustainable wastewater treatment: a short review. Proceedings. https://doi.org/10.3390/proceedings2024105046

Chen Z, Wang D, Sun M, Ngo HH, Guo W, Wu G, Jia W, Shi L, Wu Q, Guo F et al (2018) Sustainability evaluation and implication of a large-scale membrane bioreactor plant. Biores Technol 269:246–254. https://doi.org/10.1016/j.biortech.2018.07.112

Chen M, Li Y, Sun X, Dai R, Zheng J, Wang X, Wang Z (2023) Recent advances in electrochemical processes integrated with anaerobic membrane bioreactor in wastewater treatment. Chem Eng J 468:143822. https://doi.org/10.1016/j.cej.2023.143822

Corominas L, Foley J, Guest JS, Hospido A, Larsen HF, Morera S, Shaw A (2013) Life cycle assessment applied to wastewater treatment: state of the art. Water Res 47(15):5480–5492

Drews A (2010) Membrane fouling in membrane bioreactors—characterisation, contradictions, cause, and cures. J Membr Sci 363(1–2):1–28. https://doi.org/10.1016/j.memsci.2010.06.046

El-Ghorab HM, Abdel Azeem MM, Badawy MI, Gad-Allah TA, Abdel-Wahed MS, Momen M (2025) Navigating the membrane bioreactors operational parameters in eliminating emerging contaminants from wastewater: a review. Egypt J Chem 68(9):183–204. https://doi.org/10.21608/ejchem.2025.332819.10710

Ensano BM, Borea L, Naddeo V, Belgiorno V, De Luna MD, Ballesteros FC Jr (2016) Combination of electrochemical processes with membrane bioreactors for wastewater treatment and fouling control: a review. Front Environ Sci 4:57. https://doi.org/10.3389/fenvs.2016.00057

Erkan HS, Turan NB, Engin GO (2020) Fouling control in MBR in a sustainable perspective. In: Current developments in biotechnology and bioengineering—advanced membrane separation processes for sustainable water and wastewater management—case studies and sustainability analysis. Elsevier, pp 21–57

Firozjaee TT, Kootenaei FG, Hasanlou H, Abdi J (2024) Water recycling, reuse, and sustainable development goals. In: Water, the environment, and the sustainable development goals. Elsevier, pp 107–125. https://doi.org/10.1016/B978-0-12-819852-0.00010-5

Frontistis Z, Lykogiannis G, Sarmpanis A (2023) Artificial neural networks in membrane bioreactors: a comprehensive review—overcoming challenges and future perspectives. https://doi.org/10.3390/sci5030031

Galinha CF, Sanches S, Crespo JG (2018) Membrane bioreactors. In: Fundamental modelling of membrane systems. Elsevier, pp 209–249

Gao T, Xiao K, Zhang J, Zhang X, Wang X, Liang S, Sun J, Meng F, Huang X (2021) Cost-benefit analysis and technical efficiency evaluation of full-scale membrane bioreactors for wastewater treatment using economic approaches. J Clean Prod 301:126984. https://doi.org/10.1016/j.jclepro.2021.126984

Glover CJ, Phillips JA, Marchand E, Hiibel SR (2022) Modeling and life cycle assessment of a membrane bioreactor–membrane distillation wastewater treatment system for potable reuse. Separations 9(6):151. https://doi.org/10.3390/separations9060151

Gupta R, Sharma A, Prakash S (2020) Treatment of landfill leachate using HF and RO hybrid membrane bioreactor in Delhi. Waste Manage 102:293–301

Hai FI, Yamamoto K, Lee CH (eds) (2018) Membrane biological reactors: theory, modeling, design, management and applications to wastewater reuse. IWA Publishing

Hu Y, Hu Y, Cheng H, Ji J, Li YY (2020) A review of anaerobic membrane bioreactors for municipal wastewater treatment with a focus on multicomponent biogas and membrane fouling control. Environ Sci Water Res Technol 6(10):2641–2663. https://doi.org/10.1039/D0EW00528B

Issa L, El Kik O, Katuri KP, Saikaly PE, Alameddine I, Abou El-Fadel MH (2022) A novel anaerobic fluidized membrane bioreactor system: improving process performance and fouling control. Environ Technol Innov 28:102821. https://doi.org/10.1016/j.eti.2022.102821

Jalilnejad E, Sadeghpour P, Ghasemzadeh K (2020) Advances in membrane bioreactor technology. Elsevier, pp 1–29. https://doi.org/10.1016/B978-0-12-816822-6.00001-X

Jijingi HE, Yazdia SK, Abakr YA, Etim E (2024) Evaluation of membrane bioreactor (MBR) technology for industrial wastewater treatment and its application in developing countries: a review. Case Stud Chem Environ Eng 100886. https://doi.org/10.1016/j.cscee.2024.100886

Jiménez-Benítez A, Vázquez JR, Seco A, Serralta J, Rogalla F, Robles A (2023) Life cycle assessment of AnMBR technology for urban wastewater treatment: a case study based on a demo-scale AnMBR system. J Environ Chem Eng 11(6):111141

Jolis RP (2015) Environmental and economic sustainability of submerged anaerobic membrane bioreactors treating urban wastewater. https://doi.org/10.4995/THESIS/10251/58864

Judd S (2004) A review of fouling of membrane bioreactors in sewage treatment. Water Sci Technol 49(2):229–235. https://doi.org/10.2166/wst.2004.0121

Judd S (2017) The MBR book: principles and applications of membrane bioreactors for water and wastewater treatment. Elsevier

Khan NA, Singh S, López-Maldonado EA, Pavithra NS, Méndez-Herrera P, López-López JR, Baig U, Ramamurthy PC, Mujawar Mubarak N, Karri RR, Aljundi IH (2023) Emerging membrane

technology and hybrid treatment systems for the removal of micropollutants from wastewater. Desalination. https://doi.org/10.1016/j.desal.2023.116873
Kim J, Bae E, Park H, Park H-J, Shah SSA, Lee K, Lee J, Oh H-S, Park P-K, Moon H, Naddeo V, Choo K-H (2023) Membrane reciprocation and quorum quenching: an innovative combination for fouling control and energy saving in membrane bioreactors. Water Res. https://doi.org/10.1016/j.watres.2023.121035
Kim J, Wu B, Jeong S, Jeong S, Kim M (2024) Recent advances of membrane-based hybrid membrane bioreactors for wastewater reclamation. Front Membr Sci Technol. https://doi.org/10.3389/frmst.2024.1361433
Kimura K (2022) Control of membrane fouling occurring in membrane bioreactors (MBRs). Jpn J Multiph Flow 36(2):180–186. https://doi.org/10.3811/jjmf.2022.t006
Kitanou S, Elmidaoui A, Taky M, Hafida A, Soukaina B, Mahi M, Taky M, Elmidaoui A (2019) Performance of an ultrafiltration membrane bioreactor (UF-MBR) in wastewater treatment. Desalin Water Treat 157:393–398. https://doi.org/10.5004/DWT.2019.24258
Kösem E, Ersöz ÖN, Yılmaz RN, Babaoğlu D, Guvenc SY, Can-Güven E, Varank G (2024) Sequential electrocoagulation and Fenton/photo-Fenton processes for the treatment of membrane bioreactor effluent of landfill leachate. Colloids Surf, A 689:133740. https://doi.org/10.1016/j.colsurfa.2024.133740
Krzeminski P, van der Graaf JHJM, van Lier JB (2017) Membrane bioreactors for municipal wastewater treatment—a review. Water Res 45(4):1489–1509. https://doi.org/10.1016/j.watres.2017.01.005
Kumar A, Reddy V (2022) Petrochemical wastewater treatment using ceramic UF-MBR: a case study from Jamnagar, India. Sep Purif Technol 285:120430
Le-Clech P (2010) Membrane bioreactors and their uses in wastewater treatments. Appl Microbiol Biotechnol 88(6):1253–1260. https://doi.org/10.1007/s00253-010-2856-3
Li J, Feng J, Jiang Z, Li J, He C, Luo Y, Wei L (2022) Anaerobic membrane bioreactors for wastewater treatment: mechanisms, fouling control, novel configurations, and future perspectives. Environ Eng Res 28(1):210575. https://doi.org/10.4491/eer.2021.575
Lopez-Herguedas N, Irazola M, Alvarez-Mora I, Mijangos L, Bilbao D, Etxebarria N, Zuloaga O, Olivares M, Prieto A (2023) Evaluating membrane bioreactor treatment for the elimination of emerging contaminants using different analytical methods. J Hazard Mater. https://doi.org/10.1016/j.jhazmat.2023.132833
Mao Z, Liu H, Niu B, Bhagat WA, Fan W, Liang D, Yang L, Zhao Q, Meng S (2024) Mitigation of fouling problem and optimization of treatment effect in the polyvinylidene fluoride (PVDF)-based electrochemical membrane bioreactor (EMBR). Sep Purif Technol. https://doi.org/10.1016/j.seppur.2024.126340
Mehta S, Verma R (2022) Sustainable treatment of food and beverage wastewater using membrane bioreactors. Water Sci Technol 85(6):1325–1338
Meng F, Zhang H, Yang F, Zhang S (2017) Strategies for membrane fouling control in MBR systems: a review. Water Res 43(6):1489–1509. https://doi.org/10.1016/j.watres.2017.02.005
Millanar-Marfa JMJ, Corpuz MVA, Borea L, Cabreros C, De Luna MDG, Ballesteros F, Vigliotta G, Zarra T, Hasan SW, Korshin GV, Buonerba A, Belgiorno V, Naddeo V (2022) Advanced wastewater treatment and membrane fouling control by electro-encapsulated self-forming dynamic membrane bioreactor. NPJ Clean Water 5(1):1–13. https://doi.org/10.1038/s41545-022-00184-z
Millanar-Marfa JMJ, Borea L, Hasan SW, de Luna MDG, Belgiorno V, Naddeo V (2020) Advanced membrane bioreactors for emerging contaminant removal and quorum sensing control. Elsevier 117–147. https://doi.org/10.1016/B978-0-12-819854-4.00006-X
Namdeti R, Lakkimsetty NR, Rao GB, Thandlam AK, Karu CV, Doddamani D, Al Kathiri DS, Prema PM (2025) Membrane bioreactors: integration of biological processes with membrane technology for wastewater treatment. J Membr Sci Res 11(2). https://doi.org/10.22079/jmsr.2025.12091.1097
Nguyen ML, Nakhjiri AT, Kamal M, Mohamed A, Algarni M, Yu ST, Wang FM, Su CH (2022) State-of-the-art review on the application of membrane bioreactors for molecular micro-contaminant

removal from the aquatic environment. Membranes 12(4):429. https://doi.org/10.3390/membranes12040429
Nguyen T-B, Bui X-T, Vo T-D-H, Cao N-D-T, Dang B-T, Tra V-T, Tran H-T, Tran L-L, Ngo HH (2020) Anaerobic membrane bioreactors for industrial wastewater treatment. Elsevier, pp 167–196. https://doi.org/10.1016/B978-0-12-819852-0.00007-5
Ni L, Wang P, Westerhoff P, Luo J, Wang K, Wang Y (2024) Mechanisms and strategies of advanced oxidation processes for membrane fouling control in MBRs: membrane-foulant removal versus mixed-liquor improvement. Environ Sci Technol. https://doi.org/10.1021/acs.est.4c02659
Osman AI, Chen Z, Elgarahy AM, Farghali M, Mohamed IM, Priya AK, Hawash HB, Yap PS (2024) Membrane technology for energy saving: principles, techniques, applications, challenges, and prospects. Adv Energy Sustain Res 5(5):2400011. https://doi.org/10.1002/aesr.202400011
Pacheco MJ, Vences LJ, Moreno H, Pacheco JO, Valdivia R, Hernández C (2021) Mixed-matrix membranes with CNT for CO_2 separation processes. Membranes 11(6):457. https://doi.org/10.3390/membranes11060457
Pandey A, Singh RK (2014) Industrial wastewater treatment by membrane bioreactor system. Elixir Chem Eng 70:23772–23777
Patel K, Desai A (2020) Advancements in industrial wastewater treatment using membrane bioreactors: a case study from Gujarat. J Water Process Eng 36:101572
Pathak N, Tran VH, Phuntsho S, Shon HY (2020) Membrane bioreactors for the removal of micropollutants. Elsevier, pp 231–252. https://doi.org/10.1016/B978-0-12-819594-9.00010-3
Pervez MN, Balakrishnan M, Hasan SW, Choo KH, Zhao Y, Cai Y, Zarra T, Belgiorno V, Naddeo V (2020) A critical review on nanomaterials membrane bioreactor (NMs-MBR) for wastewater treatment. NPJ Clean Water 3(1):43. https://doi.org/10.1038/s41545-020-00085-3
Quan S, Li SW, Xiao YC, Shao L (2017) CO_2-selective mixed matrix membranes (MMMs) containing graphene oxide (GO) for enhancing sustainable CO_2 capture. Int J Greenhouse Gas Control 56:22–29. https://doi.org/10.1016/j.ijggc.2017.10.001
Racar M, Dolar D, Karadakić K, Čavarović N, Glumac N, Ašperger D, Košutić K (2020) Challenges of municipal wastewater reclamation for irrigation by MBR and NF/RO: physico-chemical and microbiological parameters, and emerging contaminants. Sci Total Environ 722:137959. https://doi.org/10.1016/j.scitotenv.2020.137959
Rahman TU, Roy H, Islam MR, Tahmid M, Fariha A, Mazumder A, Tasnim N, Pervez MN, Cai Y, Naddeo V, Islam MS (2023) The advancement in membrane bioreactor (MBR) technology toward sustainable industrial wastewater management. Membranes 13(2):181. https://doi.org/10.3390/membranes13020181
Ramesh P, Kumar S, Iyer V (2019) Treatment of textile wastewater using HF-MBR and UF membrane technology: case study from Tiruppur. J Membr Sci 580:298–306
Rao D, Choudhary R, Rajan V (2018) Removal of pharmaceutical micropollutants from wastewater using hybrid MBR-NF system. Desalin Water Treat 129:220–230
Razman KK, Hanafiah MM, Mohammad AW (2022) An overview of LCA applied to various membrane technologies: progress, challenges, and harmonization. Environ Technol Innov 27:102803. https://doi.org/10.1016/j.eti.2022.102803
Ryu T-Y, Won J, Jung H, Im H (2024) A critical review on advanced membrane bioreactors for wastewater treatment: fouling reduction and energy demand. J Korean Soc Environ Eng. https://doi.org/10.4491/ksee.2024.46.10.629
Samin G, Pavlova M, Arif MI, Postema CP, Damborsky J, Janssen DB (2014) A Pseudomonas putida strain genetically engineered for 1,2,3-trichloropropane bioremediation. Appl Environ Microbiol 80(17):5467–5476. https://doi.org/10.1128/AEM.01620-14
Sari DAP, Rahmah A, Sasongko NA (2023) Life cycle assessment (LCA) in palm oil plantation and mill with impact categories global warming potential, acidification, and eutrophication. IJMST 10(2):797–807
Sharma R, Singh N, Gupta P (2021) Membrane bioreactor applications for municipal wastewater treatment in India: performance and challenges. Environ Technol Innov 22:101482

Singh K, Raj D, Patel P (2023) Application of HF-MBR for dairy wastewater treatment in India: a performance analysis. J Environ Chem Eng 11(1):108320

Song W, Xie B, Huang S, Zhao F, Shi X (2020) Aerobic membrane bioreactors for industrial wastewater treatment. Curr Dev Biotechnol Bioeng 129–145. https://doi.org/10.1016/B978-0-12-819594-9.00010-3

Srivastava P, Jain M, Singh H (2021) PVDF-based UF membranes for hospital wastewater treatment: a study on pharmaceutical and pathogen removal. J Water Reuse Desalination 11(2):144–157

Stuckey DC (2024) Anaerobic membrane bioreactors. UWA Publishing, pp 51–92. https://doi.org/10.2166/9781789063479_0051

Tian C, Chen J, Bai Z, Wang X, Dai R, Wang Z (2023) Recycling of end-of-life polymeric membranes for water treatment: closing the loop. J Membr Sci Lett 3(2):100063. https://doi.org/10.1016/j.memsci.2023.100063

Unal BO, Dizge N, Keskinler B, Karagunduz A (2021) Can electro-cleaning solve the problem of membrane biofouling for membrane bioreactor? Desalin Water Treat 211:382–390. https://doi.org/10.5004/DWT.2021.26364

Viotti P, Tatti F, Bongirolami S, Romano R, Mancini G, Serini F, Azizi M, Croce L (2024) Life cycle assessment methodology applied to a wastewater treatment plant. Water 16(8):1177. https://doi.org/10.3390/water16081177

Visvanathan C, Aim RB, Parameshwaran K (2000) Membrane separation bioreactors for wastewater treatment. Crit Rev Environ Sci Technol 30(1):1–48. https://doi.org/10.1080/10643380091184177

Yogarathinam LT, Ismail AF, Goh PS (2024) Membrane bioreactor for sewage treatment. Springer, pp 217–226. https://doi.org/10.1007/978-3-031-48228-1_14

Yuvaraju M, Deena D (2024) Enhanced performance of membrane bioreactor in sewage treatment through integrated control strategy. Global NEST J 26(8). https://doi.org/10.30955/GNJ.06204

Zambare RS, Nemade PR (2021) Polymer nanocomposite membranes for wastewater treatment. In: Handbook of nanomaterials for wastewater treatment. Elsevier, pp 605–672. https://doi.org/10.1016/B978-0-12-821100-7.00019-3

Zhao Y, Qiu Y, Mamrol N, Ren L, Li X, Shao J, Yang X, Van der Bruggen B (2021) Membrane bioreactors for hospital wastewater treatment: recent advancements in membranes and processes. Front Chem Eng China 1–27.https://doi.org/10.1007/S11705-021-2107-1

Zielinska M, Ojo A (2023) Anaerobic membrane bioreactors (AnMBRs) for wastewater treatment: recovery of nutrients and energy, and management of fouling. Energies 16(6):2829. https://doi.org/10.3390/en16062829

Zilling I, Lestander K (2019) Environmental impact of wastewater treatment—a study of membrane bioreactor and Iso-Disc technologies from a life cycle perspective

Chapter 2
Biological Aspects of Wastewater Treatment

Ashok Kumar Joshi Puja, Irudhayaraj Peatrise Geofferina, and Shobana Sugumar

Abstract Microorganisms play a crucial part in the treatment of wastewater, aiding in its purification and treatment to reduce environmental harm. Microorganisms seem as inconspicuous laborers, conscientiously organizing the transformation of polluted water into an asset fit for recycling. In order for microbes to perform their functions, the proper conditions for pH, temperature, oxygen levels, and nutrients must be met. By employing biological wastewater treatment, suspended particles, oils, grease, and fats can be removed from wastewater. Moreover, this treatment aids in the reduction of biological and chemical oxygen demands (BOD and COD) that can be utilized in other industries for commercial purposes. Many advantages have been reported by the incorporation of microorganisms into wastewater treatment plants, including highly effective processes and environmental sustainability. Integrating microorganisms in wastewater treatment for future developments plants is promising, varying from operating microbial fuel cells to generate power to designing microbial sensors for real-time monitoring. This chapter deals with biological aspects of microbes involved in the treatment of wastewater.

Keywords Biochemical oxygen demand · Bioremediation · Filtration techniques · Microorganisms · Photobioreactors · Wastewater Treatment

Abbreviations

WWT	Waste water treatment
WWTP'S	Waste water treatment plants

A. K. J. Puja · I. P. Geofferina
Department of Biotechnology, College of Engineering and Technology, SRM Institute of Science and Technology, Kattankulathur, Tamil Nadu, India

S. Sugumar (✉)
Department of Genetic Engineering, College of Engineering and Technology, SRM Institute of Science and Technology, Kattankulathur, Chennai, India
e-mail: shobanas@srmist.edu.in

A. Mandpe et al. (eds.), *Membrane Filtration and Membrane Bioreactors for Wastewater Treatment*, Advances in Wastewater Research,
https://doi.org/10.1007/978-981-95-6706-5_2

COD	Chemical oxygen demand
TOC	Total organic carbon
DO	Dissolved oxygen
BOD	Biological oxygen demand
AD	Anaerobic digestion
MP	Micro-plastics
HRT	Hydraulic retention time
SRT	Solid retention time
RAS	Return activated sludge
MLR	Mixed liquor recirculation
OS	Reverse osmosis
FO	Forward osmosis
MF	Microfiltration
UF	Ultrafiltration
NF	Nanofiltration
PBR'S	Photobioreactors
EPS	Extracellular polymeric substances
MBBR	Moving bed biofilm reactor

2.1 Introduction

In the last few decades, the quantity of water consumed worldwide has increased two-fold. Water is promptly running out due to its ample use in agriculture, industry, transportation, and residential purposes, as well as the consequences of climate change (Ahmed et al. 2021). Industrial advancement, sophisticated lifestyles, and agricultural expansion driven by growing demands are the main causes of the increased water contamination. Industries are constantly releasing polluted wastewater into water bodies. Both organic and inorganic materials, including heavy metals such as copper and cadmium, are present in the effluents. Agricultural wastes, electronic debris, and runoff water from storms elevate the level of pollutants in water (Kumar and Shah 2021).

Wastewater treatment's foremost aims are to: (i) remove organic materials that decompose naturally; (ii) remove different nutrients, like phosphates; (iii) destroy pathogens; and (iv) prevent water pollution to protect aquatic life. The main drawbacks of wastewater treatment, however, are sludge treatment and disposal, low efficiency, emerging pollutants, maintenance and monitoring, and low efficiency. These factors also raise the overall cost of wastewater treatment. To clean up contaminated water, conventional remediation techniques (chemical, physical, and biological) can be performed off- or on-site. In terms of eliminating harmful substances and nutrients, biosorption using microbes, bacteria (microbe-mediated remediation), fungi (mycoremediation), and algae (phytoremediation) has demonstrated encouraging outcomes (Koul et al. 2022). Several approaches are used in conventional wastewater treatment to eliminate the desired contaminants. However, because of their high energy requirements and associated costs, these stated technologies are not entirely appropriate for eliminating wastewater pollutants. Traditional wastewater treatment techniques need an extensive area of land and can't remove all contaminants. The biological techniques commonly used in wastewater treatment can be implemented in this situation to lower treatment costs and improve removal efficiency. Algae, fungi, and bacteria are frequently employed in biological treatment to eliminate contaminants (Bhatt et al. 2022a, b).

The most significant way of treating wastewater is biological digestion, which utilizes complex microbial communities to convert pollutants into low or harmless substances. The efficiency of the treatment is mostly influenced by the microbial community. The two categories of biological mechanisms used in wastewater treatment are the attached growth process and the suspended growth process (Karia et al. 2023). When wastewater is treated biologically, microbes use the pollutants in the wastewater as food and break down the organic substrate to produce simpler molecules like CO_2 and water. Various microorganisms release distinct types of enzymes, and these enzymes can break down chemical compounds that are harmful to the environment. Oxidoreductases are enzymes that are mostly used in environmental remediation processes to eliminate dangerous metals and phenolic chemicals (Ali et al. 2023).

The release of lipids from industrial processes leads to the formation of an oil-like film on the surface of water, obstructing the entry of sunlight and oxygen and negatively impacting the aquatic environment. Additionally, droplets of oil congregate with suspended materials in the wastewater, clog sewer pipes, and disrupt the flow of wastewater and the treatment process. Lipids in wastewater not only endanger the environment but also carry a risk of cancer to humans. Lipase and other hydrolytic enzymes are formed by various bacteria and fungi and are useful in wastewater treatment (Nimkande and Bafana 2022).

An all-biological method for lowering environmental pollution is microbial bioremediation, which uses both aerobic and anaerobic microbial species. When it comes to environmental benefits, cost-effectiveness, innovation, and promise, the use of microbial enzymes to refine persistent organic pollutants (POPs) is found to be far better than other bioremediation technologies. Artificial intelligence (AI) techniques have become increasingly popular in the recent past for the thorough examination of incredibly complicated data and the presentation of the data in an easily comprehensible style which is to investigate secondary effects of effluents from readily measurable soil parameters (Haripriyan et al. 2022).

Biological wastewater treatment approaches are used to eliminate toxins from various industrial wastewater to safeguard the long-term sustainability of the ecosystem and the well-being of living creatures (Sharma et al. 2023). This book concentrates on illustrating how functionally distinct microbes can be employed in wastewater treatment to accomplish environmental sustainability and economic feasibility (Fig. 2.1).

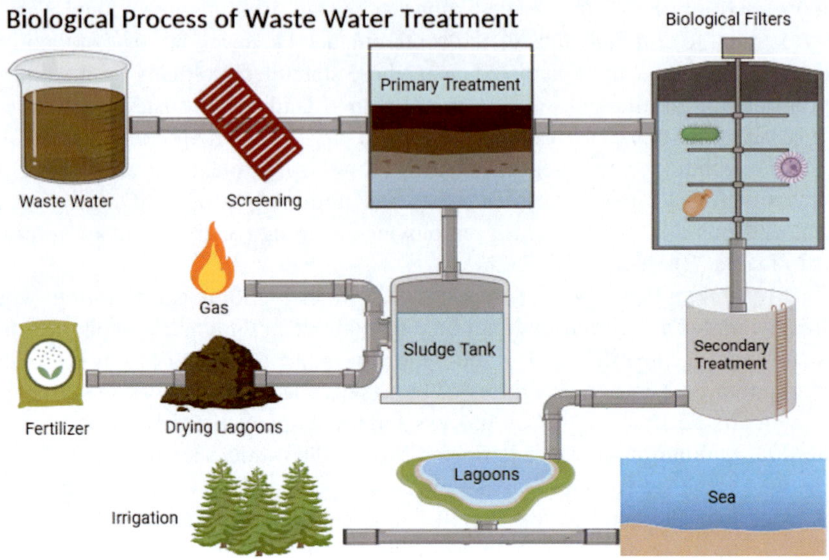

Fig. 2.1 Biological process of wastewater treatment

2.2 Characteristics of Wastewater Treatment

2.2.1 Physical Characteristics of Wastewater

The physical attributes of wastewater include turbidity, color, odor, total solids, and temperature.

- **Temperature**: The processed effluents from several industries in WW have high temperatures. Its discharge into various bodies of water endangers the aquatic life of living things.
- **Odor and Color**: wastewater has a foul smell and a soiled, dark look. Wastewater can be made odor-free utilizing a variety of methods, such as the adsorption process with activated carbon.
- **Turbidity**: The presence of suspended particles, clay, and other organic and inorganic pollutants causes wastewater to have a very high turbidity. Wastewater treatment lowers the turbidity to a tolerable level.
- **Total Dissolved Solids**: Toxic compounds that pose harm to aquatic life can be found in wastewater as part of certain solid components or their traces. Wastewater that has been treated has less suspended solids and is safe to release into the environment.

2.2.2 Chemical Characteristics Due to Chemical Impurities

The various chemical components that characterize wastewater are Total Organic Carbon (TOC), Chemical Oxygen Demand (COD), chlorides, nitrogen, sulfates, phosphorus, pH, trace elements, heavy metals, etc.

- **pH**: Chemical and petrochemical industry effluents have high pH levels, making them unsuitable for aquatic life since marine life cannot survive at higher pH levels.
- **BOD and COD**: The biochemical and chemical oxygen demands of WW both rise. The elevated levels of BOD and COD are inappropriate for aquatic life because they cause a reduction in DO, which is necessary for the survival of living things.
- **DO**: As waste increases, dissolved oxygen in water bodies diminishes. Because there is not enough oxygen in the water, living things suffocate and eventually perish. Following wastewater treatment, DO levels rise but stay below the danger threshold (Nishat et al. 2023).

2.2.3 Biological Characteristics of Contaminants

- **Biochemical Oxygen Demand (BOD)**: BOD estimates how much oxygen is required to use microorganisms to stabilize organic materials. Oil and Grease: Petroleum products and food waste are the sources of oil and grease. Microbial life in wastewater: The following microorganisms can be found in wastewater: Protozoa

  a. Bacteria
  b. Fungi
  c. Virus
  d. Nematode
  e. Seaweed
  f. Rotifers (Chen et al. 2020).

2.3 Biological Living

Microbiological activities play an essential part in wastewater and drinking water treatment. The perspectives provide useful insight into the microbial ecosystems utilized in wastewater and drinking water treatment. Enhancing our understanding of these ecosystems will greatly enhance our capacity to improve the effectiveness of current treatment processes and capabilities, as well as develop novel methods to meet the needs of a rapidly expanding global population. Furthermore, the retrieval or repurposing of resources holds great significance, such as the retrieval of phosphorus or its transformation of trash into bioplastic (Nielsen and Loosdrecht 2010). In 1930, Tausz and Donath, proposed the use of microbes to deal with both organic and inorganic contaminants (Tausz and Donath 1930). Since then, it has been extensively employed for the breakdown, transformation, and adsorption of harmful substances, where the toxic pollutants serve as an energy source as well as a carbon resource for the microbial breakdown of wastewater (Raj Deena et al. 2022).

2.3.1 Fungi

In regions with lax environmental regulations, brewery effluent is typically dumped into rivers, lakes, or the ocean, or straight into municipal wastewater treatment plants. Pre-treatment is often necessary for breweries that discharge to municipal wastewater treatment plants to ensure compliance with regulations and lessen the strain on the treatment facility (Simate et al. 2011). Treatment of brewery waste streams by fungi may be more advantageous than wastewater treatment methods based on bacteria. One benefit of using fungal-based treatments is developing mycelium which transforms organic materials and nutrients in the wastewater into simply fungal biofuel, reducing the organic load compared to little value slurry obtained via bacterial processing (Zheng et al. 2015). The fungal species most frequently utilized in wastewater treatment are *Agaricus bisporus*, *Lentinula edodes*, *Pleurotus ostreatus*, *Trichoderma harzianum*, and *Trametes versicolor* (Shao et al. 2008). The wastewater streams that are highest in nutrients can therefore be isolated and gathered separately; because of the decreased organic load, the residual wastewater would subsequently be less problematic for the municipal wastewater treatment facility, and this strategy could also result in lower waste disposal costs (Damle and Shukla 2010).

2.3.2 Yeast

As research has advanced, new technologies focused on yeast have been used more in the treatment of water. These technologies have demonstrated enormously promising futures. It has been discovered that yeast produces glycolipids and lipids. For this reason, it has a high usage value and is frequently employed in the treatment of wastewater with high-level concentrations of organic materials, wastewater containing heavy metal ions, and household discharge. With a high rate of wastewater treatment efficacy, yeast can transform the majority of organic debris into nutrient-dense, non-toxic single-cell protein. Oily wastewater comes from a variety of sources, and its COD and BOD values are comparatively high. It is often engaged in the sludge treatment with increased concentrations of oil (Wang et al. 2018) (Table 2.1).

Table 2.1 Role of yeast in wastewater treatment

Species	Sources of wastewater	Treatments	Removal efficiency	References
Kluyveromyces marxianus, *Saccharomyces cerevisiae*, and *Pichia anomala*	Synthetic wastewater	Bioremediation	COD 70%, nitrate 97%, nitrite 80%, phosphate 93% sulfate ions 70%, Pb^{2+}43 mg/L, Cd^{2+} ions 39 mg/L	Nicula et al. (2023)
Saccharomyces cerevisiae	Baker's yeast wastewater	Membrane Separation	COD (73% – 93%)	Igwegbe et al. (2022)

2.3.3 Bacteria

Due to the upsurge of antibiotic resistance, bacteriophage therapy has been opted as a means of inactivating bacteria that are resistant to antibiotics (Karthikeyan and Meyer 2006). Furthermore, WWTPs contain antibiotics, which allow the pathogenic microbes to develop resistance against antibiotics. More pathogenicity was shown by antibiotic-resistant bacteria; hence, different strains of bacteria were developed that released enzymes. These enzymes could break down the antibiotics. (Mancuso et al. 2021). It has been demonstrated that eradication of viral infections from wastewater and wastewater treatment plants has been achieved by bacteriophages (Dias et al. 2018). The biofilm of the planktonic bacteria is often vulnerable to bacteriophages that lyse their cells (Harper and Enright 2011). The biofilm-forming bacteria *Pseudomonas aeruginosa* can clog the filters of drinking water treatment plants; therefore, high-priced chlorination and flushing treatments are required to eliminate these biofilms formed on the filters (Jassim et al. 2016). Both chlorination and bacteriophage methods were successful in eliminating *P. aeruginosa* biofilm by up to 96% (Zhang and Hu 2013). The most significant class of autotrophic organisms in bioprocess sewage treatment is nitrifiers. Nitrification is caused by many bacterial taxa, including *Nitrosomonas*, *Nitrobacter*, and *Nitrospira*. *Nitrosomonas* oxidizes ammonia to nitrite (NO_2^-); the latter is changed to nitrate (NO_3^-) by *Nitrobacter* and *Nitrospira*. Nitrifying bacteria are delicate creatures with limited optimal pH and temperature ranges. Inhibitors also affect them adversely. It has been noticed that certain medications reduce the power of ammonia-oxidizing bacteria to function Wang and Gunsch (2011) (Table 2.2).

Table 2.2 Role of bacteria in wastewater treatment

Species	Origin of wastewater	Type of biological treatment	Removal efficiency	References
Prototheca zopfii	Meat processing wastewater	Up-flow anaerobic sludge blanket (UASB) reactor	BOD: total 56%, 85% TSS, 58% Oil and Grease	Sanghamitra et al. (2021), Yamaguchi et al. (1999)
Prototheca zopfii	Meat processing wastewater	UASB followed by rotating biological contactors (RBC)	94% BOD 96% TSS 91% Oil and grease	Wahaab and El-Awady (1999), Sanghamitra et al. (2021)
Prototheca zopfii	Petrochemical refinery wastewater	UASB reactor	85% BOD 90% TOC 91.1% Oil 95.6% Sulfide	Wei et al. (2000), Sanghamitra et al. (2021)
Prototheca zopfii	Petrochemical oil refinery wastewater	Anaerobic–aerobic system reactor	BOD5: 94%–99%, TOC: 93.4%, oil: 96.3%	Shon et al. (2002), Sanghamitra et al. (2021)
Acinetobacter sp. KUL 8	Bakery wastewater	Biodegradation reactor	Fat and oil: 70%	Shon et al. (2002), Sanghamitra et al. (2021)
Acinetobacter sp. KUL 8	Wastewater from Palm oil industry	Biodegradation reactor	87.7% fat and oil removal	Mukherji et al. (2004), Sanghamitra et al. (2021)
Bacillecs sp. KUL 39	Bakery wastewater	Biodegradation reactor	Fat and oil: 64%	Shon et al. (2002), Sanghamitra et al. (2021)
Bacillecs sp. KUL 39	Palm oil industry wastewater	Biodegradation reactor	Fat and oil: 80.6%	Shon et al. (2002), Sanghamitra et al. (2021)
Pseudomonas sp. strain D2D3	Diesel oil wastewater	Biodegradation reactor	Fats, oils, and Grease (FOG): 42%	Mukherji et al. (2004), Sanghamitra et al. (2021)
Streptomyces sp. B350	Oil field wastewater	Biological aerated filter (BAF) reactor	TOC: 64%, oil: 86%, Polycyclic aromatic hydrocarbons: 84%	Zhao et al. (2006), Sanghamitra et al. (2021)

(continued)

Table 2.2 (continued)

Species	Origin of wastewater	Type of biological treatment	Removal efficiency	References
Streptomyces sp. B350	Pet food industrial wastewater	PBR-UASB (hybrid packed bed reactor) combined with a UASB blanket	Oil and grease removal: 90%	Jeganathan et al. (2006), Sanghamitra et al. (2021)
Phototrophic microorganism-Cyanobacteria	Wastewater carrying diesel oil	RBC	TPH at N:P ratio of 98.6%	Chavan and Mukherji (2008), Sanghamitra et al. (2021)
Burkholderia cepacia	Wastewater carrying diesel oil	RBC	TPH removal: 99.4%	Fadile et al. (2011), Sanghamitra et al. (2021)
Cyanobacteria	Synthetic and restaurant wastewater	Aerobic biodegradation reactor	Grease removal 93.3% for adapted biomass	Fadile et al. (2011), Sanghamitra et al. (2021)
Burkholderia cepacia	Petrochemical wastewater	Moving bed biofilm reactor (MBBR)	Diesel removal: more than 80%	Ghorbanian et al. (2014), Sanghamitra et al. (2021)
Burkholderia cepacia	Wastewater carrying diesel oil	Upflow anoxic fixed-bed bioreactor (UAnFB)	99.9% TPH	Ghorbanian et al. (2014), Sanghamitra et al. (2021)
Burkholderia cepacia	Petrochemical wastewater	Sequencing anoxic batch reactor (SAnBR),	99.7% TPH	Emadian et al. (2015), Sanghamitra et al. (2021)
Pseudomonas strain	Bilge water	Hybrid upflow anaerobic sludge blanket (HUASB) reactor	95% Oil	Sanghamitra et al. (2020), Sanghamitra et al. (2021)
Pseudomonas strain	Petroleum derived synthetic oily wastewater	Aerobic batch reactor	54.6–80% Oil	Sanghamitra et al. (2020), Sanghamitra et al. (2021)

2.3.4 Algae

Algae have currently come into focus as a potentially economical way to remove harmful chemicals from the environment by using the processes of biosorption, bioaccumulation, and intracellular breakdown. Therefore, it is possible to use more effective methods for environmental resource recovery and sustainable development before releasing wastewater into the natural surroundings. More significantly, algae have the potential to be used as a feedstock for several industrial processes, including the synthesis of biofuel. Scientists are currently working on using algae as a source of food additives, biofuels, medications, and fertilizers (Bhatt et al. 2022a, b). The majority of algae are aquatic, autotrophic creatures without actual stems, roots, or leaves. They often fall into two categories: macroalgae and microalgae. Algae consist of single-celled organisms (microalgae) like Prototheca, Chlorella, and diatoms, or multicelled organisms (macroalgae) like brown and green seaweeds. Algae can be grown industrially in dual methods: (i) in pond system and (ii) in an enclosed photobioreactor system. Because it is more difficult to regulate the nutrients for development in an open pond system, photobioreactor-based algae culture is a more convenient option (Nguyen et al. 2022).

2.3.4.1 Methods for Growing Algae-Based Wastewater Treatment

Utilizing algae-based wastewater treatment depends on the growing technique, which affects downstream processes, biomass growth, and nutrient removal.

2.3.4.1.1 Phototrophic Cultivation of Algae

Open systems culture is one of the most widely used methods for producing phototrophic microorganisms. A substantial risk exists for contamination by predators and parasites of microalgae, including rotifers or ciliates, which can quickly deplete microalgal biomass and result in significant productivity losses. Closed photobioreactors (PBRs), on the other hand, have a larger initial investment, ongoing maintenance expenses, and a lower danger of contamination. However, they provide good control over cultural conditions. Since gas exchange with the atmosphere happens through sterilized filters, photobioreactors are typically defined as closed, lit vessels where the environment and culture are not directly in contact, and the risk of contamination is greatly reduced. Moreover, this configuration results in reduced CO_2 and loss of water throughout the process, and overall productivity is significantly higher than with an open system (Chanquia et al. 2022).

2.3.4.1.2 Heterotrophic Cultivation of Algae

Heterotrophic production of algae can be made attainable by repeated and bulk production of high-density biomass. Heterotrophic algae are grown in fermenters that get a regular or constant carbon source supply. For the growth of heterotrophic microalgae, glycerol, acetate, and glucose are the most common carbon sources. Significant amounts of docosahexaenoic acid (DHA) can be produced by the heterotrophic microalgae such as *Schizochytrium mangrovei, Schizochytrium limacinum, and Crypthecodinium cohnii* using waste materials. This substance is a long-chain fatty acid that is beneficial to man's health in the cure of numerous conditions, including psoriasis, Alzheimer's, rheumatoid arthritis, cancer, and atherosclerosis. Ammonium from high-nitrogen wastewater stimulates the pentose phosphate pathway. Heterotrophic cultivation may therefore be chosen as a high-nitrogen wastewater source. Compared to phototrophic algae, heterotrophic culture removes more organic carbon from wastewater and overcomes restrictions related to light and dissolved CO_2 (Silva et al. 2021).

2.3.4.1.3 Mixotrophic Cultivation of Algae

Algae grown by mixotrophic use both phototrophic and heterotrophic metabolisms, with light and both organic and inorganic carbon serving as sources of energy and carbon. While organic carbon and dissolved CO_2 can be employed for metabolisms, light limitation, and inhibition are less effective in mixotrophic culture. Thus, mixotrophic cultivation offers increased biomass productivity, flexibility, and efficiency in the removal of nutrients. The most prevalent organic molecules have historically been glycerol, acetate, and carbohydrates (such as glucose). The elimination of nitrogen and phosphorus was more effective when the mixotrophic culture of *Chlorella vulgaris* with glucose was added, as opposed to sodium acetate or protein. Wastewater with a high organic content is often suitable for mixotrophic cultivation. One benefit of mixotrophic cultivation is the capacity to employ organic carbon in waste to develop biomass with a high lipid content (Giuliano Dragone 2022) (Table 2.3).

2.4 Biological Methods of Wastewater Treatment

The significance of wastewater treatment by biological approaches is increasing due to the retrieval of biosolids, heavy metals, valuable plant nutrients and bioenergy resources. Pure cultures or mixed microbial consortia can simultaneously eliminate several types of solid pollutants present in effluents. Microbial biomass determines the classification of biological wastewater treatment systems into a pair of methods: suspended growth system and attached growth system. Biological wastewater treatment techniques can also be used for the extraction of valuable materials (Machineni 2019).

Table 2.3 Role of algae in wastewater treatment

Species	Origins of wastewater	Type of biological treatment	Removal efficiency	References
Chlorella variabilis	Domestic waste eater	Closed photobioreactors and open reactor system	COD 74.8–89.8%, total nitrogen 93.8–96.1% and total phosphorous 97.1–99.9%	Tran et al. (2021)
Chlorella vulgaris	Synthetic municipal wastewater	Activated sludge	81% (NH_3), 39% (PO_4^{3-}), and 98% (COD)	Corpuz et al. (2021)
Scenedesmus abundans	Domestic wastewater	Hybrid loop airlift photobioreactor	COD 80.19%, Dissolved inorganic nitrogen 90.73%, and dissolved inorganic phosphorus 86.31%	SundarRajan et al. (2020)
Chlorella sorokiniana	Wastewater	Activated sludge	Total nitrogen 65.3% Total phosphorus 42.6%	Qv et al. (2023)
Chlorella vulgaris, Scenedesmus sp., & Chlorococcum sp.	Tannery wastewater	Co-culture	COD 89.9% TKN % 80.9%	Nagabalaji et al. (2023)
*Synechocysti strain (*ΔSphU)	Shrimp wastewater	Photobioreactor	Phosphate 71.5% Nitrate 42.7%, Ammonium 48.5% Nitrite 92.9%, Harmful bacteria belonging to Cellvibrionaceae and Pseudomonadaceae families were recorded at 5.85% and 3.18%	Krasaesueb et al. (2023)
Macroalgae				
Caulerpa lentillifera	Marine aquaculture wastewater	Membrane separation	COD 48.82%, NH_4^+ 9.14% and NO_2^- 40.74%	Ye et al. (2023)
Chaetomorpha maxima)	Marine aquaculture wastewater	Moving bed bioreactor	Total Nitrogen 5.5% Total Phosphorous 7.7%	Li et al. (2020)
Padina pavonica	Agriculture wastewater	Batch biosorption	Co(II) 84.3%	Aloufi et al. (2024)
Gracilaria verrucose, Enteromorpha intestinalis	Textile wastewater	Batch biosorption	Malachite Green *G. verrucose* 97.49% *E. intestinalis* 93.37%	Amreen et al. (2024)

(continued)

Table 2.3 (continued)

Species	Origins of wastewater	Type of biological treatment	Removal efficiency	References
Chara corallina	Wastewater	Membrane separation	Cd^{2+} 98.20%, Ni^{2+} 95.15%, and Pb^{2+} ions 93.80%	Yousif et al. (2024)
Sargassum fusiforme, Enteromorpha prolifera	Wastewater	Phytoremediation	*S. fusiforme*—Mn 95.73% Pb 99.46% *E. prolifera*—Mn 98.17% Pb 100.00%	Diaconu et al. (2023)

2.4.1 Suspended Growth

In a suspended-growth system like activated sludge operations, aerated lagoons, and membrane separation, the pollutants passes through and around free-floating microorganisms, aggregating into flocks of biological matter that eventually settle out of the wastewater (Fig. 2.2).

2.4.1.1 Activated Sludge

Many organic and inorganic pollutants, including pesticides, medications, heavy metals, pathogens, bacterial biomass, surfactants, microplastics (MPs), and per-

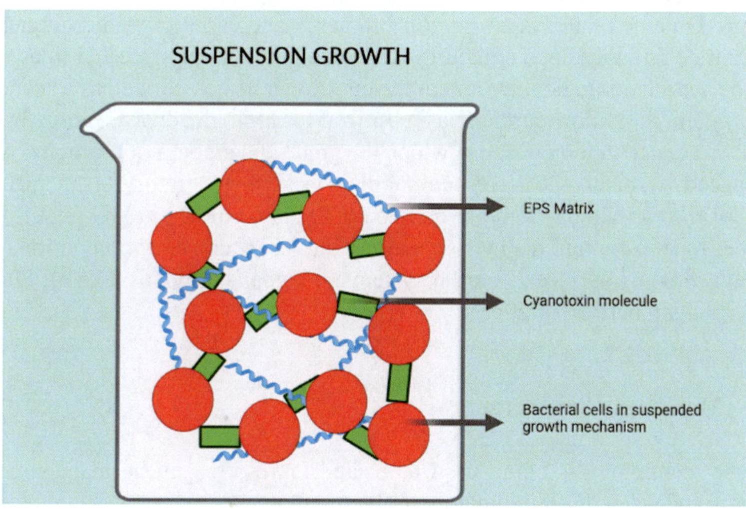

Fig. 2.2 Suspended growth system in wastewater treatment

and polyfluoroalkyl compounds, are present in the biosolid. Biodegradation of compounds can be accelerated by adding an appropriate rate of activated sludge that is enriched in protein and carbohydrates. Anaerobic digestion (AD) of the biosolid is an additional technique. Biogas is produced when the activated sludge undergoes anaerobic digestion. Phosphorus (P) and nitrogen (N), two organic nutrients, have been shown to be abundant in activated sludge (Awasthi et al. 2023). Anaerobic, anoxic, and aerobic system performance is significantly impacted by influent wastewater features like total N, COD and total P as well as operating factors like HRT, SRT, the mixed liquor recirculation (MLR), Dissolved oxygen (DO), and return activated sludge (RAS) flow rates. In suspended growth systems within the combined fluid of the reactor, the microbial mass is held in place as flocs. These flocs are kept in suspension through agitation by gas injection or mechanical mixers. In anaerobic systems, biogas will often be used instead of air in aerobic operations. The intimate connection between the biomass and the substrate is facilitated by agitation. In biofilm systems, microorganisms are bonded to the supporting medium in a thin film. The latter may be either a stationary or movable bed. While moving beds might have sand or granular activated carbon, stationary beds are often formed of plastic or gravel. During reactor operation, these support medium beds may be immersed in a mixed liquor or can be subjected to air and wastewater. On the other hand, suspended-growth biofilms can occasionally be incorporated into these aerobic systems (Musa and Idrus 2021).

2.4.1.2 Aerated Lagoons

In the aerobic wastewater treatment process, substrates or natural elements present in wastewater are broken down by very potent bacteria that thrive in the presence of oxygen. Organic matter is a reservoir of energy and carbon for microorganisms. If its substrate and microbes remain in suspension, commonly referred to as mixed liquor, the aerobic biological treatment technique can be categorized as a suspended growth system. A shallow earthen basin utilized for biological wave energy transfer (WWT), an aerated lagoon is one in which the grade and amount of biological WWT are enhanced by mechanical oxygen supply via surface aerators or air diffusers. The bed of the earthen basin might be lined with clay, asphalt, or any impermeable substances to stop contaminants from leaking into it. After a retention period of one to ten days, floating surfaces aeration systems in lagoons or ponds take off 80–90% of BOD (Sangamnere et al. 2023).

2.4.1.3 Membrane Separation

Membrane technology is being used more and more often in the management of wastewater and water due to regulatory problems about meeting the stringent requirements for water quality. Reverse osmosis (OS), ultrafiltration (UF), microfiltration (MF) and nanofiltration are the main membrane technologies used to purify water

(NF). Based on the pore size, membranes can filter out colloids, particles, and macromolecules (Musa and Idrus 2021). Membranes that are utilized in wastewater treatment function as barriers since they let water through but keep impurities from doing so. Due to differential pressure on both sides of the membrane, the pollutants reside on one side of the membrane while the water flows through the other side. To increase the performance of the wastewater treatment, an advanced technology namely forward osmosis membrane was developed. The two important advantages of the forward osmosis such as long-term membrane fouling and vigorous retention capacity led to a successful process for wastewater treatment. The membrane-based technique depends on the differences in osmotic pressure between the membrane surfaces (Ahmed et al. 2022).

2.4.1.3.1 Microfiltration (MF)

Initially, MF membranes were employed in the wastewater purification to decrease the fouling and blockages. Later, they were designed with a pressure-driven permeable membranes specifically to eradicate pathogens and colloidal particles suspended in wastewater (Acarer 2023).

2.4.1.3.2 Reverse Osmosis (RO)

The inadvertent movement of water molecules across a semipermeable membrane from a solution with a small amount of solute and reduced osmotic pressure to one with a substantial amount of solute and elevated osmotic pressure is the organic process referred to as osmosis. When an extremely concentrated solution (feed solution) is subjected to atmospheric pressure, the rapid movement of water ions can be interrupted or turned. The water molecules are compelled to flow against the typical osmotic phenomenon if the pressure being applied difference is larger than the osmotic gradient across the membrane. Reverse osmosis is the process that has taken place in this instance (Xiang et al. 2022). Of all the pressure-driven processes, reverse osmosis operation demands the highest hydraulic pressure. With a pore size of about 0.1 nm, a RO membrane can purify water by eliminating nearly all types of impurities.

2.4.1.3.3 Nanofiltration (NF)

Nanofiltration membrane pore sizes are typically in the range of 0.1–1 nm, which translates to a MWCO of 0.1–5 kDa. NF can remove almost all viruses, divalent salts, and the majority of naturally occurring organic materials when it is run at pressures between 5 and 35 bars.

2.4.1.3.4 Ultrafiltration (UF)

Ultrafiltration (UF) is a widely-recognized method for treating wastewater. UF exhibits good permeate flow and can achieve high rejection of macromolecules. The primary mechanism of UF membranes, which have pores with sizes between 0.005 and 0.1 μm, is size exclusion; however, additional parameters that can significantly impact rejection efficiency include molecule structure, charge, and hydrodynamic circumstances (Goh et al. 2022).

2.4.2 *Attached Growth*

Although attached-growth biomass, sometimes referred to as biofilm, has benefits over conventional treatment technologies, its use in wastewater treatment has expanded recently (Muhamad et al. 2015). The attached growth mechanism creates an ideal surface for the growth of the bacteria, allowing them to take in food and nutrients within and surrounding the specific area close to their micro-colonies. In this way, it maintains a certain equilibrium between the area in which the bacteria proliferate and the ease with which they may carry out their metabolic function (Kumar et al. 2019). A bio filter, a trickling sand filter, and a swim-bed reactor are components of the biological NH 4-N removal system used to clean groundwater and drinking water. But to effectively remove NH 4-N, all of them need constant aeration, which raises energy and financial concerns. Three primary components comprise the attached growth reactors: a lab-scale reactor designed to investigate the impact of reactor layout on the removal of NH 4-N; a lab-scale reactor developed to analyze the impact of raw underground water standard on the removal of NH 4-N; in addition, a pilot-scale attached growth reactor incorporated into the Chyasal drinking water supply system (Khanitchaidecha et al. 2012) (Fig. 2.3).

2.4.2.1 Fixed Biofilms

A coating of extracellular polymeric substances (EPS) that functions as glue surrounds the micro-colonies of the various species that build up biofilms—fungi, bacteria, algae, and yeast—creating something known as the "biofilm cells' house" (Dang and Lovell 2016). One of the primary building blocks of biofilms, EPS is made up of lipids, proteins, nucleic acids, and polysaccharides. Biofilm wettability varies due to the varied characterizations that EPS might have in water, such as hydrophilic or hydrophobic properties (Tanaka et al. 2019). The use of biofilm systems for wastewater treatment is growing in popularity. Their growing popularity is a result of their endearing qualities. They have demonstrated efficacy concerning economic and environmental factors (Mahto and Das 2022). When it comes to fixed-film wastewater treatment devices, rotating biological contactors (RBC) and trickling filters were the most commonly utilized options. In a study published in 2011,

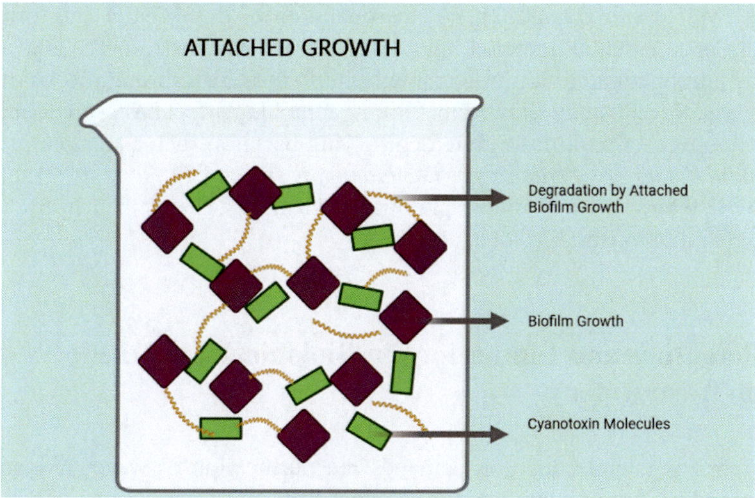

Fig. 2.3 Attached growth system in wastewater treatment

Boelee et al. (2011) demonstrated the effectiveness of microalgal biofilms in both phosphorus elimination and nitrogen removal from domestic wastewater contaminants. Upon adhering metal ions to the EPS matrix, a biofilm can remove harmful metals from the effluent, such as zinc, lead, and copper (Teitzel and Parsek 2003). Wastewater purification technology driven by microbial biofilms is an effective and less expensive for a limited number of apartments fit for single-family dwellings (Mononen et al. 2010). Rubber latex wastewater is treated with different inorganic anions using bacteria based on biofilms. The primary component of biofilms is extracellular polymeric substances (EPS), which have many chemical functional groups and are rich in lipids and polysaccharides. The bacteria *Rhodococcus rhodochrous* has shown a great potency to remove NO_3 (95%) and PO_9^- (75%) and was the most effective candidate to treat wastewater produced from rubber latex (Kim et al. 2021).

2.4.2.2 Moving Bed Biofilms Reactor

The discovery of moving bed biofilm reactor in the early 1980s particularly to clean domestic and industrial wastewater has gained a lot of attention. A higher biomass concentration can be maintained over a long period of time by fixing the microbes onto the solid biofilm surfaces to promote sluggish growth of microbes. These microbes efficiently degrade the contaminants available in the wastewater thereby stabilizing the function of the bioreactor for a long run (Raj Deena et al. 2022). In recent years, the wastewaters from sources such as agriculture and municipalities were treated by two kinds of aerobic biofilm reactors, namely moving bed biofilm reactors and aerobic submerged fixed bed reactors. These bioreactors possess many advantages which include increased reaction rates, low costs, reusability and firmness (Mahto and Das 2022). For the operation of this MBBR, the formation of biofilm is an essential factor (Liu et al. 2022). Insufficient resource is available about the introduction of biofilm-forming bacteria in MBBR through bioaugmentation process. Recent study showed that nitrate removing effect has been doubled by using immobilizing biofilm-forming denitrifying bacteria, such as *Diaphorobacter sp.*, *Pannonibacter sp.*, *Thauera sp.*, *Pseudomonas sp.*, and *Thauera sp.* (Patel et al. 2021). MBBR has proven to remove pharmacological waste better than the traditional sludge separation methods (Ooi et al. 2017).

2.5 Potentials and Limitations in Biological Treatment of Wastewater

Present environmental condition demands the purification of water, reduction in chemical and biological impurities can be achieved by latest waste water management methods. It is important to investigate microbial interactions in aquatic environments. The influence of geo-environmental elements connected to adsorption mechanisms requires further investigation (Younas et al. 2021).

Numerous microbes, including bacteria and yeasts, have well-established metabolic engineering pathways that allow for the customization of these organisms for particular uses. To address some of the difficulties in microalgal bioremediation of wastewater, greater study in the field of microalgal metabolic engineering would be necessary. Combining microalgae with other appropriate microorganisms could improve the removal efficiency of such undesirable pollutants from wastewater (Al-Jabri et al. 2020).

There are still a number of restrictions in place despite several improvements in the efficacy of the created biologic treatment techniques. In general, it would be advised to combine many modalities of treatment to overcome technological challenges (Sathya et al. 2023).

Most biotreatment processes are extremely vulnerable to different environmental parameters (such as pH value, temperature, and wastewater components) because functional microbes grow slowly. This results in an excessively long start-up time, which lowers the economic benefits and limits the practical applications (Yi et al. 2019).

Because the biological approach makes use of bioremediation systems, microbial breakdown, fungal decolorization, and adsorption by living or dead microbial biomass, it is typically more cost-effective. The microbes have the potential for breaking down the contaminants present in the wastewater. However, the use of microorganisms is typically confined due to the technological restraints. The various drawbacks of wastewater treatment using biological methods are vast area, toxic chemicals, diurnal variation, design obstructions, and inflexible operation (Hanafi and Sapawe 2020).

Even though the microbes play an important role in the environment and industries, the changes in various factors like pH, temperature, and seasonal diversions, their applications are often limited (Singh et al. 2022).

2.6 Conclusion

The process of bioremediation depends on assembly of microorganisms that clears and regains wastewater and minimizes environmental impact. In order to achieve sustainable wastewater treatment, microbes such as fungi, protozoa, cyanobacteria and bacteria have gained a lot of attention in sewage management. Organic and inorganic wastewater contains nitrogen (N), carbon (C) and phosphorus (P) in high concentration, the above microorganisms utilize these elements for their growth and survival hence they are found to be attractive alternatives. As an effect of these microbes in WWT, the concentration of harmful chemicals in water is significantly reduced (Singh et al. 2022). The traditional chemical coagulants method is least effective in treating wastewater than the bio-coagulants. The sludge produced by the action of bio-coagulants can be reused as they are biodegradable. The slurry can be digested by anaerobic method which further yields biogas which has various commercial applications. The activity of microorganisms in anaerobic digestion are not affected as the bio-coagulants are harmless. The current scenario also reveals that the sludge generated from wastewater can be used as a fertilizer (Kurniawan et al. 2020).

References

Acarer S (2023) A review of microplastic removal from water and wastewater by membrane technologies. Water Sci Technol 88(1):199–219

Ahmed SF, Mofijur M, Nuzhat S, Chowdhury AT, Rafa N, Uddin MA, Show PL (2021) Recent developments in physical, biological, chemical, and hybrid treatment techniques for removing emerging contaminants from wastewater. J Hazard Mater 416:125912

Ahmed SF, Mehejabin F, Momtahin A, Tasannum N, Faria NT, Mofijur M, Mahlia TMI (2022) Strategies to improve membrane performance in wastewater treatment. Chemosphere 306:135527

Ali S, Sharma B, Deoli K, Saini D, Bisht M (2023) An overview on bioremediation strategies for waste water treatment and environmental sustainability. Appl Ecol Environ Sci 11(2):64–70

Al-Jabri H, Das P, Khan S, Thaher M, AbdulQuadir M (2020) Treatment of wastewaters by microalgae and the potential applications of the produced biomass—a review. Water 13(1):27

Aloufi AS, Al Riyami B, Fawzy MA, Al-Yasi HM, Koutb M, Hassan SH (2024) Model-assisted optimization of cobalt biosorption on macroalgae Padina pavonica for wastewater treatment. Water 16(6):887

Amreen S, Dash SK, Naik B, Dash AK, Verma AK, Pradhan A (2024) Marine macro algae-derived activated carbon an efficient bio-sorbent for Malachite green removal from wastewater. Environ Qual Manag

Awasthi MK, Ganeshan P, Gohil N, Kumar V, Singh V, Rajendran K, Taherzadeh MJ (2023) Advanced approaches for resource recovery from wastewater and activated sludge: a review. Biores Technol 384:129250

Bhatt P, Bhandari G, Turco RF, Aminikhoei Z, Bhatt K, Simsek H (2022b) Algae in wastewater treatment, mechanism, and application of biomass for production of value-added product. Environ Pollut 309:119688

Bhatt P, Bhandari G, Turco RF, Aminikhoei Z, Bhatt K, Simsek H (2022) Algae in wastewater treatment, mechanism, and application of biomass for production of value-added product. Environ Pollut (Barking, Essex: 1987) 309:119688. https://doi.org/10.1016/j.envpol.2022.119688

Boelee NC, Temmink H, Janssen M, Buisman CJN, Wijffels RH (2011) Nitrogen and phosphorus removal from municipal wastewater effluent using microalgal biofilms. Water Res 45(18):5925–5933

Chanquia SN, Vernet G, Kara S (2022) Photobioreactors for cultivation and synthesis: Specifications, challenges, and perspectives. Eng Life Sci 22(12):712–724

Chavan A, Mukherji S (2008) Treatment of hydrocarbon-rich wastewater using oil degrading bacteria and phototrophic microorganisms in rotating biological contactor: effect of N: P ratio. J Hazard Mater 154(1–3):63–72

Chen GH, van Loosdrecht MC, Ekama GA, Brdjanovic D (eds) (2020) Biological wastewater treatment: principles, modeling and design. IWA Publishing

Corpuz MVA, Borea L, Senatore V, Castrogiovanni F, Buonerba A, Oliva G, Naddeo V (2021) Wastewater treatment and fouling control in an electro algae-activated sludge membrane bioreactor. Sci Total Environ 786:147475

Damle AJ, Shukla SR (2010) Production of laccase from Coriolus versicolor and its application in dye decolorization in combination with UV/H_2O_2 technique. Clean-Soil, Air, Water 38(7):663–669

Dang H, Lovell CR (2016) Microbial surface colonization and biofilm development in marine environments. Microbiol Mol Biol Rev 80(1):91–138

Diaconu LI, Covaliu-Mierlă CI, Păunescu O, Covaliu LD, Iovu H, Paraschiv G (2023) Phytoremediation of wastewater containing lead and manganese ions using algae. Biology 12(6):773

Dias E, Ebdon J, Taylor H (2018) The application of bacteriophages as novel indicators of viral pathogens in wastewater treatment systems. Water Res 129:172–179

Dragone G (2022) Challenges and opportunities to increase economic feasibility and sustainability of mixotrophic cultivation of green microalgae of the genus Chlorella. Renew Sustain Energy Rev 160:112284

Emadian SM, Hosseini M, Rahimnejad M, Shahavi MH, Khoshandam B (2015) Treatment of a low-strength bilge water of Caspian Sea ships by HUASB technique. Ecol Eng 82:272–275

Fadile A, El Hassani FZ, Aissam H, Merzouki M, Benlemlih M (2011) Aerobic treatment of lipid-rich wastewater by a bacterial consortium. Afr J Microbiol Res 5(30):5333–5342

Ghorbanian M, Moussavi G, Farzadkia M (2014) Investigating the performance of an up-flow anoxic fixed-bed bioreactor and a sequencing anoxic batch reactor for the biodegradation of hydrocarbons in petroleum-contaminated saline water. Int Biodeterior Biodegradation 90:106–114

Goh PS, Wong KC, Ismail AF (2022) Membrane technology: a versatile tool for saline wastewater treatment and resource recovery. Desalination 521:115377

Hanafi MF, Sapawe N (2020) A review on the current techniques and technologies of organic pollutants removal from water/wastewater. Mater Today Proc 31:A158–A165

Haripriyan U, Gopinath KP, Arun J, Govarthanan M (2022) Bioremediation of organic pollutants: a mini review on current and critical strategies for wastewater treatment. Arch Microbiol 204(5):286. https://doi.org/10.1007/s00203-022-02907-9

Harper DR, Enright MC (2011) Bacteriophages for the treatment of Pseudomonas aeruginosa infections. J Appl Microbiol 111(1):1–7

Igwegbe CA, Obiora-Okafo IA, Iwuozor KO, Ghosh S, Kurniawan SB, Rangabhashiyam S, Ighalo JO (2022) Treatment technologies for bakers' yeast production wastewater. Environ Sci Pollut Res 29(8):11004–11026

Jassim SA, Limoges RG, El-Cheikh H (2016) Bacteriophage biocontrol in wastewater treatment. World J Microbiol Biotechnol 32(4):70

Jeganathan J, Nakhla G, Bassi A (2006) Long-term performance of high-rate anaerobic reactors for the treatment of oily wastewater. Environ Sci Technol 40(20):6466–6472

Karia GL, Christian RA, Jariwala ND (2023) Wastewater treatment: concepts and design approach. PHI Learning Pvt. Ltd

Karthikeyan KG, Meyer MT (2006) Occurrence of antibiotics in wastewater treatment facilities in Wisconsin, USA. Sci Total Environ 361(1–3):196–207

Khanitchaidecha W, Shakya M, Nakano Y, Tanaka Y, Kazama F (2012) Development of an attached growth reactor for NH 4-N removal at a drinking water supply system in Kathmandu Valley, Nepal. J Environ Sci Health Part A 47(5):734–743

Kim ES, Ha JH, Choi J (2021) Biological fixed-film systems. Water Environ Res Res Publ Water Environ Fed 93(4):491–501. https://doi.org/10.1002/wer.1445

Koul B, Yadav D, Singh S, Kumar M, Song M (2022) Insights into the domestic wastewater treatment (DWWT) regimes: a review. Water 14(21):3542

Krasaesueb N, Boonnorat J, Maneeruttanarungroj C, Khetkorn W (2023) Highly effective reduction of phosphate and harmful bacterial community in shrimp wastewater using short-term biological treatment with immobilized engineering microalgae. J Environ Manage 325:116452

Kumar P, Hegde K, Brar SK, Cledon M, Kermanshahi-Pour A (2019) Potential of biological approaches for cyanotoxin removal from drinking water: a review. Ecotoxicol Environ Saf 172:488–503

Kumar V, Shah MP (2021) Advanced oxidation processes for complex wastewater treatment. In: Advanced oxidation processes for effluent treatment plants. Elsevier, pp 1–31

Kurniawan SB, Abdullah SRS, Imron MF, Said NSM, Ismail NI, Hasan HA, Purwanti IF (2020) Challenges and opportunities of biocoagulant/bioflocculant application for drinking water and wastewater treatment and its potential for sludge recovery. Int J Environ Res Public Health 17(24):9312

Li X, Deng Y, Li X, Ma X, Wang J, Li J (2020) Integration of marine macroalgae (Chaetomorpha maxima) with a moving bed bioreactor for nutrient removal from maricultural wastewater. Archaea 2020(1):8848120

Liu X, Tang P, Liu Y, Xie W, Chen C, Li T, Liu B (2022) Efficient removal of organic compounds from shale gas wastewater by coupled ozonation and moving-bed-biofilm submerged membrane bioreactor. Biores Technol 344:126191

Machineni L (2019) Review on biological wastewater treatment and resources recovery: attached and suspended growth systems. Water Sci Technol 80(11):2013–2026

Mahto KU, Das S (2022) Bacterial biofilm and extracellular polymeric substances in the moving bed biofilm reactor for wastewater treatment: a review. Biores Technol 345:126476

Mancuso G, Midiri A, Gerace E, Biondo C (2021) Bacterial antibiotic resistance: the most critical pathogens. Pathogens 10(10):1310

Mononen T, Coloma S, Lehtonen A, Kapanen A, Vikman M, Itävaara M, Romantschuk M (2010) Biodegration of BPA and HHCB from waste water by multi-stage biofilm processes. Linnaeus Eco-Tech, 156–163

Muhamad MH, Abdullah SRS, Hasan HA, Rahim RAA (2015) Comparison of the efficiencies of attached-versus suspended-growth SBR systems in the treatment of recycled paper mill wastewater. J Environ Manage 163:115–124

Mukherji S, Jagadevan S, Mohapatra G, Vijay A (2004) Biodegradation of diesel oil by an Arabian Sea sediment culture isolated from the vicinity of an oil field. Biores Technol 95(3):281–286

Musa MA, Idrus S (2021) Physical and biological treatment technologies of slaughterhouse wastewater: a review. Sustainability 13(9):4656

Nagabalaji V, Maharaja P, Nishanthi R, Sathish G, Suthanthararajan R, Srinivasan SV (2023) Effect of co-culturing bacteria and microalgae and influence of inoculum ratio during the biological treatment of tannery wastewater. J Environ Manage 341:118008

Nguyen LN, Aditya L, Vu HP, Johir AH, Bennar L, Ralph P, Nghiem LD (2022) Nutrient removal by algae-based wastewater treatment. Current Pollution Reports 8(4):369–383

Nicula NO, Lungulescu EM, Rîmbu GA, Marinescu V, Corbu VM, Csutak O (2023) Bioremediation of wastewater using yeast strains: an assessment of contaminant removal efficiency. Int J Environ Res Public Health 20(6):4795. https://doi.org/10.3390/ijerph20064795

Nielsen PH, van Loosdrecht MC (2010) Microbial ecology of drinking water and waste water treatment processes

Nimkande VD, Bafana A (2022) A review on the utility of microbial lipases in wastewater treatment. J Water Process Eng

Nishat A, Yusuf M, Qadir A, Ezaier Y, Vambol V, Khan MI, Eldin SM (2023) Wastewater treatment: a short assessment on available techniques. Alex Eng J 76:505–516

Ooi GTH, Escola Casas M, Andersen HR, Bester K (2017) Transformation products of clindamycin in moving bed biofilm reactor (MBBR). Water Res 113:139–148. https://doi.org/10.1016/j.watres.2017.01.058

Patel RJ, Patel UD, Nerurkar AS (2021) Moving bed biofilm reactor developed with special microbial seed for denitrification of high nitrate containing wastewater. World J Microbiol Biotechnol 37(4):68. https://doi.org/10.1007/s11274-021-03035-0

Qv M, Dai D, Liu D, Wu Q, Tang C, Li S, Zhu L (2023) Towards advanced nutrient removal by microalgae-bacteria symbiosis system for wastewater treatment. Biores Technol 370:128574

Raj Deena S, Kumar G, Vickram AS, Rani Singhania R, Dong CD, Rohini K, Anbarasu K, Thanigaivel S, Ponnusamy VK (2022) Efficiency of various biofilm carriers and microbial interactions with substrate in moving bed-biofilm reactor for environmental wastewater treatment. Biores Technol 359:127421. https://doi.org/10.1016/j.biortech.2022.127421

Sangamnere R, Misra T, Bherwani H, Kapley A, Kumar R (2023) A critical review of conventional and emerging wastewater treatment technologies. Sustain Water Resour Manag 9(2):58

Sanghamitra P, Mazumder D, Mukherjee S (2020) A study on aerobic biodegradation of oil and grease containing wastewater. J Indian Chem Soc 97(5):819–822

Sanghamitra P, Mazumder D, Mukherjee S (2021) Treatment of wastewater containing oil and grease by biological method- a review. J Environ Sci Health A Tox Hazard Subst Environ Eng 56(4):394–412. https://doi.org/10.1080/10934529.2021.1884468

Sathya R, Arasu MV, Al-Dhabi NA, Vijayaraghavan P, Ilavenil S, Rejiniemon TS (2023) Towards sustainable wastewater treatment by biological methods–a challenges and advantages of recent technologies. Urban Climate 47:101378

Shao X, Peng D, Teng Z, Ju X (2008) Treatment of brewery wastewater using anaerobic sequencing batch reactor (ASBR). Biores Technol 99(8):3182–3186

Sharma M, Agarwal S, Agarwal Malik R, Kumar G, Pal DB, Mandal M, Sarkar A, Bantun F, Haque S, Singh P, Srivastava N, Gupta VK (2023) Recent advances in microbial engineering approaches for wastewater treatment: a review. Bioengineered 14(1):2184518. https://doi.org/10.1080/21655979.2023.2184518

Shon HK, Tian D, Kwon DY, Jin CS, Lee TJ, Chung WJ (2002) Degradation of fat, oil, and grease (FOGs) by lipase-producing bacterium Pseudomonas sp. strain D2D3. J Microbiol Biotechnol 12(4):583–591

Silva TLD, Moniz P, Silva C, Reis A (2021) The role of heterotrophic microalgae in waste conversion to biofuels and bioproducts. Processes 9(7):1090

Simate GS, Cluett J, Iyuke SE, Musapatika ET, Ndlovu S, Walubita LF, Alvarez AE (2011) The treatment of brewery wastewater for reuse: state of the art. Desalination 273(2–3):235–247

Singh D, Goswami RK, Agrawal K, Chaturvedi V, Verma P (2022) Bio-inspired remediation of wastewater: a contemporary approach for environmental clean-up. Curr Res Green Sustain Chem 5:100261

SundarRajan P, Gopinath KP, Arun J, GracePavithra K, Pavendan K, AdithyaJoseph A (2020) An insight into carbon balance of product streams from hydrothermal liquefaction of Scenedesmus abundans biomass. Renewable Energy 151:79–87

Tanaka N, Kogo T, Hirai N, Ogawa A, Kanematsu H, Takahara J, Tanaka Y (2019) In-situ detection based on the biofilm hydrophilicity for environmental biofilm formation. Sci Rep 9(1):8070

Teitzel GM, Parsek MR (2003) Heavy metal resistance of biofilm and planktonic Pseudomonas aeruginosa. Appl Environ Microbiol 69(4):2313–2320

Tran DT, Van Do TC, Nguyen QT, Le TG (2021) Simultaneous removal of pollutants and high value biomaterials production by Chlorella variabilis TH03 from domestic wastewater. Clean Technol Environ Policy 23:3–17

Wahaab RA, El-Awady MH (1999) Anaerobic/aerobic treatment of meat processing wastewater. Environmentalist 19:61–65

Wang Y, Qiu L, Hu M (2018) Application of yeast in the wastewater treatment. In: E3S web of conferences, vol 53. EDP Sciences, p 04025

Wang S, Gunsch CK (2011) Effects of selected pharmaceutically active compounds on treatment performance in sequencing batch reactors mimicking wastewater treatment plants operations. Water Res 45(11):3398–3406

Wei SQ, Xue L, Jun HZ (2000) Petrochemical wastewater treatment by biological process

Xiang H, Min X, Tang CJ, Sillanpää M, Zhao F (2022) Recent advances in membrane filtration for heavy metal removal from wastewater: a mini review. J Water Process Eng 49:103023

Yamaguchi T, Ishida M, Suzuki T (1999) Biodegradation of hydrocarbons by Prototheca zopfii in rotating biological contactors. Process Biochem 35(3–4):403–409

Ye T, Li M, Lin Y, Su Z (2023) An effective biological treatment method for marine aquaculture wastewater: combined treatment of immobilized degradation bacteria modified by chitosan-based aerogel and macroalgae (Caulerpa lentillifera). Aquaculture 570:739392

Yi H, Li M, Huo X, Zeng G, Lai C, Huang D, Zhang M (2019) Recent development of advanced biotechnology for wastewater treatment. Crit Rev Biotechnol 40(1):99–118. https://doi.org/10.1080/07388551.2019.1682964

Younas F, Mustafa A, Farooqi ZUR, Wang X, Younas S, Mohy-Ud-Din W, Hussain MM (2021) Current and emerging adsorbent technologies for wastewater treatment: trends, limitations, and environmental implications. Water 13(2):215

Yousif Y, Hassan O, Ibraheem IJ (2024) Removal of heavy metal ions from water using nanocellulose-based membranes derived from macroalgae Chara corallina. J Degrad Min Lands Manag 11(3):5793–5803

Zhang Y, Hu Z (2013) Combined treatment of Pseudomonas aeruginosa biofilms with bacteriophages and chlorine. Biotechnol Bioeng 110(1):286–295

Zhao X, Wang Y, Ye Z, Borthwick AG, Ni J (2006) Oil field wastewater treatment in biological aerated filter by immobilized microorganisms. Process Biochem 41(7):1475–1483

Zheng T, Li P, Wang Q, Li X, Ai H, Gan K, Sharavsambuu A (2015) Pilot-scale experiments on brewery wastewater treatment and sludge reduction based on food chain predation. Desalin Water Treat 55(5):1142–1151

Chapter 3
Biological Treatment Pathways for Cleaner Water Systems

Simranjeet Singh, Etisha Arun, Khushhal Rathore, Daljeet Singh Dhanjal, Ruby Angurana, Joginder Singh, and Praveen C. Ramamurthy

Abstract Wastewater is a complex mixture of contaminants and poses a significant threat to the environment and human health. Effective treatment is crucial to protect water resources and public well-being. Biological treatment, particularly activated sludge, is a common method but faces challenges due to the complexity and variability of wastewater. To address these issues, the wastewater treatment industry is exploring innovative approaches. These include advanced technologies like membrane filtration and nanotechnology, as well as optimizing existing methods through automation and the use of engineered microorganisms. The goal is to enhance treatment efficiency, reduce costs, and recover valuable resources. Ultimately, the successful management of wastewater requires a multifaceted approach, considering factors such as wastewater composition, treatment technology, and environmental impact. By advancing in research and development, we can create more sustainable and effective wastewater treatment solutions for the future.

Keywords Wastewater · Biological treatment · Microbial consortia · Techniques

S. Singh · P. C. Ramamurthy (✉)
Interdisciplinary Centre for Water Research (ICWaR), Indian Institute of Science, Bengaluru, Karnataka, India
e-mail: onegroupb203@gmail.com

S. Singh
Ministry of Environment, Forest and Climate Change, Indira Paryavaran Bhawan, Jor Bagh Road, New Delhi 110003, India

E. Arun · K. Rathore
Department of Biosciences and Bioengineering, Indian Institute of Technology Roorkee, Roorkee, India

D. S. Dhanjal · R. Angurana
Department of Bioengineering and Biosciences, LPU, Phagwara, Punjab, India

J. Singh
Microbial Biotechnology Laboratory, Department of Botany, Nagaland University, Lumami, Nagaland, India

A. Mandpe et al. (eds.), *Membrane Filtration and Membrane Bioreactors for Wastewater Treatment*, Advances in Wastewater Research,
https://doi.org/10.1007/978-981-95-6706-5_3

3.1 Introduction

Water is an indispensable fluid for most biological, natural, and human processes and an omnipotent solvent. To ensure the continuous availability of water on Earth, wastewater treatment must be optimized to remove the maximum number of contaminants that have accumulated through the action of microorganisms or chemical contamination (Saravanan et al. 2021). The water must undergo this treatment prior to being recycled or reintroduced into bodies of water; doing so will significantly mitigate its detrimental effects on the environment. Currently, biological treatment is a highly popular method for treating effluent (Ahmed et al. 2021; Ahmed et al. 2021). This characteristic sets it apart due to its wide range of uses, ability to adjust to different situations, cost-effectiveness, and lack of harmful effects. There are two commonly used biological methods in wastewater treatment: the conventional activated sludge method and the biological contact oxidation method (Saravanathamizhan and Perarasu 2021). At present, the activated sludge method is the most used conventional biological treatment approach in wastewater treatment. Activated sludge is a type of aerobic biological sewage treatment that can eliminate some phosphorus and nitrogen from sewage. It can also remove dissolved and colloidal biochemical organic matter, suspended particles, and other substances that can be absorbed by activated sludge. This treatment method is ideal for handling effluent that has strict water quality standards and maintains consistent water quality. Its efficiency in removing contaminants is remarkably high (Seviour et al. 2003).

In the field of wastewater treatment plant (WWTP) design, construction, and operation, ensuring that the effluent consistently meets legal standards of quality is the principal objective (Hakanen et al. 2013). Wastewater treatment facilities are commonly characterized by substantial financial outlays due to the costly infrastructure and machinery, in addition to the ongoing need for supervision, upkeep, and functioning under typical operational circumstances (Drewnowski et al. 2018). Moreover, effluent treatment facilities typically necessitate regular operation. It exerts strain on both the social and economic spheres, even in nations that are designated as developed (Silva 2023). Engineering professionals, water utility companies, and plant operating experts are in pursuit of environmentally sustainable, technologically advanced, cost-efficient, and innovative solutions (Macêdo-Júnior et al. 2023). The primary objective of operational operations is to guarantee that a wastewater treatment plant (WWTP) produces treated water of suitable quantity and quality, while also adhering to the established regulations (Serdarevic and Dzubur 2019). Conversely, maintenance consists of activities that guarantee the equipment's consistent and efficient operation with the intention of achieving the objectives of sustainable operation (Calais et al. 2017). For example, a compact, straightforward, and initially inexpensive treatment facility might accrue substantial recurring costs, potentially leading to a greater overall expenditure in comparison with an alternative technology. Following the testing phase, water treatment plants (WWTPs) situated in developing and impoverished nations often encounter operational and maintenance-related obstacles (Mara 2003; Pryce et al. 2022). Presently, it is imperative for the water utility industry and society to amass adequate funds to cover the immense costs associated with operation. Currently, the process of reducing the expenses and operational procedures associated with operation and maintenance can be initiated (Blokus-Dziula et al. 2023). In regards to the expenses associated with damage restoration, the ramifications are often substantial and significantly influence the overall cost escalation. To further compound the situation, the restoration of regular operations at a wastewater treatment plant (WWTP) necessitates a substantial investment of effort (Obaideen et al. 2022). Wastewater has emerged as a key contributor to pollution because of the rapid growth of industrialization. This has a substantial impact on the quality of the environment as well as the stability of ecosystems (Zheng et al. 2015). Distinct features of this kind of wastewater include a complex makeup including a great variety of suspended solids and high levels of industrial effluent. If the formation of organic pollutants is not controlled properly, it may have significant ecological effects (Sathya et al. 2022; Tariq et al. 2006). Some of the classical physicochemical treatment technologies can remove certain contaminants; however, their application in the treatment of wastewater is limited since they are expensive, require a lot of energy, and result in residues of waste (Patel and Bhatt 2022). Closely related to this has been increased focus on the application of biological therapy technologies. In the aspect of environmental protection, the biological treatment of wastewater is a process in which microorganisms decompose organic pollutants and remove suspended solids and inorganic pollutants, converting them into inoffensive by-products (Howard et al. 2004). It reduces environmental contamination due to biological treatment, which is the breakdown of organic molecules available in the wastewater from industrial processes by bacteria. This finally results in the production of simplified compounds (Kanaujiya et al. 2019). Also, based on the principles of sustainable development, biological treatment has the positive aspect of minimizing costs and energy utilization in addition to decreasing the generation of waste. This complies with the tenets of sustainable development. Due to the development in the fields of biology, microbiology, and bioengineering, wastewater biological treatment systems have continuously innovated and improved in the past few years (Silva 2023).

Applied technologies include activated sludge, biofilm, biosorption, and plant wetlands, among others, which have proved to work excellently Murnane et al. (2019). Nevertheless, due to the fact that wastewater is a complex, multifaceted natural substance, the different technologies used for wastewater treatment differ in effectiveness and applicability according to the nature of the input. This is due to the variability and complexity of the wastewater itself. It is precisely due to this that identification and assessment of effective strategies in the biological treatment of industrial process waste effluent is still very critical at this point (Crini and Lichtfouse 2019; Singh et al. 2023a, b).

3.2 Composition of Wastewater

Wastewater is a general term that includes sewage and other liquid waste by-products disposed of because of manufacturing. Components of wastewater include contaminants, intermediate materials, final products, and compounds used throughout the process of production (Devda et al. 2021). Among the other characteristics are that it has a complex structure, high concentration and level of pollution, changes in temperature, unpredictability, and chances of holding some other contaminants that are not otherwise found elsewhere (Adejumoke et al. 2018). The different and wide chemical composition of wastewater makes it difficult to define and categorize the many forms of effluent. A wide range of the digestion of waste depends on the microorganism present (Asthana et al. 2017). The wastewater can, therefore, be categorized based on the components changed and end products from the different industries. These ranges include metal working, papermaking, chemical, dyeing, tanning, pesticide, textile printing and dyeing, among many other processes (Yadu et al. 2021). Suppose the classification of wastewater is based on the primary pollutants it contains, then such classification is possible. For example, acidic wastewater, alkaline wastewater, cyanide wastewater, chromium wastewater, mercury wastewater, phenol wastewater, oil wastewater, organo-phosphorus wastewater, and many other related substances in wastewater are classified based on pH. It is important to have a classification system based on real-world occurrences (Karri et al. 2021). This has resulted in a tremendous increase in both the quantity and type of wastewater due to rapid industrialization. From such a perspective, pollution in aquatic ecosystems, causing the prevalence of risks linked to human health and general human safety, is growing far beyond acceptable boundaries. Nothing is more crucial in preserving the environment other than the treatment of wastewater originating from municipal sources. Most critical is the control of wastewater from industrial sources (Kloepfer et al. 2005). While this is a very vast history of wastewater treatment beginning in the late nineteenth century and has been refinished through many trials and approaches, there have so far remained technological challenges (Lofrano and Brown 2010). This problem occurs mainly due to the complex nature of wastewater and its consistently variable chemical composition. Thus, treating sewers in the industrial setting involves different characteristics and creates various issues in comparison with treating wastewater in the municipal setting (Soares 2020).

3.3 Importance of Wastewater Treatment

It not only decreases the amount of excess pollutants produced but also the fact that it offers significant cost reductions that are of a larger magnitude. This chapter is aimed at presenting the complete panorama of the significance, methodology, and mechanism of biological alteration of biochar along with its application in the process of environmental remediation process in the context of environmental remediation

(Wei et al. 2003). The biologically modified biochar can be achieved using several techniques. These may be ex situ technologies, in which anaerobically digested material is pyrolyzed, or in situ technologies, wherein extracellular enzymes are used (Azad et al. 2022). This process is recognized to be one through which biofilms are produced by microorganisms on the surface of biochar; colonization and attachment of microbes to the biochar surface have been identified as culprits. Biochar, having undergone biological changes, can metabolize organic and inorganic pollutants, catalyzing adsorption, biodegradation, and biosorption at the same time (Daims et al. 2006). The linkage of wastewater to the rapid fluctuations in climate could be well established.

This Sustainable Development Goal aims to avert the adverse impacts of climate change by regulating and developing renewable energies. Since water can feasibly be a resource for renewable energy, it is achievable to recycle and subject part of the wastewater to chemical processes to develop inexpensive energy that is equally eco-friendly (Smit and Pilifosova 2003). The treatment efficiency of wastewater is expected to increase, mainly in developing nations, to combat issues brought about by climate change (Zarei 2020). Mixed with the other sources, the chemically treated water is free of many water contaminants and chemicals to provide energy, thus reducing the intake of fresh water and water from underneath the ground (Ahmed et al. 2017). The potential growth of the renewable sector is so promising in drastically reducing pollution and providing a safe, healthy environment for both human beings and animals using renewable energy (Algarni et al. 2023). To give light to the additional environmental issues associated with desalination, it is shown in the research that the recycling of wastewater is favored over the desalination process of saltwater. In the near future, the intake of fresh water from rivers and lakes is likely to diminish because of the invention of renewable energy resources, and this is the result of the significance of wastewater treatment plants (Dhakal et al. 2022; Miller et al. 2015). Furthermore, the treated wastewater can serve multiple functions beyond its initial purpose of being released back into the environment. It can be used for activities like washing, cleaning, and even in agriculture (Jhansi and Mishra 2013) (Fig. 3.1).

Within the context of the ongoing expansion of industry and the projection that the global population will reach 10 or 11 billion people in the twenty-first century, wastewater treatment is an extremely important component in the protection of both the environment and the health of the human population (Gomes da Silva and Gouveia 2020). In the end, contamination that is created by households, industries, and agricultural practices ends up in the environment or the waste stream, where it may be processed in wastewater treatment plants, provided that such facilities are available. Even though they might have relatively large operational costs, modern wastewater treatment plants are quite efficient (Zheng et al. 2013). Furthermore, they consume a substantial amount of electricity, which can be a big contributor to the overall electricity expenses of a municipality to a significant degree. It is possible that a substantial step forward might be taken by investigating novel technologies that can make use of the potential of sewage (Obileke et al. 2021). Sewage contains both organic and inorganic contaminants, each of which contains useful energy that could

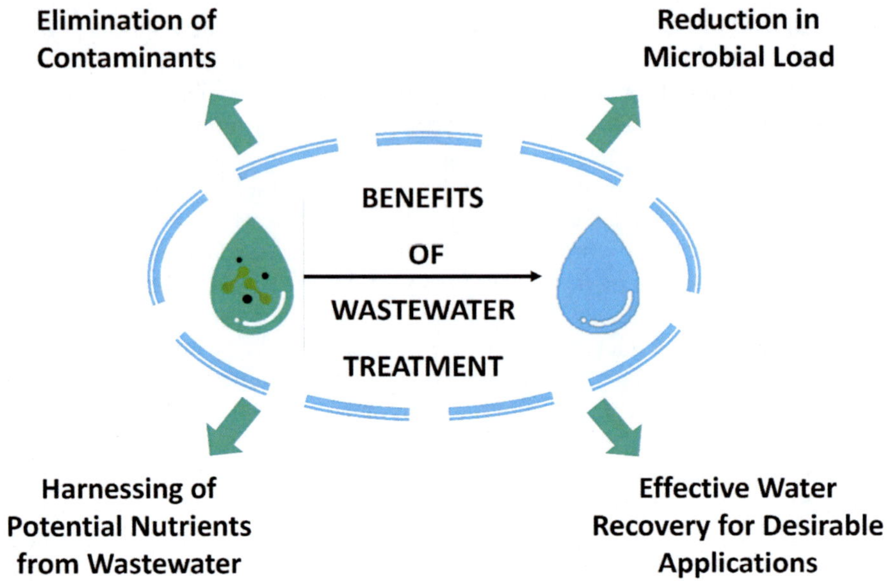

Fig. 3.1 Various benefits of wastewater treatment

be recovered in an effective manner throughout the cleansing process (Topare et al. 2011). While this is the case, the rapidly developing technology of microbial fuel cells shows promise for the future as a technique for treating wastewater that is environmentally friendly. It presents the possibility of simultaneously removing pollution while concurrently turning the chemical energy of pollutants into power that is less harmful to the environment (Roy et al. 2023). In addition, we investigate the foundations of the Microbial Fuel Cell and the ways in which it might be utilized in the environmentally responsible treatment of wastewater (Topare et al. 2011).

3.4 Advantages and Limitations of Biological Treatment

When it comes to the treatment of wastewater from industrial processes, challenges that include the presence of complex organics and vast quantities of wastewater are among the most difficult to overcome. Wastewater is generated by many industries, ranging from that produced by the textile industry, chemical, and pharmaceutical industries to that produced by the food processing business (Samaei et al. 2018). There are enormous varieties of wastewater produced from industrial activities (Zieliński et al. 2023). The long list of organic compounds present in wastewaters of concern includes organic solvents, oils and fats, phenolics, and heavy metal ions, to mention just a few. There have been many cases where a single biological treatment is not enough to satisfactorily provide the results expected (Zheng et al. 2013).

The reason for this is that the drug possesses not only a high concentration but also complex degrading properties. Because of this, the removal of organic matter through the biological treatment process is either inefficient or unsuccessful. This is because high concentrations of organic matter inhibit the growth and metabolic activities of microorganisms, which ultimately leads to the removal of organic matter (Khan et al. 2023a, b; Molinuevo-Salces et al. 2019). Microorganisms have a tough time breaking down certain complicated organic matter because of its convoluted structure. On the other hand, "some organic matter is complex," (Parmar et al. 2022). When this factor is put into consideration, the biodegradation of hazardous substances can be said to be a difficult process to achieve. The structural complexity of toxic substances and suppression of microbial activity by toxic substances are the main obstacles that must be conquered to achieve the biodegradation of toxic compounds as indicated by (Cesaro et al. 2013). The most general sense of the structural complexity of hazardous substances lies in the fact that it hampers the ability of bacteria to break down such hazardous substances into their core components, which are the first step in the process. Due to the fact that these molecules may have a few different functional groups, identification of the enzymes in the microbial metabolic pathway that are responsible for breaking down these toxins is relatively difficult (Forrez et al. 2011). Additionally, some hazardous materials are persistent such that it becomes difficult for them to break down in an organism present in the natural environment (Sharma et al. 2023). Microorganisms are hampered in their ability to carry out their functions when toxic chemicals are present. The growth of microorganisms along with the metabolic activity of organisms can be disrupted by significant quantities of harmful compounds, which can lead to a reduction in degradation or even a complete cessation of degradation. This can be a consequence of the disruption of the growth by harmful compounds (Shahid et al. 2020). Some toxicants tend to cause damage to the genetic material of bacteria, which might further slowdown the extent to which biological systems deteriorate. Therapy is frequently required in conjunction with other approaches, such as physicochemical methods, to be successful in overcoming the problems (Okoye et al. 2022). Since wastewater produced by a variety of industries and processes has variability, which includes the composition of wastewater, concentration of wastewater, pH of wastewater, temperature of wastewater, and so on, the technological adaptability and stability of biological treatment is also a comparable challenge. This is because of the many factors that contribute to the variability of wastewater (Nishat et al. 2023). The selection of biological treatment technologies that are suitable for the characteristics of certain industrial effluents is therefore of the utmost importance from a technical standpoint (Leitão et al. 2006). For example, the treatment of wastewater having a high content of organic matter may include aerobic and anaerobic bioreactors such as MBR, SBR, UASB, etc. (Osman and Hodaifa 2023). In contrast, as cited by Razzak et al. (2022), the process may involve a combination of biological and chemical treatments in the case of heavy metal-contaminated wastewater (Razzak et al. 2022). Therefore, the selection of the applied biological treatment technologies needs to be considered in relation to Wastewater characteristics and the existing situation for a technology to be adaptable. This is to ensure that the technology is able to handle wastewater (Prasse et al.

2015). The most important factor, however, is the stability of the biological treatment of wastewater, which is by and large essential for the smooth running of the treatment system in ensuring contaminant removal efficiency. Wastewater happens to be a complex and rather unpredictable substance that, by nature, may tend to have detrimental effects on the biological treatment system, thus reducing its capacity or even leading to failure (Mishra et al. 2023). Operational parameters of the technology can be modified to generate optimal environmental conditions that promote microbial activity and stability. In this regard, temperature, pH, and aeration, among others, can be adjusted in order to make the technology more stable (Schultz 2005). Consequently, the generation of an artificial ecosystem with multiple species living together possesses several advantages. This might increase the system's capacity to degrade a wide array of pollutants and its effectiveness in preventing the occurrence of any kind of disturbance (Ahmed et al. 2022). So that any sudden shocks could less influence the process, it needs to be controlled and monitored, with the immediate necessity for periodic evaluation of significant parameters and the treatment process for any quick changes. The parameters monitored include dissolved oxygen, COD in wastewater, and ammonia nitrogen, among others. Adjustments should be made where necessary in relation to the monitoring results. The biological stage of water treatment forms an important part of the wastewater treatment process. It caters to wastewater resulting from a very diversified basis like commercial and domestic wastewater. The term "secondary treatment" is applied to the removal of impurities that may remain after the successful completion of the first treatment (Mohamad Faudzi et al. 2023).The chemical treatment of wastewater refers to the use of chemicals that react with the contaminants present in the wastewater (Borgaonkar and Marhaba 2021). On the other hand, biological treatment of wastewater involves the use of microorganisms to break down the pollution in wastewater. This mode of treatment transforms the unstable organic wastes into stable inorganic forms through the action of bacteria, nematodes, algae, fungi, protozoa, and rotifers (Jung and Pauly 2011; Djadouni 2019). Secondary treatment has been defined as being "treatment of wastewater by a process involving biological treatment with a secondary sedimentation," according to Samer (2015). Biology serves as the foundation for the technique of secondary therapy. The settled effluent is fed to a specifically designed bioreactor where the organic matter is digested by aerobic or anaerobic microorganisms such as bacteria, algae, and fungi (Khan et al. 2024). In the bioreactor, a very ideal bioenvironment would be provided for the microorganisms to grow and make use of the soluble organic materials as a supply of energy for sustenance purposes (Zhang et al. 2018).

The biological oxidation process of the dissolved organic matter will continue so long as the bacteria gain access to food and oxygen in the form of settling wastewater. This will give them the ability to go on with their process. A vast majority of responsibility for the biological process rests with the bacteria, forming the basic tropic level that a bioreactor can be visualized to, which is the position an organism occupies in a food chain (Hussain et al. 2021). The bioconversion of dissolved organic molecules can result in the production of thick bacterial biomass, which has the potential to greatly cleanse wastewater. In the subsequent step, the microbial biomass that was present in the treated wastewater is eliminated by the process of sedimentation (Goffin et al. 2018; Samer 2015). The only difference between this secondary sedimentation and primary sedimentation is that the sludge contains bacterial cells rather than fecal particles. This secondary sedimentation is considered to be practically identical to primary sedimentation (Cervantes et al. 2006). Microorganisms, primarily heterotrophic bacteria but also occasionally somewhat unusual fungi, are responsible for the biological removal of organic compounds from wastewater that has settled because of the process of settling (Rahimi et al. 2020). Biosynthesis and biological oxidation are the two distinct biological mechanisms that bacteria use to break down organic molecules (Nguyen et al. 2021). These mechanisms are utilized by bacteria in order to degrade such compounds. In addition to the effluent, minerals and other end products of biological oxidation are still present in the solution (3.1). Biosynthesis is the process by which new cells are created from organic matter that has been dissolved and brought into the colloidal structure. It is possible to remove the dense biomass that these cells create by the process of sedimentation (3.2).

3.5 Oxidation:

$$\text{COHNS} + O_2\,(\text{Organic matter}) + \text{Bacteria} \rightarrow CO_2 + NH_3 + \text{Energy} + \text{Other end products} \quad (3.1)$$

3.6 Biosynthesis:

$$\text{COHNS} + O_2\,(\text{Organic matter}) + \text{Bacteria} \rightarrow C_5H_7NO_2\,(\text{New cells}) \quad (3.2)$$

Several bioenvironmental variables have an impact on the rate of biochemical processes and bacterial activity. The most critical variables are temperature, pH, dissolved oxygen, nutrition concentration, and dangerous chemicals. Controlling all these parameters in a bioreactor or biological treatment system ensures the preservation of microbial growth in an optimal bioenvironment (Kanaujiya et al. 2019). The optimal performance of most biological treatment systems occurs within the mesophilic temperature range, often between 20 °C and 40 °C. The temperature of the wastewater in aeration tanks and percolating filters usually ranges from 12 °C to 25 °C. However, the efficiency of heat dissipation percolating filters is significantly influenced by the rate of ventilation and the temperature of the ambient air (Chowdhury et al. 2010). High temperatures increase biological activity and metabolism—the rate of substrate removal. However, the raised metabolic rates under higher temperatures may give rise to issues related to adequate availability of oxygen supplies (Dammak et al. 2023).

Another essential topic has to do with the utilization of both autotrophs and heterotrophs in the bioremediation of wastewater (Grandclément et al. 2017). In this context, a heterotroph can be considered an organism that uses organic carbon for growth because it cannot synthesize needed organic compounds independently from carbon dioxide (Van Aken 2011). The heterotrophic division is determined by the energy sources used. A photoheterotroph is an organism deriving its energy from light (Sepúlveda-Muñoz et al. 2023) A chemoheterotroph obtains energy from organic compounds as well as from inorganic ones (Nirmalakhandan et al. 2019; Samer 2015). Photoautotrophs are organisms, particularly plants and algae, which can make use of solar radiation for energy requirements (Sepúlveda-Muñoz et al. 2023). A lithoautotroph is an organism that utilizes inorganic carbon dioxide to synthesize organic compounds, including proteins, lipids, and carbohydrates (Oller et al. 2011) Autotrophs receive an energy source from the reduced carbon molecules, while heterotrophs receive their main energy source from the food which they consume (Kim et al. 2020). Heterotropic organisms comprise about 95 percent of the total number of species (Calow 1977).

3.7 Types of Biological Wastewater Treatment Process:

Wastewater treatment systems can be broadly classified into two classes based on biological processes. It includes aerobic and anaerobic treatment.

3.7.1 Aerobic Wastewater Treatment

Aerobic Wastewater Treatment is a biological process that takes place when there is oxygen present. According to Loukidou and Zouboulis (2001), this method is the most efficient and rapid way to treat biological waste, eliminating about 98% of organic contaminants. This method efficiently decomposes organic pollutants, leading to a purer water discharge compared to anaerobic treatment. Some examples of aerobic biological treatment methods are activated sludge, trickling filters, aerated lagoons, and oxidation ponds. The activated sludge method is widely used for both domestic and industrial wastewater treatment. Aerobic biological treatment will consistently demonstrate high efficiency and stability regardless of the prevailing conditions (Loukidou and Zouboulis 2001).

Aeration can effectively remove trace organic volatile compounds (VOCs) from water, making it free of such contaminants. The method referred to as "gas adsorption" or "oxidation" involves the oxidation of iron and/or manganese to transfer a chemical, such as oxygen, from air or a gas phase into water. Aeration also offers the advantage of releasing dissolved gases such as CO2 and H2S. In addition to other compounds present in water and volatile odors, air stripping has proven to be an effective method for removing NH_3 from wastewater (Kurniawan et al. 2010).

3.7.1.1 Activated Sludge Process

The activated sludge technique is the most widely used method for treating biological waste in the secondary stage of wastewater treatment. An activated sludge process refers to a multi-chamber reactor unit that employs concentrated microorganisms to decompose organic matter and eliminate nutrients from wastewater, resulting in the production of superior-quality effluent (Benner et al. 2013). The sewage, which contains organic waste and germs, is aerated in an aeration tank using a mechanical aerator. This method accelerates the decomposition of waste. The process of aeration in an activated sludge system involves the introduction of air into a tank, which promotes the growth of microorganisms in the wastewater. The sludge, which refers to the effluent from the aeration tank that contains the flocculent microbiological material, is separated into a settling tank. This settling tank is also known as a clarifier or secondary settler (Cesaro et al. 2013).

Mixed liquor is formed by the combination of wastewater and microbial substances. After a specific duration in the aeration tank, the mixed liquor is transferred to a settling tank, which is also referred to as a clarifier. In this tank, the biomass or sludge present in the treated wastewater settles down due to the force of gravity. The waste generated by this continuous process might be directed to publicly owned treatment works (POTW) or subjected to further treatment (Bouwer and Crowe 1988). The activated sludge process, initially developed in the early 1900s for treating domestic wastewater, has now been adapted to effectively eliminate naturally decomposing organic substances from industrial wastewater. Currently, it is widely employed for the purpose of purifying wastewater originating from both industrial and municipal sources (Orhon 2015).

Activated sludge treatment is an effective method for facilities that handle waste organic compounds. These include municipal sewage treatment plants, chemical manufacturing facilities, paper and pulp mills, oil refineries, and food processing companies (Rittmann et al. 1989). The activated sludge method, when developed correctly, offers effective treatment for waste that contains biodegradable elements. Typically, it is employed for the treatment of biodegradable organic wastewaters that range from mild concentrations (10–1000 ppm) to high concentrations (>1000 ppm). The performance is contingent upon the particular compounds and their respective concentrations (Sustarsic 2009). The activated sludge treatment process is not designed to remove metals or other dissolved or suspended pollutants. In order for things to work properly, there must be a relatively high concentration of organic compounds, specifically above 100 parts per million (ppm) (Dignac et al. 2000).

3.7.1.2 Tracking Filters

The aerobic treatment method, also known as percolating or sprinkling filters, is the second most often used type. Following the initial treatment, these filters are frequently employed to remove pollutants, such as ammonia, from water. Once it has reached a state of stability, the treated wastewater will be fed into a digestion system (Paździor et al. 2019). Biological trickling filters, commonly referred to as BTFs, have been employed for almost sixty years to purify wastewater (Butler and Boltz 2014). The first iterations of BTFs were developed using a pragmatic comprehension of biological wastewater treatment. Currently, there is an increasing trend in the popularity of attached growth (biofilm) systems for wastewater treatment. These systems entail the use of biomass growth on support media. Attached growth systems provide various benefits in comparison with conventional activated sludge processes, such as lower energy consumption, reduced biomass generation, and faster treatment durations (Ganzenko et al. 2014). Biological trickling filters (BTFs) have been used in wastewater treatment for more than sixty years, with their original designs relying on practical knowledge of biological wastewater treatment. The use of attached growth (biofilm) systems for wastewater treatment, which involves the development of biomass on support media, is currently becoming widely popular (Wik 2003). Attached growth systems provide several advantages over typical activated sludge processes, including lower energy usage, decreased biomass generation, and shorter treatment times (Naz et al. 2015).

3.7.1.3 Oxidation Ponds

These are shallow, man-made bodies of water that are used to treat wastewater.

The bacteria, algae, and other species that break down the organic waste in the primary effluent participate in interactions within the ponds. According to Mantzavinos and Kalogerakis (2005), the wastewater produced by these ponds can be used for several reasons, making them productive. Significant amounts of land are necessary, and the procedure is usually slow (Mantzavinos and Kalogerakis 2005). Oxidation ponds are frequently used in regions characterized by sparse populations and abundant land resources (Butler et al. 2017). Oxidation ponds are aerobic habitats where the oxygen required by heterotrophic bacteria, which are organisms incapable of producing their carbon and depend on organic carbon for growth, is supplied by photosynthetic algae and atmospheric transfer (Johnson et al. 2020). The algae reside solely in the euphotic zone, sometimes referred to as the sunlight zone, which typically has a depth of only a few millimeters. Ponds are intentionally constructed with a depth ranging from 1.2 to 1.8 m to maximize the penetration of sunlight into the water. Ponds appear to have a dark green color due to the extensive growth of algae (Mantzavinos and Kalogerakis 2005; Nageshwari et al. 2021).

3.7.2 Anaerobic Treatment

Typically, wastewater with a high concentration of pollutants is processed to a degree that permits it to be released into a local sewer system. This treatment technique is effective in the treatment of high-strength wastewater and employs anaerobic organisms capable of surviving in the absence of oxygen (Grady et al. 2011). Compared to aerobic treatment, the amount of sludge generated in this situation is less. Anaerobic treatment occurs in multiple stages and is a lengthy procedure (Fricke et al. 2005; Kassab et al. 2010). Wastewater treatment facilities utilize anaerobic digestion, a biological process, to stabilize and decompose sludge. Once the procedure is completed, the wastewater may undergo a range of further treatments. This method is approved because it can stabilize the water with a minimum creation of biomass. Biogas is produced by the anaerobic process when microorganisms eat biodegradable material. The process converts around 40% to 60% of the organic materials into carbon dioxide and methane (CH_4) (Tran et al. 2016). The food industry and agriculture sectors are the main contributors to potent organic wastewaters that contain substantial quantities of easily decomposable substances. Due to the challenges and expenses involved in fulfilling the increased oxygen requirements for maintaining aerobic conditions, treating these wastewaters aerobically is difficult (Samer 2015). Anaerobic degradation occurs in the absence of oxygen. The anaerobic treatment of highly organic wastewaters offers numerous advantages, despite its time-consuming nature. The benefits of this process include a high level of purification, the capacity to handle huge amounts of organic matter, the production of methane, which is a non-reactive flammable gas, as the final product, and the generation of small quantities of highly stable sludge (Zieliński et al. 2023). Livestock manure, food processing effluents, petroleum wastes (as long as the toxicity is regulated), and wastes from canning and dyeing operations that use soluble organic components in their treatment are all considered appropriate examples of wastewaters (Karić et al. 2022; Samer 2015). Anaerobic reactions, such as solids fermentation, mostly occur within two distinct temperature ranges: mesophilic and thermophilic. The temperature ranges are 30–38 °C and 38–50 °C, respectively (Steiniger et al. 2023). In contrast to aerobic systems, anaerobic environments do not allow for the complete stability of organic components. Consequently, it is typically essential to employ aerobic treatment for the anaerobic effluents afterward (Chipasa and Mędrzycka 2006). Due to the presence of solubilized organic matter that can be easily treated aerobically, it is possible to install a combination of anaerobic and aerobic units for the treatment of the final waste matter (Zheng et al. 2013).

3.7.2.1 Up-Flow Anaerobic Sludge Blanket (UASB) Reactors

A dense population of bacteria suspended in aerobic wastewater forms the basis of the activated sludge process. High rates of bacterial growth and respiration are possible in environments with abundant oxygen and nutrients. In the process, the organic matter that already exists is either used to create more microbes or transformed into oxidized by-products like CO_2, NO_3^-, SO_4^{2-}, and PO_4^{3-} (Hussain et al. 2021; Samer 2015). According to Van Rijn (1996), there are five interconnected parts that make up the activated sludge process: the sludge itself, the aeration and mixing system, the sedimentation tank, the bioreactor, and the returned sludge (van Rijn 1996). When it comes to treating wastewater, using activated sludge in the biological process is a great option because it is effective and doesn't cost much to run. The oversized quantity of excess sludge, however, that WWTPs produce creates major problems in terms of environmental impact and cost. Given the high water content, excess sludge is challenging to clean. The by-products of the WWTP are first dried and dewatered before being burned into ash (Lofrano et al. 2013).

The biological treatment process is the most widely adopted process for the treatment of domestic wastewater. This process is driven by bacterial populations with unique sedimentation characteristics (Machineni 2019). Live microorganisms added to the polluted wastewater can break down organic materials and consume biodegradable waste particles, improving the quality of the contaminated water (Parmar et al. 2022).

3.7.2.2 Combined Aerobic Anaerobic Processes

According to Liu et al. (2019), the most common application of anaerobic and aerobic systems is streams that have a considerable number of organic contaminants (Liu et al. 2019). The operation in this kind of set-up overview includes anaerobic treatment, a process that minimizes organic pollutants; and aerobic treatment, which will polish and reduce further the amount of biological oxygen demand and total suspended solid. Anaerobic systems, on the other hand, should be designed in such a manner that they hold the biomass sludge in them and do not allow the sludge to come into contact with air. This is ensured in the UASB system or in an airtight enclosed digester—a configuration more commonly used for batch treatment cycles—by the preservation of a biomat below the treated effluent as it rises up toward the surface of the tank. The oxidation tends to form nitrate in the second-stage aerobic treatment step. In effect, the use of both technologies in conjunction normally proves a better and enhanced treatment than the use of the aerobic system alone (Lodha and Chaudhari 2007). This aspect manifests itself more so profoundly when only an anaerobic treatment approach is utilized without the administration of oxygen, as determined by Chan et al. (2009) and Zieliński et al. (2023) (Chan et al. 2009; Zieliński et al. 2023). Due to the nature of the pollutants, concentration levels, temperatures, and other characteristics of the wastewater stream, one of the methods may be more effective for the treatment of biological components than the other (Dionisi 2017).

There are several ways through which anaerobic and aerobic systems handle biological components. Generally, however, the aerobic treatment systems, which are also able to remove phosphate and nitrogen, are best suitable for streams with a low BOD and COD. On the contrary, according to Hussain et al. (2021), anaerobic systems are generally employed for the treatment of either warm wastewater streams or waste streams that primarily consist of large quantities of organic pollutants (Hussain et al. 2021). It is known in the study by Velis et al. (2009) that the type of waste generated for treatment by anaerobic and aerobic technologies varies from one another. This difference can be the purpose for the specific application of one method over the other. In this respect, anaerobic treatment systems become better than aerobic systems in terms of valuable by-products produced and sludge production, respectively (Velis et al. 2009). Anaerobic by-products, comprising energy-rich biogas and stable, safe sludge for use as agricultural fertilizer, are often treated, reused, and commercialized by facilities (Cortez et al. 2008; Diener et al. 2011). These by-products are often marketed to agricultural producers. While these benefits exist, there remains the paramount importance of mitigating the unpleasant odors often generated by anaerobic treatment systems. Anaerobic systems generally have a few design advantages over aerobic ones with respect to energy and operational costs. This also involves high efficiency. On the flip side, the systems are typically more costly and require a longer period for successful completion (de Souza et al. 2010; Lettinga 1996).

3.8 Factors Affecting the Efficiency of Biological Treatment

Various methods are being employed to treat wastewater and make it safe enough to be released into water bodies. Every method comes with a few limitations and their efficacy is governed by several factors:

I. Influent wastewater characteristics
II. Operational parameters
III. Microbial community composition and activity
IV. Reactor design and configuration

3.8.1 Influent Wastewater Characteristics

For efficient denitrification of the wastewater, the ratio of Chemical Oxygen Demand to total Nitrogen (COD/N) of the water is of critical importance (Chen et al. 2015). The nitrate recycling ratio (R) also affects the removal of nitrogen as well as phosphorus from untreated water (Li et al. 2023). Chen et al. (2015) have shown that the COD/N ratio and R significantly affect aerobic and anaerobic wastewater treatment using biological aerated systems (BAF). The efficiency of denitrification at low COD/N ratio and high R decreased, whereas at COD/N ratios at moderate level, the nitrogen depletion efficiency increased with increasing R but only up to a certain level (Chen et al. 2015). The efficiency was highest at a very high COD/N ratio with increasing R-values. The COD/N ratio also depends on the age of the sludge. The sludge age of more than 20 days and low organic content of sludge cause issues in the settling of sludge (Ahn and Park 2008). A significant improvement in denitrification was observed in the addition of a volatile fatty acid source as an additional carbon source. The higher ratio of volatile fatty acid to nitrogen (VFA/N) shortened the denitrification time (Komorowska-Kaufman et al. 2006). Thus, the amount of carbon source that microorganisms can easily utilize is vital to enhance the wastewater treatment process. Optimizing the size of the anoxic zone also enhances the biological treatment of wastewater. The quality of wastewater also determines the microbiota and their preferred metabolic pathway that would govern the treatment process (Plaza et al. 1991).

3.8.2 *Operational Parameters*

Apart from the quality of wastewater, the pH of the treatment conditions also affects the efficiency of the process. The release of phosphorus from wastewater increased when pH was around the neutral range in aerobic–anaerobic cycles of batch reactors. The uptake of short-chain fatty acids increases when pH is raised to a slightly alkaline range of 8.0 (Liu et al. 2007). For phosphorus accumulating organisms (PAOs), the change in pH in the range of 6.8–8.0 had an insignificant impact on the uptake of acetate, whereas an increase in pH decreased the acetate uptake capability of glycogen accumulating organisms (GAOs) (Filipe et al. 2001a, b).

Similarly, an optimum temperature of 16–21 °C is proven best for denitrification. Temperature lower than 15 °C is suitable for sludge age of 20 days or above (Komorowska-Kaufman et al. 2006). High temperatures of 31–40 °C increase the effluent suspended solids and reduce the soluble chemical oxygen demand, thus lowering the quality of sludge. Changes in temperature during the treatment process result in differences in the diversity of microorganisms. A shift to higher temperatures reduces the population of metazoans and protozoans (Morgan-Sagastume and Allen 2003). The aerobic process of wastewater treatment entirely depends on the levels of dissolved oxygen maintained by aeration of the sludge. Microorganisms use the dissolved oxygen of water to utilize the organic contaminants as a source of nitrogen and carbon essential for their metabolic processes. To ensure proper treatment of water contaminated with high organic content, a sludge system supplying pure oxygen is preferable (Zupančič and Roš 2008). Denitrification is an anaerobic reaction and is directly affected by the dissolved oxygen content. The enzymes governing the denitrification process are inhibited in the presence of oxygen (Grießmeier et al. 2019). Thus, maintaining oxygen levels at different stages of wastewater treatment is essential. Oxygen concentration also impacts the diversity of microorganisms present and active to carry out the biological treatment of wastewater (Abdelfattah et al. 2023).

3.8.3 *Microbial Community Composition and Activity*

Biological wastewater treatment mostly depends on the quality of microorganisms. Bacteria form a major group of microbiota involved in wastewater treatment (Daims et al. 2006). Many species of *Pseudomonas, Bacillus, Lactobacillus,* and *Alcaligenes* are major bacteria that are involved in biodegradation and decolorization of the wastewater (Chattopadhyay et al. 2022). Anaerobic bacteria do not require continuous aeration and thus consume less energy compared to aerobes (Kumar et al. 2008). Contaminated wastewater with high molasses content needs fungal species like *Aspergillus* to efficiently carry out biodegradation. The use of algae for bioremediation is better as algae can utilize nitrate, phosphorus, and ammonia from water as an energy source. This reduces the load of artificially added carbon sources (Kushkevych 2021). The diversity of the microbial population also impacts the pH of wastewater governing the treatment process. The diversity of microorganisms is also determined by the type of substrate available in the wastewater for them to utilize in the metabolic pathways (Szymanski and Patterson 2003).

3.8.4 Reactor Design and Configuration

In the case of aerobic bioreactors, the number of stirred tanks affects the aeration and results in suspended microorganism growth (Metcalf et al. 1991). This can be improved by using multiple stirrers in series to avoid the settlement of microbes, enhance aeration, and reduce the operation cost of the bioreactor (Narayanan and Narayan 2019). For anaerobic bioreactors, the rate of digestion can be increased by the addition of multiple substrates like paper waste (Yen and Brune 2007). These then can be digested by introducing microalgae to the sludge which improves the digestion rate as well as the biogas production rate (Olsson et al. 2014). The trickle bed biofilm reactor can be made open or closed based on the sensitivity of the products to be retrieved (Olalere et al. 2022). To enhance the efficient removal of nitrogen and organic matter circulating fluidized-bed bioreactor (CFBBR) was developed at the University of Western Ontario which has a circulating loop that increases the combustion. CFBBRs are used to treat wastewater from both municipal and industrial sources. It has better mixing ability and enhanced mass transfer. CFBBRs have been proved to increase the denitrification and removal of organic matter in wastewater by as much as 80% (Nelson et al. 2017). Aerobic, anoxic, and anaerobic conditions can be alternatively maintained to select the quality and diversity of microbiota in the reactor (Wang et al. 2022).

3.9 Advantages and Disadvantages of Different Biological Processes

Biological methods of wastewater treatment have been very widely used, as they are highly efficient and environmentally sustainable with lower treatment costs compared to chemical methods (Sathya et al. 2023). Both aerobic and anaerobic methods are used widely based on the quality of wastewater. Anaerobic treatment is preferred for water with high Chemical Oxygen Demand (COD) of more than 4000 mg/L, whereas low COD of 1000 mg/L is treated using aerobic methods. Anaerobic treatments do not require constant aeration for the microbial population and hence have lower energy requirements (Chan et al. 2009). Industrial wastewaters have high COD and are thus treated anaerobically. Anaerobic treatment has some other advantages over aerobic as it produces valuable by-products which can be utilized as biofuels. Aerobic treatment on the other hand produces high-quality effluent and is more efficient in degrading organic matter (Grady et al. 2011) (Table 3.1).

Recently, combinations of both aerobic and anaerobic processes have been utilized for enhanced treatment of wastewater. The wastewater is first subjected to anaerobic treatment to produce biofuels. The low-quality anaerobic effluent is then treated in aerobic conditions to improve the quality of the effluent generated. The effluent thus produced after an anaerobic–aerobic combination can easily be disposed of. Anaerobic pretreatment decreases the aeration requirement thus making the process cost-effective (Khalidi-Idrissi et al. 2023; Li and Yu 2016). Volatile components are also removed in anaerobic treatment so volatilization during the aerobic process is minimized. The use of both systems in a single treatment process is both economical and environmentally sustainable with better by-products (Cervantes et al. 2006; Ros and Zupančič 2004).

The activated sludge process is cost-effective and can easily be installed in space-deficit areas. It requires less space and produces high-quality sludge. It is convenient to operate and hygienic as the process results in less odor (Waqas et al. 2020). Though the initial installation cost is low, the overall high running cost of activated sludge bioreactors makes it difficult to continue. If the water quality is very poor, the effluent produced also suffers the same fate. Industrial wastewater with very high COD cannot be treated using this method. It is a durable process with low power needs (Oulebsir et al. 2020).

Tracking filters are also simple technology with less area requirement. This is a reliable process which does not require expertise or prior handling experience. It is very efficient for wastewater with high organic content, removes nitrogen, and reduces the Biological Oxygen Demand (BOD) of water (Daigger and Boltz 2011). The effluent produced by tracking filters might need secondary treatment before being discharged into water bodies. They also have a major issue of biomass accumulation

Table 3.1 Comparison between aerobic and anaerobic biological treatment of wastewater

Property	Aerobic	Anaerobic
COD concentration	Low	High
Effluent quality	Good	Poor
Efficiency of organic removal	High	High
Sludge production	High	Low
Biofuel production	No	Yes
Energy requirement	High	Low
Nutrient requirement	High	Low
Temperature sensitivity	Low	High

Modified from Chan et al. (2009)

causing clogging and impaired performance. The process is less flexible compared to the activated sludge process and produces significant odor (Mehrnaz et al. 2020).

Oxidation pond is a natural method for the degradation of wastewater and is most suitable for villages with sufficient sunlight. These reduce the pollutants from wastewater at low cost and do not require much maintenance (Butler et al. 2017). Oxidation ponds do not rely on chemicals and thus are safer if proper maintenance is carried out. They may act as a breeding ground for mosquitoes and produce foul smell. Oxidation ponds cannot operate in low sunlight conditions and might serve as a habitat for disease-carrying vectors and weeds. They require large land for construction and thus cannot be easily installed (Mtethiwa et al. 2008).

3.10 Recent Advances and Future Directions

Old methods used for wastewater treatment are good but their efficacy is quite low. To increase the efficacy and lower the cost multiple different methods are being invented, like using engineering strains of microorganisms, using consortia of different organisms. Also, the use of different technology (i.e., nanotechnology, membrane filtration) increased over time. Here, we will discuss more about these new technologies and future directions (Fig. 3.2).

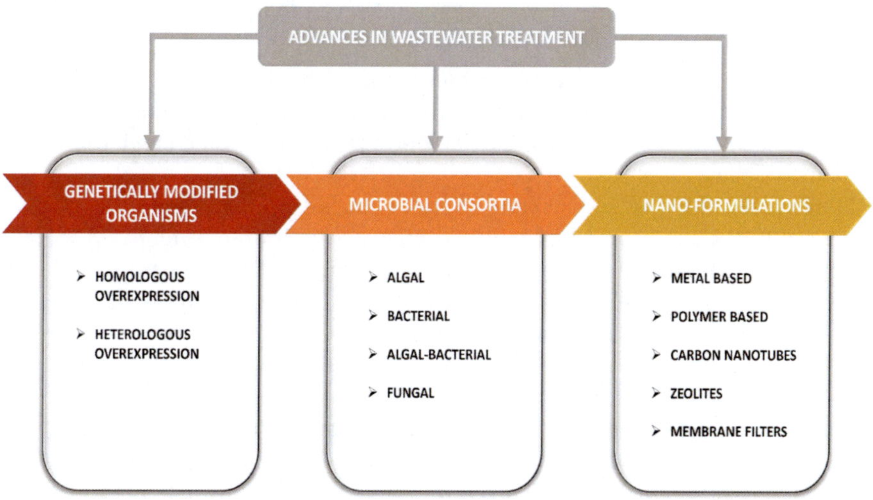

Fig. 3.2 Advances in wastewater treatment

3.10.1 Application of Engineered Strains and Microbial Consortia

The utilization of genetically modified organisms (GMOs) in waste management is called bioremediation. Overexpression of in situ protein or insertion of outside protein for overexpression is the most common way of genetically modifying microorganisms. Engineered proteins are also expressed because they are more stable and more effective. The wastewater contains multiple pollutants like heavy metals, xenobiotics, and more dangerously radioactive compounds (Rafeeq et al. 2023). There are studies that show genetically modified microorganisms removing heavy metals from wastewater than their respective wild types. Genetically modified *Saccharomyces cerevisiae, Aspergillus fumigates, Penicillium canescens* remove three times more heavy metals than their wild strains (Ghosh et al. 2023), while *Escherichia coli, Staphylococcus aureus* and *Enterococcus faecium* were genetically modified to increase their heavy metal removal capacity from their respective wild types (Ebah et al. 2022). The industrial revolution introduced some chemicals that were not natural, natural or very rare to be found in nature. These chemicals were named "Xenobiotic," which means "foreign to life." These xenobiotics are very stable in nature, and there are almost null natural microorganisms that can degrade these xenobiotics (Štefanac et al. 2021). One example of xenobiotics is synthetic pesticides (i.e., organochlorines, pyrethroids). Many reports show pyrethroid hydrolases in multiple species of bacteria and them effectively degrading the pesticides (Ambaye et al. 2022; Shehata et al. 2023; Singh et al. 2023a, b). Genetically engineered strain MC4-5222 of Pseudomonas putida with a dehalogenase gene (*dhaA31*) cloned behind the constitutive *dhlA* promoter was used for degradation of 1,2,3-trichloropropane bioremediation (Samin et al. 2014). One of the most useful characteristics of these organisms is their survival in high pH and temperature ranges, biotic–abiotic stress and withstand high toxicity of pollutants (Ebah et al. 2022).

In nature it is hard to find a single micro-organism species as fully functioning, they usually work in a group of multiple species or with other micro-organisms, i.e., cheese, curd, soy sauce, etc. This strong feature of natural microbial groups, consortium, made the scientist interested and laid the foundation of synthetic consortia (Jiang et al. 2024; Kumara Behera and Varma 2017; Song et al. 2014). Over time several studies reported the plausible applications of microbial consortia (microalgal, bacterial, microalgal–bacterial, fungal, etc.) that include bioremediation of organic material from wastewater (Ramanan et al. 2016). The use of multiple micro-organisms (polyculture) for bioremediation can be very beneficial considering these micro-organisms are acclimatized to different environmental conditions and have different metabolic activities that can help to facilitate the growth of biological systems which is capable of functioning under different conditions (Boonma et al. 2015). Different microbial consortia behave differently. In microalgal–bacterial consortia, both microalgae and bacteria play important roles. Microalgae recover inorganic nutrients (nitrogen, phosphorus) and produce oxygen and photosynthates (sugars) by photosynthesis that helps in the growth of aerobic bacteria present and synthesis of high-value compounds (i.e., lipids, fats, etc.), while the bacteria degrade organic material using photosynthates provided by microalgae, produces CO_2 that helps in growth of algae, and also helps in regulation of algal metabolism to enhance the assimilation power of pollutants (Jiang et al. 2021). Since many new possible combinations are still to be verified or identified, the use of consortia in wastewater treatment is still very restricted (Mu et al. 2021).

3.10.2 Integration of Advanced Technologies (Membrane Filtration, Nanotechnology)

Nanotechnology is one of the rising technologies that are working in all fields of science (Singh et al. 2019; Singh et al. 2022a, b). The nanoformulations (i.e., nanoparticles, nanotubes, quantum dots, etc.) are very useful in drug delivery, catalysis, adsorption, and much more. Properties of these nanomaterials like high surface area, high catalysis activity, antimicrobial, and other electronic and optical properties make them beneficial for wastewater treatment (Malik et al. 2023). One of the chief techniques for the separation of pollutants from water is adsorption. It is also one of the preferred methods for wastewater management because of it being high efficiency, flexibility, low cost, and high potency of a large range of pollutants (Kumari et al. 2019). There are multiple nanomaterials that can be used for adsorption of pollutants. Metal-based, carbon nanotubes, zeolites, and polymer-based nanoabsorbents are the most used nanoformulations for the removal of pollutants (Jain et al. 2021). All of these nanoformulations have their own benefits. Carbon nanotubes have high specific areas, mechanical flexibility, and large pore volumes (Li et al. 2018), while nanometal oxides offer a small size and large surface area that helps in lowering the diffusion distance (El-sayed 2020). Polymeric nanoformulations show exceptional stability, improved selectivity toward pollutants, and high surface area for quick decontamination (Berber 2020), while zeolite-based nanomaterials became center of attraction due to their compatibility in the natural environment, stability in water and high surface area with low cost of production (Rahman et al. 2022).

Another nanoformulation that became quite famous for wastewater treatment in recent years is nanomembrane. The basic principle of nanoformulation is to remove the water contamination using different-sized pores. It uses a pressure of 5–20 bars and a pore size of 0.5 to 2 nm that shows higher rejection than ultrafiltration, consumes low energy, and shows higher flux than reverse osmosis (Abdel-Fatah 2018; Mulyanti and Susanto 2018). Major reasons for acceptance of this method are its flexible design, eco-friendly, and cost-effectiveness with high-quality water processing even after occupying significantly less space (Guo et al. 2016; Khan et al. 2023a, b; Zhou et al. 2014).

3.10.3 Development of Resource Recovery Systems from Wastewater

Wastewater contains many nutrients (e.g., potassium, sulfur, magnesium, phosphorus, etc.) and different organic matter (nitrogen-containing organic compounds) that can further be used for other purposes (fertilizers); also excess nutrient removal slows the eutrophication process and stops the water body destruction. Nutrient removal can be done using different methods like biological, physical, chemical, or they can be used in mixed format (Al-Ghouti et al. 2019; Kim and Chung 2014). Water, energy, and biosolids recovery are very useful in decreasing waste and increasing the efficiency and sustainability of resources. Energy recovery is one of the most useful and important aspects of resource recovery from wastewater. Energy recovery can be done in the form of biogas from anaerobic bioreactors (Manyuchi et al. 2018), sunlight-driven splitting of wastewater sludge for hydrogen production (Sekar et al. 2020), and other biofuel production (i.e., biodiesel, biocrude oil, bioethanol) (Yadav et al. 2019). Besides these, different excreta (mostly urine) of domesticated animals can be used in microbial fuel cells to generate power (electricity) (Lu et al. 2019; Yadav et al. 2020). One of the recent problems that humanity is facing, fresh water scarcity, can also be decreased by water recovery from wastewater. Recovered water can be used in different fields like irrigation and drinking water after treating it accordingly (Guo et al. 2014).

3.10.4 Automation and Optimization of Treatment Processes

The wastewater treatment process consists of multiple steps, each step is very crucial for a good quality treatment. To keep it that way, each step has to be observed. In past years usage of automatic sensors for measurement of dissolved oxygen, chemical oxygen demand, biochemical oxygen demand, pH, electrical conductivity and oxidation–reduction potential has increased (Zhang et al. 2020). These sensors automatically maintain a level of these characteristics like the dissolved oxygen sensor maintains the oxygen levels by controlling blowers (Keskitalo and Leiviskä 2012). Optical, ion-selective electrodes, biosensors are some of the majorly used sensors. Optical sensors are durable sensors that are based on the absorbance and fluorescence principle. Durability and minimal maintenance requirements make these optical biosensors common and one of the most used sensors for the automation of wastewater treatment process (Ferrari et al. 1996; Goffin et al. 2018). One of the emerging techniques, artificial intelligence, is also used in the automation process of wastewater treatment plants. It uses different AI algorithms for automation of the process (Nakkasunchi et al. 2021).

3.10.5 Case Studies: Examples of Successful Biological Wastewater Treatment Plants

Research is done for humanity, and it is important to integrate it into day-to-day life. These case studies are examples of the integration of research, common knowledge, and wastewater treatment. One of the most famous case studies is the Arcata Wastewater treatment facility, in California. It is the perfect blend of artificial and natural processes. In the first step, sedimentation, bleaching, and other primary treatments were done that remove large pieces. In the second step, different plants and microorganisms including bacteria, fungi, and algae are used to remediate the heavy metals and organic compounds. Here, it can be seen that in the first step, it is a chemical and physical method for the removal of pollutants, and in the second step, it is biological. The organic and inorganic matter are used as fertilizers by plants and other microorganisms (Allen and Gearheart 1978). Recently, in the same wastewater treatment facility, microplastic removal was studied on different microbes and chemical processes that enhanced the wastewater treatment efficacy and improved the water quality (Connor 2023; Singh et al. 2022a, b).

Another case study is from Curauma, Chile. In this, they used integrated biological methods for resource recovery (Vergara-Araya et al. 2020). Water from different water sources is collected and purification is done using physical methods. The water is then sent for treatment; that water is either used for irrigation or further treated for another purpose like drinking or household cleaning. All the waste from household consumption and water sources is collected and used in anaerobic digestion for biogas production (Han et al. 2016). That biogas is then sent to a combined heat and power plant for electricity and heat generation that completes the demand of the city. This case study can be considered a more modern solution for the cities where a project like the Arcata wastewater treatment plant cannot be established because of confined land space. The study also shows that the integrated system is more efficient than the conventional method (Han et al. 2016; Vergara-Araya et al. 2020).

One more small-scale case study was done in Vadodara, India, for wastewater treatment using biofilters. In this study, a small wastewater treatment plant was assembled in the campus of Gharda Institute of Technology, by M/s Transchem Agritech Ltd, Vadodara, Gujrat. They used earthworms as vermifilter and biological degradation of large organic materials for the collection of vermicompost that can be used as fertilizers in farms. Studies for chemical oxygen demand, biological oxygen demand, total dispersed sediments, and dissolved oxygen were performed. The addition of vermifilter (earthworms) for preliminary treatment decreased the amount of solid waste and also COD, and BOD of wastewater. The vermifilter is efficient, cost-effective, and economic. This small plant is advantageous because it is very easy to use, needs low energy, and there is no formation of sludge or odor because it is an aerobic process. It also provides profitable organic fertilizer without involving a large amount of steps. This type of project can be useful for household wastewater or farm waste from cattle (Garkal et al. 2015).

3.11 Conclusions

Over the years, different methods and techniques emerged for wastewater treatment and the use of conventional methods was decreased. These methods like the use of genetically engineered microbes, the use of microbial consortia even the nanotechnology that is used for wastewater treatment nowadays. These methods are efficient and provide more resource recovery chances than conventional methods. To ease the manual work, automation of wastewater treatment is also in process. Automation uses different techniques including the use of artificial intelligence. Automation of optimization is also one of the recent advances in wastewater treatment with application sensors, especially biosensors. The case studies tell us about the on-field implications of all the research going on in the wastewater treatment area. Overall, recent advancement in wastewater treatment is in good time.

Acknowledgements Dr. Simranjeet Singh acknowledges the DBT HRD Project and Management Unit at the Regional Center for Biotechnology, NCR Biotech Science Cluster, Faridabad, Haryana, for the Research Associateship (DBT-RA) and associated funding as per award letter No DBT-RA/2022/July/N/2044 dated January 12, 2023. The authors are grateful to the Ministry of Education (MoE) for their support through the grant MoE-STARS/STARS-2/2023-0714, dated September 26, 2022.

References

Abdel-Fatah MA (2018) Nanofiltration systems and applications in wastewater treatment: review article. Ain Shams Eng J 9(4):3077–3092. https://doi.org/10.1016/j.asej.2018.08.001

Abdelfattah A, Ramadan H, Elsamahy T, Eltawab R, Mostafa S, Zhou X, Cheng L (2023) Multifaced features and sustainability of using pure oxygen in biological wastewater treatment: a review. J Water Process Eng 53:103883. https://doi.org/10.1016/j.jwpe.2023.103883

Ahmed MB, Zhou JL, Ngo HH, Guo W, Thomaidis NS, Xu J (2017) Progress in the biological and chemical treatment technologies for emerging contaminant removal from wastewater: a critical review. J Hazard Mater 323:274–298. https://doi.org/10.1016/j.jhazmat.2016.04.045

Ahmed M, Mavukkandy MO, Giwa A, Elektorowicz M, Katsou E, Khelifi O, Naddeo V, Hasan SW (2022) Recent developments in hazardous pollutants removal from wastewater and water reuse within a circular economy. Npj Clean Water 5(1):12. https://doi.org/10.1038/s41545-022-00154-5

Ahmed SF, Mofijur M, Nuzhat S, Chowdhury AT, Rafa N, Uddin MdA, Inayat A, Mahlia TMI, Ong HC, Chia WY, Show PL (2021) Recent developments in physical, biological, chemical, and hybrid treatment techniques for removing emerging contaminants from wastewater. J Hazard Mater 416:125912. https://doi.org/10.1016/j.jhazmat.2021.125912

Ahn CH, Park JK (2008) Critical factors affecting biological phosphorus removal in dairy wastewater treatment plants. KSCE J Civ Eng 12(2):99–107. https://doi.org/10.1007/s12205-008-0099-8

Algarni S, Tirth V, Alqahtani T, Alshehery S, Kshirsagar P (2023) Contribution of renewable energy sources to the environmental impacts and economic benefits for sustainable development. Sustain Energy Technol Assess 56:103098. https://doi.org/10.1016/j.seta.2023.103098

Al-Ghouti MA, Al-Kaabi MA, Ashfaq MY, Da'na DA (2019) Produced water characteristics, treatment and reuse: a review. J Water Process Eng 28:222–239. https://doi.org/10.1016/j.jwpe.2019.02.001

Allen GH, Gearheart RA (1978) Arcata integrated wastewater treatment, reclamation, and Salmon Ranching Project. https://repository.library.noaa.gov/view/noaa/34738%0A

Ambaye TG, Chebbi A, Formicola F, Prasad S, Gomez FH, Franzetti A, Vaccari M (2022) Remediation of soil polluted with petroleum hydrocarbons and its reuse for agriculture: recent progress, challenges, and perspectives. Chemosphere 293:133572. https://doi.org/10.1016/j.chemosphere.2022.133572

Asthana M, Kumar A, Sharma BS (2017) Wastewater treatment. In: Singh RL (ed) Principles and applications of environmental biotechnology for a sustainable future. Springer Singapore, pp 173–232. https://doi.org/10.1007/978-981-10-1866-4_6

Azad D, Pateriya RN, Arya R, Sharma RK (2022) Biological treatment for biochar modification: opportunities, limitations, and advantages. In: Ramola S, Mohan D, Masek O, Méndez A, Tsubota T (eds) Engineered biochar: fundamentals, preparation, characterization and applications. Springer Nature Singapore, pp 85–104. https://doi.org/10.1007/978-981-19-2488-0_6

Benner J, Helbling DE, Kohler H-PE, Wittebol J, Kaiser E, Prasse C, Ternes TA, Albers CN, Aamand J, Horemans B, Springael D, Walravens E, Boon N (2013) Is biological treatment a viable alternative for micropollutant removal in drinking water treatment processes? Water Res 47(16):5955–5976. https://doi.org/10.1016/j.watres.2013.07.015

Berber MR (2020) Current advances of polymer composites for water treatment and desalination. J Chem 2020(1):7608423. https://doi.org/10.1155/2020/7608423

Blokus-Dziula A, Dziula P, Kamedulski B, Michalak P (2023) Operation and maintenance cost of water management systems: analysis and optimization. Water 15(17):3053. https://doi.org/10.3390/w15173053

Boonma S, Chaiklangmuang S, Chaiwongsar S, Pekkoh J, Pumas C, Ungsethaphand T, Tongsiri S, Peerapornpisal Y (2015) Enhanced carbon dioxide fixation and bio-oil production of a microalgal consortium. CLEAN Soil Air Water 43(5):761–766. https://doi.org/10.1002/clen.201400171

Borgaonkar AD, Marhaba TF (2021) Chapter 16—Evaluation of sustainability strategies—a water quantity and quality perspective. In: Ahuja E (ed) Handbook of water purity and quality, 2nd edn. Academic Press, pp 409–454. https://doi.org/10.1016/B978-0-12-821057-4.00011-2

Bouwer EJ, Crowe PB (1988) Biological processes in drinking water treatment. J Am Water Works Ass 80(9):82–93

Butler CS, Boltz JP (2014) Biofilm processes and control in water and wastewater treatment. In: Ahuja P (ed) Comprehensive water quality and purification. Elsevier, pp 90–107. https://doi.org/10.1016/B978-0-12-382182-9.00083-9

Butler E, Hung Y-T, Suleiman Al Ahmad M, Yeh RYL, Liu RLH, Fu YP (2017) Oxidation pond for municipal wastewater treatment. Appl Water Sci 7(1):31–51. https://doi.org/10.1007/s13201-015-0285-z

Calais JFLG, Azevedo SG, Matias JCDO, Catalão JPDS (2017) Management and operations maintenance for a water treatment and supply company. Int J Ind Syst Eng 25(3):360–382

Calow P (1977) Conversion efficiencies in heterotrophic organisms. Biol Rev 52(3):385–409. https://doi.org/10.1111/j.1469-185X.1977.tb00840.x

Cervantes FJ, Pavlostathis SG, van Haandel AC (2006) Advanced biological treatment processes for industrial wastewaters, 1st edn. IWA Publishing. https://doi.org/10.2166/9781780402345

Cesaro A, Naddeo V, Belgiorno V (2013) Wastewater treatment by combination of advanced oxidation processes and conventional biological systems. J Biorem Biodegrad 4(8):1–8

Chan YJ, Chong MF, Law CL, Hassell DG (2009) A review on anaerobic–aerobic treatment of industrial and municipal wastewater. Chem Eng J 155(1):1–18. https://doi.org/10.1016/j.cej.2009.06.041

Chattopadhyay IJ, Usman TMM, Varjani S (2022) Exploring the role of microbial biofilm for industrial effluents treatment. Bioengineered 13(3):6420–6440. https://doi.org/10.1080/21655979.2022.2044250

Chen Y, Li B, Ye L, Peng Y (2015) The combined effects of COD/N ratio and nitrate recycling ratio on nitrogen and phosphorus removal in anaerobic/anoxic/aerobic (A2/O)-biological aerated filter (BAF) systems. Biochem Eng J 93:235–242. https://doi.org/10.1016/j.bej.2014.10.005

Chipasa KB, Mędrzycka K (2006) Behavior of lipids in biological wastewater treatment processes. J Ind Microbiol Biotechnol 33(8):635–645. https://doi.org/10.1007/s10295-006-0099-y

Chowdhury P, Viraraghavan T, Srinivasan A (2010) Biological treatment processes for fish processing wastewater—a review. Biores Technol 101(2):439–449. https://doi.org/10.1016/j.biortech.2009.08.065

Connor F (2023) Investigating the effectiveness of microplastic removal and the relationship between microplastics, bacterial communities, and nitrogen cycling in the natural treatment systems at the Arcata Wastewater Treatment Facility [Humboldt State University]. https://digitalcommons.humboldt.edu/etd/650

Cortez S, Teixeira P, Oliveira R, Mota M (2008) Rotating biological contactors: a review on main factors affecting performance. Rev Environ Sci Bio/technology 7(2):155–172. https://doi.org/10.1007/s11157-008-9127-x

Crini G, Lichtfouse E (2019) Advantages and disadvantages of techniques used for wastewater treatment. Environ Chem Lett 17(1):145–155. https://doi.org/10.1007/s10311-018-0785-9

Daigger GT, Boltz JP (2011) Trickling filter and trickling filter-suspended growth process design and operation: a state-of-the-art review. Water Environ Res 83(5):388–404. https://doi.org/10.2175/106143010X12681059117210

Daims H, Taylor MW, Wagner M (2006) Wastewater treatment: a model system for microbial ecology. Trends Biotechnol 24(11):483–489. https://doi.org/10.1016/j.tibtech.2006.09.002

Dammak I, Fersi M, Hachicha R, Abdelkafi S (2023) Current insights into growing microalgae for municipal wastewater treatment and biomass generation. Resources 12(10):119. https://doi.org/10.3390/resources12100119

de Souza SM, Bonilla KAS, de Souza AAU (2010) Removal of COD and color from hydrolyzed textile azo dye by combined ozonation and biological treatment. J Hazardous Mater 179(1):35–42. https://doi.org/10.1016/j.jhazmat.2010.02.053

Devda V, Chaudhary K, Varjani S, Pathak B, Patel AK, Singhania RR, Taherzadeh MJ, Ngo HH, Wong JWC, Guo W (2021) Recovery of resources from industrial wastewater employing electrochemical technologies: status, advancements and perspectives. Bioengineered 12(1):4697–4718

Dhakal N, Salinas-Rodriguez SG, Hamdani J, Abushaban A, Sawalha H, Schippers JC, Kennedy MD (2022) Is desalination a solution to freshwater scarcity in developing countries? Membranes 12(4). https://doi.org/10.3390/membranes12040381

Diener S, Studt Solano NM, Roa Gutiérrez F, Zurbrügg C, Tockner K (2011) Biological treatment of municipal organic waste using black soldier fly larvae. Waste Biomass Valorization 2(4):357–363. https://doi.org/10.1007/s12649-011-9079-1

Dignac M-F, Ginestet P, Rybacki D, Bruchet A, Urbain V, Scribe P (2000) Fate of wastewater organic pollution during activated sludge treatment: nature of residual organic matter. Water Res 34(17):4185–4194. https://doi.org/10.1016/S0043-1354(00)00195-0

Dionisi D (2017) Biological wastewater treatment processes: mass and heat balances, 1st edn. CRC Press

Djadouni F (2019) Microorganisms based biological agents in wastewater treatment and agriculture: potential use and benefits. J Biol Environ Sci, 13(38):93–100. https://izlik.org/JA97ZF57NC

Drewnowski J, Remiszewska-Skwarek A, Fernandez-Morales FJ (2018) Model based evaluation of plant improvement at a large wastewater treatment plant (WWTP). J Environ Sci Health Part A Toxic/Hazardous Substances Environ Eng 53(7):669–675. https://doi.org/10.1080/10934529.2018.1438821

Ebah EE, Yange IA, Ohie IR, Inya OJ (2022) Application of genetically modified organisms in waste management—a review. Stamford J Microbiol 12(1 SE-Review Articles):15–20. https://doi.org/10.3329/sjm.v12i1.63338

El-sayed MEA (2020) Nanoadsorbents for water and wastewater remediation. Sci Total Environ 739:139903. https://doi.org/10.1016/j.scitotenv.2020.139903

Ferrari GM, Dowell MD, Grossi S, Targa C (1996) Relationship between the optical properties of chromophoric dissolved organic matter and total concentration of dissolved organic carbon in the southern Baltic Sea region. Mar Chem 55(3):299–316. https://doi.org/10.1016/S0304-4203(96)00061-8

Filipe CD, Daigger GT, Grady CPJ (2001a) Stoichiometry and kinetics of acetate uptake under anaerobic conditions by an enriched culture of phosphorus-accumulating organisms at different pHs. Biotechnol Bioeng 76(1):32–43. https://doi.org/10.1002/bit.1023

Filipe CDM, Daigger GT, Grady CPL Jr (2001b) A metabolic model for acetate uptake under anaerobic conditions by glycogen accumulating organisms: stoichiometry, kinetics, and the effect of pH. Biotechnol Bioeng 76(1):17–31. https://doi.org/10.1002/bit.1022

Forrez I, Boon N, Verstraete W, Carballa M (2011) Biodegradation of micropollutants and prospects for water and wastewater biotreatment. In: Moo-Young (ed) Comprehensive biotechnology, 3rd edn. Pergamon, pp 433–442. https://doi.org/10.1016/B978-0-444-64046-8.00371-2

Fricke K, Santen H, Wallmann R (2005) Comparison of selected aerobic and anaerobic procedures for MSW treatment. Waste Manage 25(8):799–810

Ganzenko O, Huguenot D, van Hullebusch ED, Esposito G, Oturan MA (2014) Electrochemical advanced oxidation and biological processes for wastewater treatment: a review of the combined approaches. Environ Sci Pollut Res 21(14):8493–8524. https://doi.org/10.1007/s11356-014-2770-6

Garkal DJ, Mapara JV, Prabhune M (2015) Domestic waste water treatment by bio-filtration: a case study. Int J Sci Environ Technol 4(1):140–145

Ghosh S, Rusyn I, Dmytruk OV, Dmytruk KV, Onyeaka H, Gryzenhout M, Gafforov Y (2023) Filamentous fungi for sustainable remediation of pharmaceutical compounds, heavy metal and oil hydrocarbons. Frontiers Bioengineering Biotechnology 11:1106973

Goffin A, Guérin S, Rocher V, Varrault G (2018) Towards a better control of the wastewater treatment process: excitation-emission matrix fluorescence spectroscopy of dissolved organic matter as a predictive tool of soluble BOD(5) in influents of six Parisian wastewater treatment plants. Environ Sci Pollut Res Int 25(9):8765–8776. https://doi.org/10.1007/s11356-018-1205-1

Gomes da Silva FJ, Gouveia RM (2020) Global population growth and industrial impact on the environment. In: Gomes da Silva FJ, Gouveia RM (eds) Cleaner production. Springer International Publishing, pp 33–75. https://doi.org/10.1007/978-3-030-23165-1_3

Grady Jr CPL, Daigger GT, Love NG, Filipe CDM (2011) Biological wastewater treatment, 3rd edn. CRC Press. https://doi.org/10.1201/b13775

Grandclément C, Seyssiecq I, Piram A, Wong-Wah-Chung P, Vanot G, Tiliacos N, Roche N, Doumenq P (2017) From the conventional biological wastewater treatment to hybrid processes, the evaluation of organic micropollutant removal: a review. Water Res 111:297–317. https://doi.org/10.1016/j.watres.2017.01.005

Grießmeier V, Leberecht K, Gescher J (2019) NO(3) (-) removal efficiency in field denitrification beds: key controlling factors and main implications. Environ Microbiol Rep 11(3):316–329. https://doi.org/10.1111/1758-2229.12758

Guo J, Zhang Q, Cai Z, Zhao K (2016) Preparation and dye filtration property of electrospun polyhydroxybutyrate–calcium alginate/carbon nanotubes composite nanofibrous filtration membrane. Sep Purif Technol 161:69–79. https://doi.org/10.1016/j.seppur.2016.01.036

Guo T, Englehardt J, Wu T (2014) Review of cost versus scale: water and wastewater treatment and reuse processes. Water Sci Technol J Int Assoc Water Pollution Res 69(2):223–234. https://doi.org/10.2166/wst.2013.734

Hakanen J, Sahlstedt K, Miettinen K (2013) Wastewater treatment plant design and operation under multiple conflicting objective functions. Environ Model Softw 46:240–249. https://doi.org/10.1016/j.envsoft.2013.03.016

Han W, Zhao Y, Chen H (2016) Study on biogas production of joint anaerobic digestion with excess sludge and kitchen waste. Procedia Environ Sci 35:756–762. https://doi.org/10.1016/j.proenv.2016.07.089

Howard I, Espigares E, Lardelli P, Martín JL, Espigares M (2004) Evaluation of microbiological and physicochemical indicators for wastewater treatment. Environ Toxicol 19(3):241–249. https://doi.org/10.1002/tox.20016

Hussain A, Kumari R, Sachan SG, Sachan A (2021) Chapter 8—Biological wastewater treatment technology: advancement and drawbacks. In: Shah M, Rodriguez-Couto (eds) Microbial ecology of wastewater treatment plants. Elsevier, pp 175–192. https://doi.org/10.1016/B978-0-12-822503-5.00002-3

Inyinbor Adejumoke A, Adebesin Babatunde O, Oluyori Abimbola P, Adelani Akande Tabitha A, Dada Adewumi O, Oreofe Toyin A (2018) Water pollution: effects, prevention, and climatic impact. In: Water challenges of an urbanizing world, vol 33, pp 33–47

Jain K, Patel AS, Pardhi VP, Flora SJS (2021) Nanotechnology in wastewater management: a new paradigm towards wastewater treatment. Molecules (Basel, Switzerland) 26(6):1797. https://doi.org/10.3390/molecules26061797

Jhansi SC, Mishra SK (2013) Wastewater treatment and reuse: sustainability options. Consilience 10:1–15

Jiang L, Li Y, Pei H (2021) Algal–bacterial consortia for bioproduct generation and wastewater treatment. Renew Sustain Energy Rev 149:111395. https://doi.org/10.1016/j.rser.2021.111395

Jiang X, Peng Z, Zhang J (2024) Starting with screening strains to construct synthetic microbial communities (SynComs) for traditional food fermentation. Food Res Int 190:114557. https://doi.org/10.1016/j.foodres.2024.114557

Johnson I, Girijan S, Tripathy BK, Ali MAS, Kumar M (2020) Algal–bacterial symbiosis and its application in wastewater treatment. In: Emerging technologies in environmental bioremediation. Elsevier, pp 341–372

Jung H, Pauly D (2011) Water in the pulp and paper industry. In: Wilderer WS (ed) Reference module in earth systems and environmental sciences: treatise on water science. Elsevier, pp 667–683. https://doi.org/10.1016/B978-0-444-53199-5.00100-7

Kanaujiya DK, Paul T, Sinharoy A, Pakshirajan K (2019) Biological treatment processes for the removal of organic micropollutants from wastewater: a review. Current Pollution Rep 5(3):112–128. https://doi.org/10.1007/s40726-019-00110-x

Karić N, Maia AS, Teodorović A, Atanasova N, Langergraber G, Crini G, Ribeiro ARL, Đolić M (2022) Bio-waste valorisation: agricultural wastes as biosorbents for removal of (in) organic pollutants in wastewater treatment. Chem Eng J Adv 9:100239

Karri RR, Ravindran G, Dehghani MH (2021) Chapter 1—Wastewater—sources, toxicity, and their consequences to human health. In: Karri RR, Ravindran G, Dehghani WM (eds) Soft computing techniques in solid waste and wastewater management. Elsevier, pp 3–33. https://doi.org/10.1016/B978-0-12-824463-0.00001-X

Kassab G, Halalsheh M, Klapwijk A, Fayyad M, Van Lier JB (2010) Sequential anaerobic–aerobic treatment for domestic wastewater—a review. Biores Technol 101(10):3299–3310

Keskitalo J, Leiviskä K (2012) Application of evolutionary optimisers in data-based calibration of activated sludge models. Expert Syst Appl 39(7):6609–6617. https://doi.org/10.1016/j.eswa.2011.12.041

Khalidi-Idrissi A, Madinzi A, Anouzla A, Pala A, Mouhir L, Kadmi Y, Souabi S (2023) Recent advances in the biological treatment of wastewater rich in emerging pollutants produced by pharmaceutical industrial discharges. Int J Environ Sci Technol 20(10):11719–11740. https://doi.org/10.1007/s13762-023-04867-z

Khan MJ, Wibowo A, Karim Z, Posoknistakul P, Matsagar BM, Wu KC-W, Sakdaronnarong C (2024) Wastewater treatment using membrane bioreactor technologies: removal of phenolic contaminants from oil and coal refineries and pharmaceutical industries. Polymers 16(3):443. https://doi.org/10.3390/polym16030443

Khan NA, López-Maldonado EA, Majumder A, Singh S, Varshney R, López JR, Méndez PF, Ramamurthy PC, Khan MA, Khan AH, Mubarak NM, Amhad W, Shamshuddin SZM, Aljundi IH (2023a) A state-of-art-review on emerging contaminants: environmental chemistry, health effect, and modern treatment methods. Chemosphere 344:140264. https://doi.org/10.1016/j.chemosphere.2023.140264

Khan NA, Singh S, López-Maldonado EA, Méndez-Herrera PF, López-López JR, Baig U, Ramamurthy PC, Mubarak NM, Karri RR, Aljundi IH (2023) Emerging membrane technology and hybrid treatment systems for the removal of micropollutants from wastewater. Desalination 565:116873. https://doi.org/10.1016/j.desal.2023.116873

Kim J-O, Chung J (2014) Implementing chemical precipitation as a pretreatment for phosphorus removal in membrane bioreactor-based municipal wastewater treatment plants. KSCE J Civ Eng 18(4):956–963. https://doi.org/10.1007/s12205-014-0070-9

Kim K, Hur JW, Kim S, Jung J-Y, Han H-S (2020) Biological wastewater treatment: comparison of heterotrophs (BFT) with autotrophs (ABFT) in aquaculture systems. Biores Technol 296:122293. https://doi.org/10.1016/j.biortech.2019.122293

Kloepfer A, Jekel M, Reemtsma T (2005) Occurrence, sources, and fate of benzothiazoles in municipal wastewater treatment plants. Environ Sci Technol 39(10):3792–3798. https://doi.org/10.1021/es048141e

Komorowska-Kaufman M, Majcherek H, Klaczyński E (2006) Factors affecting the biological nitrogen removal from wastewater. Process Biochem 41(5):1015–1021. https://doi.org/10.1016/j.procbio.2005.11.001

Kumar MA, Abdullah SS, Kumar PS, Dheeba B, Mathumitha C (2008) Comparitive study on potentiality of bacteria and fungi in bioremediatio n and decolorization of molasses spent wash. J Pure Appl Microbiol 2:393–400

Kumara Behera B, Varma A (2017) Concept of microbiological processes. In: Microbial biomass process technologies and management. Springer International Publishing, pp 1–43. https://doi.org/10.1007/978-3-319-53913-3_1

Kumari P, Alam M, Siddiqi WA (2019) Usage of nanoparticles as adsorbents for waste water treatment: an emerging trend. Sustain Mater Technol 22:e00128. https://doi.org/10.1016/j.susmat.2019.e00128

Kurniawan TA, Lo W, Chan G, Sillanpää MET (2010) Biological processes for treatment of landfill leachate. J Environ Monit 12(11):2032–2047. https://doi.org/10.1039/C0EM00076K

Kushkevych I (2021) The application of microorganisms in wastewater treatment. In: Processes, vol 9, issue 11, p 1914. https://doi.org/10.3390/pr9111914

Leitão RC, van Haandel AC, Zeeman G, Lettinga G (2006) The effects of operational and environmental variations on anaerobic wastewater treatment systems: a review. Biores Technol 97(9):1105–1118. https://doi.org/10.1016/j.biortech.2004.12.007

Lettinga G (1996) Sustainable integrated biological wastewater treatment. Water Sci Technol 33(3):85–98. https://doi.org/10.1016/0273-1223(96)00303-4

Li C, Wu K, Chen L, Liu Z, Zhao X, Li Y, Hu M, Zhao Q, Ye Z (2023) Advanced treatment of low-pollution and poor biodegradability sewage by combined process. J Clean Prod 414:137366. https://doi.org/10.1016/j.jclepro.2023.137366

Li W-W, Yu H-Q (2016) Advances in energy-producing anaerobic biotechnologies for municipal wastewater treatment. Engineering 2(4):438–446. https://doi.org/10.1016/J.ENG.2016.04.017

Li Y, Zhou B, Zheng G, Liu X, Li T, Yan C, Cheng C, Dai K, Liu C, Shen C, Guo Z (2018) Continuously prepared highly conductive and stretchable SWNT/MWNT synergistically composited electrospun thermoplastic polyurethane yarns for wearable sensing. J Mater Chem C 6(9):2258–2269. https://doi.org/10.1039/C7TC04959E

Liu Y, Chen Y, Zhou Q (2007) Effect of initial pH control on enhanced biological phosphorus removal from wastewater containing acetic and propionic acids. Chemosphere 66(1):123–129. https://doi.org/10.1016/j.chemosphere.2006.05.004

Liu Y, Ngo HH, Guo W, Peng L, Wang D, Ni B (2019) The roles of free ammonia (FA) in biological wastewater treatment processes: a review. Environ Int 123:10–19. https://doi.org/10.1016/j.envint.2018.11.039

Lodha B, Chaudhari S (2007) Optimization of Fenton-biological treatment scheme for the treatment of aqueous dye solutions. J Hazard Mater 148(1):459–466. https://doi.org/10.1016/j.jhazmat.2007.02.061

Lofrano G, Brown J (2010) Wastewater management through the ages: a history of mankind. Sci Total Environ 408(22):5254–5264. https://doi.org/10.1016/j.scitotenv.2010.07.062

Lofrano G, Meriç S, Zengin GE, Orhon D (2013) Chemical and biological treatment technologies for leather tannery chemicals and wastewaters: a review. Sci Total Environ 461–462:265–281. https://doi.org/10.1016/j.scitotenv.2013.05.004

Loukidou MX, Zouboulis AI (2001) Comparison of two biological treatment processes using attached-growth biomass for sanitary landfill leachate treatment. Environ Pollut 111(2):273–281. https://doi.org/10.1016/S0269-7491(00)00069-5

Lu S, Li H, Tan G, Wen F, Flynn MT, Zhu X (2019) Resource recovery microbial fuel cells for urine-containing wastewater treatment without external energy consumption. Chem Eng J 373:1072–1080. https://doi.org/10.1016/j.cej.2019.05.130

Macêdo-Júnior RO, Serpa FS, Santos BLP, De Vasconcelos CR, Silva GF, Ruzene DS, Silva DP (2023) Produced water treatment and its green future in the oil and gas industry: a multi-criteria decision-making study. Int J Environ Sci Technol 20(2):1369–1384

Machineni L (2019) Review on biological wastewater treatment and resources recovery: attached and suspended growth systems. Water Sci Technol 80(11):2013–2026

Malik S, Muhammad K, Waheed Y (2023) Nanotechnology: a revolution in modern industry. Molecules (Basel, Switzerland) 28(2):661. https://doi.org/10.3390/molecules28020661

Mantzavinos D, Kalogerakis N (2005) Treatment of olive mill effluents Part I. Organic matter degradation by chemical and biological processes—an overview. Environ Int 31(2):289–295. https://doi.org/10.1016/j.envint.2004.10.005

Manyuchi MM, Mbohwa C, Muzenda E (2018) Anaerobic treatment of opaque beer wastewater with enhanced biogas recovery through Acti-zyme bio augmentation. S Afr J Chem Eng 26:74–79. https://doi.org/10.1016/j.sajce.2018.10.002

Mara D (2003) Domestic wastewater treatment in developing countries, 1st edn. Routledge. https://doi.org/10.4324/9781849771023

Mehrnaz Z, Séverine T, Bruno M, Laurent A, Sandrine B, Dominique C (2020) Assessment of sampling techniques to investigate airborne transport of contaminants in treated wastewater from sprinkler irrigation fine droplet. World Environ Water Resources Congress 2020:27–41. https://doi.org/10.1061/9780784482957.004

Metcalf L, Eddy HP, Tchobanoglous G (1991) Wastewater engineering: treatment, disposal, and reuse, vol 4. McGraw-Hill, New York

Miller S, Shemer H, Semiat R (2015) Energy and environmental issues in desalination. Desalination 366:2–8. https://doi.org/10.1016/j.desal.2014.11.034

Mishra RK, Mentha SS, Misra Y, Dwivedi N (2023) Emerging pollutants of severe environmental concern in water and wastewater: a comprehensive review on current developments and future research. Water-Energy Nexus 6:74–95. https://doi.org/10.1016/j.wen.2023.08.002

Mohamad Faudzi SM, Mohd Souhkri DN, Mohd Akhir MF, Abdul Aziz H, Mohd Kasim MZ, Zakaria NA, Saad NA (2023) Total maximum daily load application using biological oxygen demand, chemical oxygen demand, and ammoniacal nitrogen: a case study for water quality assessment in the Perai River Basin, Malaysia. Water 15(6):1227. https://doi.org/10.3390/w15061227

Molinuevo-Salces B, Riaño B, Hernández D, Cruz García-González M (2019) Microalgae and wastewater treatment: advantages and disadvantages. In: Alam MA, Wang Z (eds) Microalgae biotechnology for development of biofuel and wastewater treatment. Springer Singapore, pp 505–533. https://doi.org/10.1007/978-981-13-2264-8_20

Morgan-Sagastume F, Allen DG (2003) Effects of temperature transient conditions on aerobic biological treatment of wastewater. Water Res 37(15):3590–3601. https://doi.org/10.1016/S0043-1354(03)00270-7

Mtethiwa AH, Munyenyembe A, Jere W, Nyali E (2008) Efficiency of oxidation ponds in wastewater treatment. Int J Environ Res 2(2):149–152

Mu R, Jia Y, Ma G, Liu L, Hao K, Qi F, Shao Y (2021) Advances in the use of microalgal–bacterial consortia for wastewater treatment: community structures, interactions, economic resource reclamation, and study techniques. Water Environ Res 93(8):1217–1230. https://doi.org/10.1002/wer.1496

Mulyanti R, Susanto H (2018) Wastewater treatment by nanofiltration membranes. IOP Conf Ser Earth Environ Sci 142:12017

Murnane JG, Ghanim B, O'Donoghue L, Courtney R, O'Dwyer TF, Pembroke JT (2019) Advances in metal recovery from wastewaters using selected biosorbent materials and constructed wetland systems. In: Eyvaz M (ed) Water and wastewater treatment. Intech Open. https://doi.org/10.5772/intechopen.84335

Nageshwari K, Baishali D, Unpaprom Y, Ramaraj R, Maniam GP, Govindan N, Thirugnanam A, Balasubramanian P (2021) Exploring the dynamics of microalgal diversity in high-rate algal ponds. In: The future of effluent treatment plants, pp 615–660

Nakkasunchi S, Hewitt NJ, Zoppi C, Brandoni C (2021) A review of energy optimization modelling tools for the decarbonisation of wastewater treatment plants. J Clean Prod 279:123811. https://doi.org/10.1016/j.jclepro.2020.123811

Narayanan CM, Narayan V (2019) Biological wastewater treatment and bioreactor design: a review. Sustain Environ Res 29(1):33. https://doi.org/10.1186/s42834-019-0036-1

Naz I, Saroj DP, Mumtaz S, Ali N, Ahmed S (2015) Assessment of biological trickling filter systems with various packing materials for improved wastewater treatment. Environ Technol 36(4):424–434

Nelson MJ, Nakhla G, Zhu J (2017) Fluidized-bed bioreactor applications for biological wastewater treatment: a review of research and developments. Engineering 3(3):330–342. https://doi.org/10.1016/J.ENG.2017.03.021

Nguyen PY, Carvalho G, Reis MAM, Oehmen A (2021) A review of the biotransformations of priority pharmaceuticals in biological wastewater treatment processes. Water Res 188:116446. https://doi.org/10.1016/j.watres.2020.116446

Nirmalakhandan N, Selvaratnam T, Henkanatte-Gedera SM, Tchinda D, Abeysiriwardana-Arachchige ISA, Delanka-Pedige HMK, Munasinghe-Arachchige SP, Zhang Y, Holguin FO, Lammers PJ (2019) Algal wastewater treatment: photoautotrophic vs. mixotrophic processes. Algal Res 41:101569

Nishat A, Yusuf M, Qadir A, Ezaier Y, Vambol V, Ijaz Khan M, Ben Moussa S, Kamyab H, Sehgal SS, Prakash C, Yang H-H, Ibrahim H, Eldin SM (2023) Wastewater treatment: a short assessment on available techniques. Alex Eng J 76:505–516. https://doi.org/10.1016/j.aej.2023.06.054

Obaideen K, Shehata N, Sayed ET, Abdelkareem MA, Mahmoud MS, Olabi AG (2022) The role of wastewater treatment in achieving sustainable development goals (SDGs) and sustainability guideline. Energy Nexus 7:100112. https://doi.org/10.1016/j.nexus.2022.100112

Obileke K, Onyeaka H, Meyer EL, Nwokolo N (2021) Microbial fuel cells, a renewable energy technology for bio-electricity generation: a mini-review. Electrochem Commun 125:107003. https://doi.org/10.1016/j.elecom.2021.107003

Okoye CO, Addey CI, Oderinde O, Okoro JO, Uwamungu JY, Ikechukwu CK, Okeke ES, Ejeromedoghene O, Odii EC (2022) Toxic Chemicals and persistent organic pollutants associated with micro- and nanoplastics pollution. Chem Eng J Adv 11:100310. https://doi.org/10.1016/j.ceja.2022.100310

Olalere OA, Gan C-Y, Taiwo AE, Alenezi H, Adeyi O, Adeyi AJ (2022) Chapter 3—Emerging trends in bioreactor systems for an improved wastes valorization. In: Mishra P, Singh L, Ghosh B (eds) Techno-economics and life cycle assessment of bioreactors: post-Covid19 waste management approach. Elsevier, pp 23–35. https://doi.org/10.1016/B978-0-323-89848-5.00006-8

Oller I, Malato S, Sánchez-Pérez JA (2011) Combination of advanced oxidation processes and biological treatments for wastewater decontamination—a review. Sci Total Environ 409(20):4141–4166. https://doi.org/10.1016/j.scitotenv.2010.08.061

Olsson J, Feng XM, Ascue J, Gentili FG, Shabiimam MA, Nehrenheim E, Thorin E (2014) Co-digestion of cultivated microalgae and sewage sludge from municipal waste water treatment. Biores Technol 171:203–210. https://doi.org/10.1016/j.biortech.2014.08.069
Orhon D (2015) Evolution of the activated sludge process: the first 50 years. J Chem Technol Biotechnol 90(4):608–640. https://doi.org/10.1002/jctb.4565
Osman RM, Hodaifa G (2023) An overview of anaerobic membrane bioreactors: current developments, fouling problems, and future prospects. J Environ Chem Eng 11(6):111482. https://doi.org/10.1016/j.jece.2023.111482
Oulebsir R, Lefkir A, Safri A, Bermad A (2020) Optimization of the energy consumption in activated sludge process using deep learning selective modeling. Biomass Bioenerg 132:105420. https://doi.org/10.1016/j.biombioe.2019.105420
Parmar S, Daki S, Bhattacharya S, Shrivastav A (2022) Microorganism: an ecofriendly tool for waste management and environmental safety. In: Development in wastewater treatment research and processes. Elsevier, pp 175–193
Patel DD, Bhatt S (2022) Environmental pollution, toxicity profile, and physico-chemical and biotechnological approaches for treatment of textile wastewater. Biotechnol Genet Eng Rev 38(1):33–86. https://doi.org/10.1080/02648725.2022.2048434
Paździor K, Bilińska L, Ledakowicz S (2019) A review of the existing and emerging technologies in the combination of AOPs and biological processes in industrial textile wastewater treatment. Chem Eng J 376:120597. https://doi.org/10.1016/j.cej.2018.12.057
Plaza E, Bosander J, Trela J (1991) Factors affecting biological nitrogen removal efficiency in a large wastewater treatment plant. Water Sci Technol 24(7):121–131
Prasse C, Stalter D, Schulte-Oehlmann U, Oehlmann J, Ternes TA (2015) Spoilt for choice: a critical review on the chemical and biological assessment of current wastewater treatment technologies. Water Res 87:237–270. https://doi.org/10.1016/j.watres.2015.09.023
Pryce D, Kapelan Z, Memon FA (2022) Economic evaluation of a small wastewater treatment plant under different design and operation scenarios by life cycle costing. Dev Eng 7:100103. https://doi.org/10.1016/j.deveng.2022.100103
Rafeeq H, Afsheen N, Rafique S, Arshad A, Intisar M, Hussain A, Bilal M, Iqbal HMN (2023) Genetically engineered microorganisms for environmental remediation. Chemosphere 310:136751. https://doi.org/10.1016/j.chemosphere.2022.136751
Rahimi S, Modin O, Mijakovic I (2020) Technologies for biological removal and recovery of nitrogen from wastewater. Biotechnol Adv 43:107570. https://doi.org/10.1016/j.biotechadv.2020.107570
Rahman ROA, El-Kamash AM, Hung Y-T (2022) Applications of nano-zeolite in wastewater treatment: an overview. Water 14(2):137. https://doi.org/10.3390/w14020137
Ramanan R, Kim B-H, Cho D-H, Oh H-M, Kim H-S (2016) Algae–bacteria interactions: evolution, ecology and emerging applications. Biotechnol Adv 34(1):14–29. https://doi.org/10.1016/j.biotechadv.2015.12.003
Razzak SA, Faruque MO, Alsheikh Z, Alsheikhmohamad L, Alkuroud D, Alfayez A, Hossain SMZ, Hossain MM (2022) A comprehensive review on conventional and biological-driven heavy metals removal from industrial wastewater. Environ Adv 7:100168. https://doi.org/10.1016/j.envadv.2022.100168
Rittmann BE, Huck PM, Bouwer EJ (1989) Biological treatment of public water supplies. Crit Rev Environ Sci Technol 19(2):119–184
Ros M, Zupančič GD (2004) Two-stage thermophilic anaerobic–aerobic digestion of waste-activated sludge. Environ Eng Sci 21(5):617–626
Roy H, Rahman TU, Tasnim N, Arju J, Rafid MM, Islam MR, Pervez MN, Cai Y, Naddeo V, Islam MS (2023) Microbial fuel cell construction features and application for sustainable wastewater treatment. Membranes 13(5):490. https://doi.org/10.3390/membranes13050490
Samaei SM, Gato-Trinidad S, Altaee A (2018) The application of pressure-driven ceramic membrane technology for the treatment of industrial wastewaters—a review. Sep Purif Technol 200:198–220. https://doi.org/10.1016/j.seppur.2018.02.041

Samer M (2015) Biological and chemical wastewater treatment processes. Wastewater Treatment Eng 150(10.5772):61250

Samin G, Pavlova M, Arif MI, Postema CP, Damborsky J, Janssen DB (2014) A Pseudomonas putida strain genetically engineered for 1,2,3-trichloropropane bioremediation. Appl Environ Microbiol 80(17):5467–5476. https://doi.org/10.1128/AEM.01620-14

Saravanan A, Senthil Kumar P, Jeevanantham S, Karishma S, Tajsabreen B, Yaashikaa PR, Reshma B (2021) Effective water/wastewater treatment methodologies for toxic pollutants removal: processes and applications towards sustainable development. Chemosphere 280:130595. https://doi.org/10.1016/j.chemosphere.2021.130595

Saravanathamizhan R, Perarasu VT (2021) Chapter 6—Improvement of biodegradability index of industrial wastewater using different pretreatment techniques. In: Shah MP, Sarkar A, Mandal (eds) Wastewater treatment: cutting edge molecular tools, techniques and applied aspects. Elsevier, pp 103–136. https://doi.org/10.1016/B978-0-12-821881-5.00006-4

Sathya R, Arasu MV, Al-Dhabi NA, Vijayaraghavan P, Ilavenil S, Rejiniemon TS (2023) Towards sustainable wastewater treatment by biological methods—a challenges and advantages of recent technologies. Urban Climate 47:101378. https://doi.org/10.1016/j.uclim.2022.101378

Schultz TE (2005) Biological wastewater treatment: the use of microorganisms to remove contaminants from wastewater is effective and widespread. To choose the right system from the many options offered, understand the various techniques available and evaluate them based on. Chem Eng 112(10):44–51. https://link.gale.com/apps/doc/A138059443/AONE?u=anon~492eb3f1andsid=googleScholarandxid=22c10f99

Sekar S, Lee S, Kim DY, Preethi V, Kalirajan KM, Sutha S, Saravanan S, Therli A, Roy M, Jagannathan K (2020) Biomass activated carbon-decorated spherical β-$Ni(OH)_2$ nanoparticles for enhanced hydrogen production from sulphide wastewater. J Water Process Eng 38:101669. https://doi.org/10.1016/j.jwpe.2020.101669

Sepúlveda-Muñoz CA, de Godos I, Muñoz R (2023) Wastewater treatment using photosynthetic microorganisms. Symmetry 15(2):525

Serdarevic A, Dzubur A (2019) Importance and practice of operation and maintenance of wastewater treatment plants. In: Avdaković S (ed) Advanced technologies, systems, and applications III (IAT 2018). Springer International Publishing, pp 121–137

Seviour RJ, Mino T, Onuki M (2003) The microbiology of biological phosphorus removal in activated sludge systems. FEMS Microbiol Rev 27(1):99–127. https://doi.org/10.1016/S0168-6445(03)00021-4

Shahid MJ, Al-Surhanee AA, Kouadri F, Ali S, Nawaz N, Afzal M, Rizwan M, Ali B, Soliman MH (2020) Role of microorganisms in the remediation of wastewater in floating treatment wetlands: a review. Sustainability 12(14):5559. https://doi.org/10.3390/su12145559

Sharma T, Singh A, Kumar N, Chauhan G, Singh DP, Singh A, Rana BB (2023) Emerging pollutants in the environment and ecological risks. In: Management and mitigation of emerging pollutants. Springer, pp 1–20

Shehata N, Singh S, Ramamurthy PC, Dhanjal DS, Kaul N, Romero R, Singh J (2023) Chapter 18—Removal of pesticides from wastewater by Fenton systems: enhanced degradation by chelating agents and solar irradiation. In: Singh J, Pandey A, Singh S, Garg VK, Ramamurthy P (eds) Current developments in biotechnology and bioengineering. Elsevier, pp 459–486. https://doi.org/10.1016/B978-0-323-91900-5.00006-0

Silva JA (2023) Wastewater treatment and reuse for sustainable water resources management: a systematic literature review. Sustainability 15(14):10940. https://doi.org/10.3390/su151410940

Singh BJ, Chakraborty A, Sehgal R (2023a) A systematic review of industrial wastewater management: evaluating challenges and enablers. J Environ Manage 348:119230. https://doi.org/10.1016/j.jenvman.2023.119230

Singh S, Garg VK, Ramamurthy PC, Singh J, Pandey A (2023) Chapter 1—Impact and prospects of pesticides on human and environmental health. In: Singh J, Pandey A, Singh S, Garg VK, Ramamurthy P (eds) Current developments in biotechnology and bioengineering. Elsevier, pp 1–32. https://doi.org/10.1016/B978-0-323-91900-5.00008-4

Singh S, Kumar Naik TSS, Anil AG, Dhiman J, Kumar V, Dhanjal DS, Aguilar-Marcelino L, Singh J, Ramamurthy PC (2022a) Micro (nano) plastics in wastewater: a critical review on toxicity risk assessment, behaviour, environmental impact and challenges. Chemosphere 290:133169. https://doi.org/10.1016/j.chemosphere.2021.133169

Singh S, Kumar V, Kapoor D, Dhanjal DS, Wani AB, Bhardwaj S, Ramamurthy PC, Singh J (2022) Chapter 12—Competence of nanoparticles for removal of pesticides from wastewater: an overview. In: Hadi Dehghani M, Karri RR, Anastopoulos I (eds) Pesticides remediation technologies from water and wastewater. Elsevier, pp 253–266. https://doi.org/10.1016/B978-0-323-90893-1.00012-X

Singh S, Kumar V, Romero R, Sharma K, Singh J (2019) Applications of nanoparticles in wastewater treatment. In: Prasad R, Kumar V, Kumar M, Choudhary D (eds) Nanobiotechnology in bioformulations. Springer International Publishing, pp 395–418. https://doi.org/10.1007/978-3-030-17061-5_17

Smit B, Pilifosova O (2003) Adaptation to climate change in the context of sustainable development and equity. Sustain Dev 8(9):9

Soares A (2020) Wastewater treatment in 2050: challenges ahead and future vision in a European context. Environ Sci Ecotechnology 2:100030. https://doi.org/10.1016/j.ese.2020.100030

Song H, Ding M-Z, Jia X-Q, Ma Q, Yuan Y-J (2014) Synthetic microbial consortia: from systematic analysis to construction and applications. Chem Soc Rev 43(20):6954–6981. https://doi.org/10.1039/C4CS00114A

Štefanac T, Grgas D, Landeka Dragičević T (2021) Xenobiotics-division and methods of detection: a review. J Xenobiotics 11(4):130–141. https://doi.org/10.3390/jox11040009

Steiniger B, Hupfauf S, Insam H, Schaum C (2023) Exploring anaerobic digestion from mesophilic to thermophilic temperatures—operational and microbial aspects. Fermentation 9(9):798. https://doi.org/10.3390/fermentation9090798

Sustarsic M (2009) Wastewater treatment: understanding the activated sludge process. In: Safety in ammonia plants and related facilities symposium, pp 26–29

Szymanski N, Patterson RA (2003) Effective microorganisms (EM) and wastewater systems. In: Future directions for on-site systems: best management practice, pp 347–355

Tariq M, Ali M, Shah Z (2006) Characteristics of industrial effluents and their possible impacts on quality of underground water. Soil Environ 25(1):64–69

Topare NS, Attar SJ, Manfe MM (2011) Sewage/wastewater treatment technologies: a review. Sci Rev Chem Commun 1(1):18–24

Tran NH, Chen H, Reinhard M, Mao F, Gin KY-H (2016) Occurrence and removal of multiple classes of antibiotics and antimicrobial agents in biological wastewater treatment processes. Water Res 104:461–472. https://doi.org/10.1016/j.watres.2016.08.040

Van Aken B (2011) Transgenic plants and associated bacteria for phytoremediation of organic pollutants. In: Moo-Young (ed) Comprehensive biotechnology, 2nd edn. Academic Press, pp 223–237. https://doi.org/10.1016/B978-0-08-088504-9.00330-5

van Rijn J (1996) The potential for integrated biological treatment systems in recirculating fish culture—a review. Aquaculture 139(3):181–201. https://doi.org/10.1016/0044-8486(95)01151-X

Velis CA, Longhurst PJ, Drew GH, Smith R, Pollard SJT (2009) Biodrying for mechanical–biological treatment of wastes: a review of process science and engineering. Biores Technol 100(11):2747–2761. https://doi.org/10.1016/j.biortech.2008.12.026

Vergara-Araya M, Lehn H, Poganietz W-R (2020) Integrated water, waste and energy management systems—a case study from Curauma, Chile. Resour Conserv Recycl 156:104725. https://doi.org/10.1016/j.resconrec.2020.104725

Wang Y, Liang B, Kang F, Wang Y, Zhao C, Lyu Z, Zhu T, Zhang Z (2022) An efficient anoxic/aerobic/aerobic/anoxic process for domestic sewage treatment: From feasibility to application. Front Microbiol 13:970548. https://doi.org/10.3389/fmicb.2022.970548

Waqas S, Bilad MR, Man Z, Wibisono Y, Jaafar J, Indra Mahlia TM, Khan AL, Aslam M (2020) Recent progress in integrated fixed-film activated sludge process for wastewater treatment: a review. J Environ Manage 268:110718. https://doi.org/10.1016/j.jenvman.2020.110718

Wei Y, Van Houten RT, Borger AR, Eikelboom DH, Fan Y (2003) Minimization of excess sludge production for biological wastewater treatment. Water Res 37(18):4453–4467. https://doi.org/10.1016/S0043-1354(03)00441-X

Wik T (2003) Trickling filters and biofilm reactor modelling. Rev Environ Sci Biotechnol 2(2):193–212. https://doi.org/10.1023/B:RESB.0000040470.48460.bb

Yadav G, Dash SK, Sen R (2019) A biorefinery for valorization of industrial waste-water and flue gas by microalgae for waste mitigation, carbon-dioxide sequestration and algal biomass production. Sci Total Environ 688:129–135. https://doi.org/10.1016/j.scitotenv.2019.06.024

Yadav G, Sharma I, Ghangrekar M, Sen R (2020) A live bio-cathode to enhance power output steered by bacteria-microalgae synergistic metabolism in microbial fuel cell. J Power Sources 449:227560. https://doi.org/10.1016/j.jpowsour.2019.227560

Yadu A, Sahariah BP, Anandkumar J (2021) Treatment and management strategies for industrial wastewater. In: Shah MP (ed) Biological treatment of industrial wastewater (p. 0). The Royal Society of Chemistry. https://doi.org/10.1039/9781839165399-00053

Yen H-W, Brune DE (2007) Anaerobic co-digestion of algal sludge and waste paper to produce methane. Biores Technol 98(1):130–134. https://doi.org/10.1016/j.biortech.2005.11.010

Zarei M (2020) Wastewater resources management for energy recovery from circular economy perspective. Water-Energy Nexus 3:170–185. https://doi.org/10.1016/j.wen.2020.11.001

Zhang L, Zhang B, Zhu X, Chang H, Ou S, Wang H (2018) Role of bioreactors in microbial biomass and energy conversion. In: Liao Q, Chang JS, Herrmann C, Xia A (eds) Bioreactors for microbial biomass and energy conversion. Springer, Singapore, pp 39–78. https://doi.org/10.1007/978-981-10-7677-0_2

Zhang W, Tooker NB, Mueller AV (2020) Enabling wastewater treatment process automation: leveraging innovations in real-time sensing, data analysis, and online controls. Environ Sci Water Res Technol 6(11):2973–2992. https://doi.org/10.1039/D0EW00394H

Zheng C, Zhao L, Zhou X, Fu Z, Li A (2013) Treatment technologies for organic wastewater. Water Treatment 11:250–286

Zheng T, Wang J, Wang Q, Nie C, Smale N, Shi Z, Wang X (2015) A bibliometric analysis of industrial wastewater research: current trends and future prospects. Scientometrics 105(2):863–882. https://doi.org/10.1007/s11192-015-1736-x

Zhou C, Shi Y, Sun C, Yu S, Liu M, Gao C (2014) Thin-film composite membranes formed by interfacial polymerization with natural material sericin and trimesoyl chloride for nanofiltration. J Membr Sci 471:381–391. https://doi.org/10.1016/j.memsci.2014.08.033

Zieliński M, Kazimierowicz J, Dębowski M (2023) Advantages and limitations of anaerobic wastewater treatment—technological basics, development directions, and technological innovations. Energies 16(1):83. https://doi.org/10.3390/en16010083

Zupančič GD, Roš M (2008) Aerobic and two-stage anaerobic–aerobic sludge digestion with pure oxygen and air aeration. Biores Technol 99(1):100–109. https://doi.org/10.1016/j.biortech.2006.11.054

Chapter 4
Advancements in Membrane Characterization

Deeptimayee Pal and Abhishek Kumar

Abstract Membrane filtration's sustainable operation, high nutrient recovery efficiency, and excellent treated water quality make it a promising technology for resource recovery and wastewater treatment. This chapter explores the importance of various membrane processes, including ultrafiltration, microfiltration, nanofiltration, and reverse osmosis. This chapter examines membranes' design, functionality, and manufacturing, emphasizing their diverse characteristics, including selectivity, charge, and pore size. It is essential to characterize the membrane to understand the intricate relationship between its morphological, structural, and functional characteristics. This chapter comprehensively examines various morphological characterization techniques, including SEM, ESEM, IFM, and AFM, as well as structural characterization techniques, such as FTIR and Raman. Additionally, it examines other characterization techniques, including XPS, nanoindentation, porometry, zeta potential, contact angle, and membrane autopsy. Each characterization technique's working principle, advantages, and drawbacks have been delineated. This chapter delves into the significance of these advanced characterization methods in resolving various membrane-related issues, including contamination and degradation. The membrane advancements for future research development and technology are further explored in this chapter.

Keywords Membrane · Membrane characterization · Morphological characterization · Porometry · Nanoindentation · Membrane autopsy

D. Pal (✉)
Department of Carbon Capture, Utilization, and Storage , Jindal Steel Limited, Angul, Odisha, India
e-mail: deeptimayeenitrkl@outlook.com

A. Kumar
Department of Chemical Engineering, Indian Institute of Technology Kharagpur, Kharagpur, West Bengal, India

A. Mandpe et al. (eds.), *Membrane Filtration and Membrane Bioreactors for Wastewater Treatment*, Advances in Wastewater Research,
https://doi.org/10.1007/978-981-95-6706-5_4

4.1 Introduction

Water, the universal solvent, is the most significant factor essential for human existence, but currently, the water quality is continuously degraded by contaminants. This attribute underscores the critical importance of water for human survival, which is increasingly threatened by climate change and numerous sources of contamination. It is anticipated that the release of contaminants will be affected by the present global warming phenomenon, which is expected to alter their dissolution rate, sorption balance, solubility, gas phase generation, mobilization of non-aqueous phase liquids, and biological degradation. Human activities have significantly influenced water quality degradation, affecting rivers, lakes, and seas worldwide. This has damaged the environment, human health, and the ecosystems that depend on these water sources (Akhtar et al. 2021). The unsustainable growth patterns witnessed worldwide are placing significant strain on water supplies, with projections showing a 50% surge in global water consumption shortly. The existence of organic and inorganic contaminants in water sources and their connection with recent disorders requires the development of more effective treatment methods to guarantee that the world's population has access to clean drinking water. The presence of emerging pollutants such as endocrine-disrupting compounds, pharmaceuticals and personal care products, flame retardants, surfactants, petroleum wastes, per- and polyfluoroalkyl substances (Pal and Sen 2023a, 2025a). These petroleum wastes adversely affect the environment (air, water, and soil) and are highly toxic to human health (Pal and Sen 2023b). Water filtering systems for residential usage, using materials such as wool, sponge, and charcoal, have existed since the seventeenth century. Standards for drinking water quality, such as those released by the World Health Organization, have been established and updated to safeguard public health from the dangers of chemical compounds and microbes (Figueras and Borrego 2010).

Furthermore, the Water Framework Directive of the European Union plays a crucial role in the political approach to combat water pollution. It necessitates that all member states monitor specific substances and devise sustainable strategies and measures to mitigate the impacts of contamination on aquatic ecosystems. The minimum quality criteria for water intended for human consumption, including drinking, heating, and other domestic purposes, are established by this directive. Its objective is to safeguard against any form of contamination. Due to the established inefficiencies of the current technologies employed in wastewater treatment facilities (WWTPs), various techniques and methods have been developed to accomplish sustainability and wastewater treatment (Pal and Sen 2024a). The conventional treatment methods employed in WWTPs, including filtration, clarification, aeration, disinfection, and oxidation, frequently fail to eliminate pollutants effectively (Koul et al. 2022). Many factors, including low volatility, hydrophobic properties, pKa values, size, shape, charge, and deficient levels of impurities, influence inefficiency (Boczkaj and Fernandes 2017). Although there has been significant research on physicochemical methods for removing pollutants in WWTPs, these techniques have several disadvantages. These include (i) high energy consumption, (ii) significant initial investment for operation and maintenance, (iii) production of toxic waste, (iv) use of harmful chemical agents, (v) need for trained personnel, and (vi) limited effectiveness in breaking down organic pollutants (Iqbal et al. 2024). Many alternative methods have been developed to overcome these constraints and remove harmful substances from water. These methods include ion exchange, electrochemical treatments, chemical precipitation, adsorption (Pal and Sen 2024b), photocatalysis (Kumar et al. 2022), advanced oxidation processes (AOPs), and membrane separation (Kumar and Gangawane 2023a).

Membrane processes, such as reverse osmosis (RO), microfiltration (MF), ultrafiltration (UF), nanofiltration (NF), and others, have proven to be effective from both an ecological and a technical standpoint for some time, making them valuable tools in water management, environmental engineering, and industrial production (Hube et al. 2020). Many critical components, such as the membrane material, the impact of the pretreatment process, the quality and pH of the feed, the operating pressure and temperature, and the membrane fouling qualities, influence how successful the RO process is in removing metal ions. Commercial polymeric RO membranes may be broadly divided into thin film composite (TFC) and asymmetric. TFC membranes were created in response to the shortcomings of asymmetric RO membranes, allowing for the optimization of the materials and techniques used to create selected thin layers with microporous support. Among the noteworthy TFC RO membrane materials are polyamides via polyethylenimine, sulfonated polysulfones, polypiperazine-amide, polypyrrolidine, and cross-linked completely aromatic polyamide (Hosseini et al. 2016).

The least restrictive membrane type, microfiltration, is used in membrane filtering. It may be used, for example, to remove germs and pigments and other particles of submicron particle sizes. These are porous membranes with pore diameters ranging from around 0.1 to 1.0 μm, with 0.2 μm being the most typical pore size for commercial membranes. There is a maximum working pressure of 500 kPa. Ultrafiltration membranes can remove viruses and germs, separate macromolecules like proteins and carbohydrates, and separate colloidal silica and pyrogens. The typical ranges for molecular weight cut-offs are 5000–200 000 g/mol. The typical pore size is about 0.05 μm (Pal and Sen 2025b). There is a maximum working pressure of 1000 kPa. In contrast to microfiltration or ultrafiltration, which stops ions from flowing through the membrane due to pore size, nanofiltration membranes work on the concept of solution diffusion, in which monovalent ions diffuse across the membrane. For example, nanofiltration may be used to remove color, sugar, and dye from water supplies, as well as THM precursors, hardness, and sulfate from wastewater sources like acid mine drainage. There is a maximum working pressure of 5000 kPa. The most restrictive kind of membrane currently in use is used for reverse osmosis. The organic and inorganic molecules are extracted from a feed solution through solution diffusion. Reverse osmosis membranes often separate ions and dissolved salts with molecular weights less than 200 g/mol. Applications include desalinating saltwater for drinking water production, purifying industrial wastes such as landfill leachate, and producing ultrapure water for semiconductors and pharmaceuticals. For high-pressure reverse osmosis systems, the operating pressure is typically up to 15,000 kPa and 7000 kPa (Rezakazemi et al. 2018).

As a result, RO uses a lot more energy than NF, which runs at pressures between 5 and 20 atm. The greater energy use of RO is one of its main disadvantages. Research on new nanomaterials has significantly contributed to several disciplines, including water treatment, owing to the ongoing development of innovative membrane types. Nanoparticles can be synthesized using sol–gel, hydrothermal, chemical coprecipitation (CCP) methods. Compared to other synthesis methods, the CCP method has been proven more effective for synthesizing iron-oxide nanoparticles (Kumar and Gangawane 2022a, b). These nanoparticles have potential for use in adsorption, catalysis, and sensing because of their large surface areas and high reactivity (Cai et al. 2018). Even though conventional membranes have already dominated the market for membranes used in water purification, it is difficult to pick a membrane type best suited for a certain application since trade-offs between flux, selectivity, stability, or expensive manufacturing costs limit each type. Another significant drawback is the ease with which existing polymeric and ceramic membranes might get fouled. Despite several efforts to chemically alter membrane surfaces, such as by grafting hydrophilic monomers onto the membrane, improvements are still required to achieve a sufficient decrease in membrane fouling. New artificial membranes for water filtration have been developed as a consequence of advancements in nanotechnology and nanomaterials. Promising membrane technologies include aligned nanotube membranes, tightly packed nanoparticles and nanofiber membrane, self-assembled two-dimensional layer materials, and their composites. These are only a few of the many ideas presented in recent years in nanotechnology (Ying et al. 2017).

The advent of nanotechnology in membrane research has created new opportunities for the creation of improved polymeric membranes to improve filtering performance (Kumar and Gangawane 2023b). Nonetheless, various challenges, such as fouling and the trade-off between membrane selectivity and water permeability, have hampered the development of classic polymeric membranes for practical applications. Membrane modifications have been attempted to address these concerns. The use of macromolecules for membrane modification has sparked widespread attention in recent years due to their intriguing chemical and structural features. Membranes treated with macromolecules have increased anti-fouling characteristics by altering their physiochemical properties, such as membrane shape, porosity, surface charge, wettability, and durability (Nagandran et al. 2020). Based on the structure of the membrane and the position of the nanomaterial, the nanocomposite membranes were categorized as conventional, thin-film nanocomposite (TFN), thin-film composite (TFC), and surface-located nanocomposite (Yin and Deng 2015). TFC polyamide membranes are preferred for NF, RO, and FO applications. The membranes consist of a non-porous, highly crosslinked polyamide selective layer and a porous support layer constructed of polysulfone. Regarding water and salt rejection, TFC polyamide membranes outperform asymmetric cellulose-acetate-based membranes, the original generation of RO membranes produced over 50 years ago. TFC membranes are widely used in desalination applications because of their high performance and stability over a broad pH range (pH 2–11 for continuous operation)

(Werber et al. 2016). The performance of the resultant TFN membrane may be significantly enhanced by adding nanoparticles to the polyamide layer. Building interlayers on prefabricated support membranes has been a potent method in recent years for producing ultra-permeable and extremely selective interlayered TFC membranes (Ji et al. 2021).

4.2 Membrane Separation

The scientific and technical methods related to the passage or exclusion of components, species, or substances via membranes are collectively known as membrane technology. Water treatment for both residential and commercial use, as well as other separation processes, relies on this technology. It finds widespread use in the chemical, biotechnological, pharmaceutical, culinary, and metallurgical sectors. A simplified outline of the most important uses of membrane technology is illustrated in Fig. 4.1. Because membrane separation is an environmentally friendly and energy-efficient alternative to traditional processes like filtering, distillation, ion exchange, and chemical treatment systems, it has several potential uses in industry and the environment. The membrane characteristics may be fine-tuned to get the desired result, and continuous separation under moderate circumstances is also possible. Hybrid processing can be easily scaled up. Limitations of this technique include concentration polarization, membrane fouling, short membrane lifespan, poor selectivity, and poor flux, among others. Various membrane morphologies with unique biological, chemical, and physical characteristics may be developed to circumvent these problems (Mamba et al. 2021; Siddique et al. 2021).

In general, membranes may be categorized as either anisotropic or isotropic. An isotropic membrane's physical and chemical properties remain constant throughout its cross-section. Layers of anisotropic membranes may differ in structure and/or chemical composition, and the membrane's overall shape is inconsistent throughout its cross-section. The intended shape (dense or porous) and the type of the raw material (organic or inorganic) determine the procedure option for membrane fabrication. Membranes may be manufactured using various techniques, including sintering, stretching, clamping tracks, coating solutions, and phase inversion. This allows for the construction of membranes with a wide range of features, making them a potentially attractive alternative to traditional contaminant removal methods by membrane separation (Tan and Rodrigue 2019).

4.3 Membrane Characterization

Membrane characterization is crucial for developing and enhancing the membrane to achieve its targeted performance. Various membrane characterization techniques such as scanning electron microscopy (SEM), interferential optical microscopy

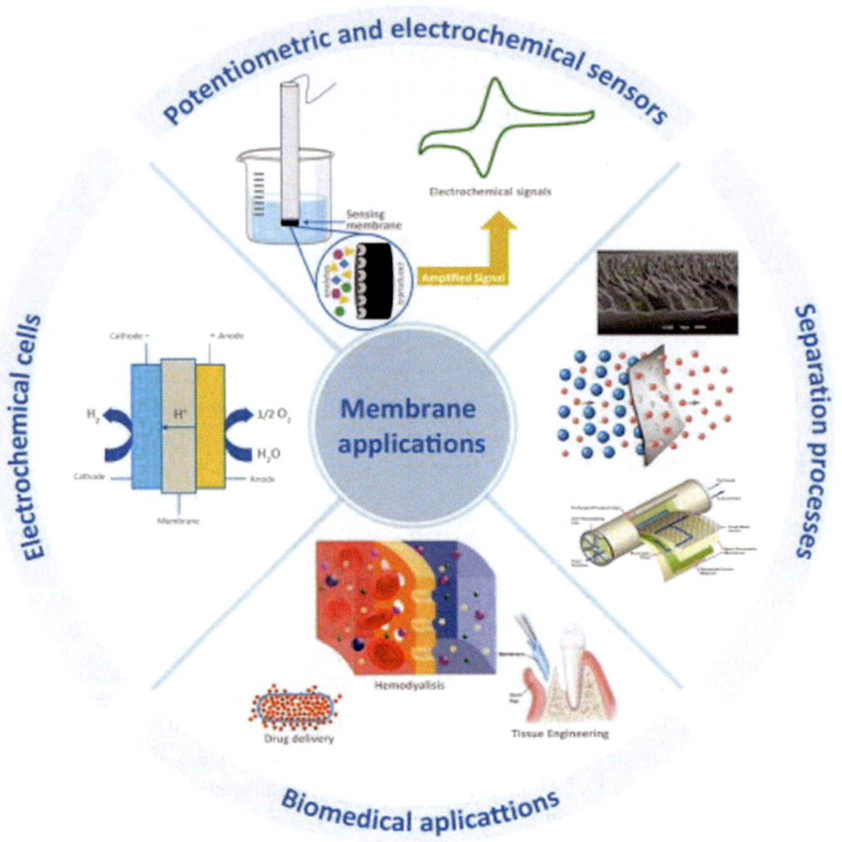

Fig. 4.1 Applications of membrane technology Cevallos-Mendoza et al. (2022)

(IFM), atomic force microscopy (AFM), porometry, nanoindentation, Fourier-transform infrared (FTIR) and Raman spectroscopy, X-ray photoelectron spectroscopy (XPS), Zeta potential, contact angle, and membrane autopsy have been discussed. The advantages and drawbacks of these characterization techniques have been listed in Table 4.1.

4.3.1 Scanning Electron Microscopy (SEM)

SEM is a helpful tool for analyzing the surfaces of materials such as membranes. It works well on any vacuum-resistant surface. The surface is first sputter-coated with a thin coating of gold metal in almost all SEMs. It includes sweeping an electron beam that has been carefully collimated over the examination specimen's surface. The beam

Table 4.1 Advantages and limitations of membrane characterization techniques Ebnesajjad (2011), Mourdikoudis et al. (2018), Titus et al. (2019)

Membrane characterization technique	Advantages	Limitations
SEM	• SEM images are exceptionally high-resolution, enabling membrane surface inspection at the micro and nanoscale • Its wide field depth enables us to see three-dimensional details and intricate surface topographies • SEM can analyze polymeric, ceramic, and composite membranes	• SEM needs careful sample preparation, including covering non-conductive objects with a thin layer of gold or platinum to prevent charging and provide high-quality pictures • Not all membrane materials can be analyzed in a vacuum, especially those susceptible to dehydration or high vacuum
IFM	• Offers resolution at the nanoscale scale, which is crucial for precise surface analysis • Allows for non-invasive analysis of material without any direct touch or harm • Provides accurate quantitative measurements of surface topography and thickness • Suitable for polymers, ceramics, and composite membranes	• Needs advanced data interpretation and processing techniques • It uses a coherent light source, which limits other lights for applications • Reflective or rough surfaces may create complex interference patterns that are hard to comprehend
AFM	• The atomic-to-nanometer resolution of AFM allows thorough surface characterization • Can image polymers, ceramics, and biological membranes • Quantifies surface roughness, mechanical characteristics, and interaction forces • It can operate in modes that protect fragile samples	• Performing high-resolution scans across wide regions may be a time-intensive process • The state of the tip and the type of the tip-sample contact may have an impact on the quality of the picture and measurements • Mainly offers surface data, whereas it does not provide access to subsurface characteristics

(continued)

Table 4.1 (continued)

Membrane characterization technique	Advantages	Limitations
Porometry	• Porometry characterizes membranes without destroying them • Provides detailed pore size and distribution data for membrane performance prediction • Characterizes polymeric, ceramic, and composite membranes • It measures quickly and accurately, making it suited for industrial research and quality monitoring	• Porometry relies on the wetting liquid thoroughly wetting the pores, which is not always true • Wetting liquid, gas type, and environmental conditions might affect results. Thus, these must be controlled and standardized • Porometry gives size and distribution data but has little information on pore structure and geometry
Nanoindentation	• Nanoindentation measures nanoscale mechanical characteristics with great accuracy and spatial resolution • It allows small-area testing of heterogeneous and composite membrane materials • Quantifies hardness, elastic modulus, and other mechanical parameters to compare and optimize membrane materials • Nanoindentation causes only localized indentations and preserves membrane integrity	• The indentation size impact affects measured characteristics by comparing the indentation size to the material's microstructure • Nanoindentation results might be difficult to understand for time-dependent or anisotropic materials • For precise measurements, thorough surface preparation is necessary. Contamination and surface roughness might impact the outcomes
FTIR	• FTIR is non-destructive and can analyze membrane samples without affecting their structure • Even minor chemical composition and molecular structure changes may be detected by FTIR, giving material information • FTIR analysis requires little sample preparation and is rapid • Compatible with polymers, ceramics, and composites	• FTIR generally measures membrane surface or near-surface characteristics, which may not reflect bulk properties • Due to overlapping absorption bands, Spectra may be difficult to discern in complicated materials • FTIR is great for qualitative analysis but difficult for quantitative analysis without calibration and reference standards

(continued)

Table 4.1 (continued)

Membrane characterization technique	Advantages	Limitations
XPS	• XPS analyzes just the top 1–10 nm, where most membrane interactions occur, making it very surface-sensitive • XPS offer chemical state details needed to understand surface chemistry and changes • Quality and quantity of elemental composition and chemical states may be measured using XPS	• XPS analysis demands a high vacuum, which may not work for membrane materials sensitive to vacuum or dehydration • Surface sensitivity is beneficial, but XPS cannot reveal membrane bulk features • Membrane samples must be properly prepared for the XPS device, and surface contamination might hinder findings
Zeta potential measurement	• It imparts information without causing any damage to the membrane • Assists in comprehending and forecasting the occurrence of fouling • Facilitates the development and enhancement of surface changes • Allows for the adjustment of operating conditions to enhance performance	• Several factors may complicate and alter the link between membrane performance and zeta potential • Accurate measurements and interpretation need specialized equipment and expertise • To get complete data, zeta potential must be tested extensively since it may alter dramatically with variations in pH, ionic strength, and other environmental factors

(continued)

Table 4.1 (continued)

Membrane characterization technique	Advantages	Limitations
Contact angle measurement	• The method is easy to implement and yields visible outcomes rapidly • The membrane remains intact, paving the way for more applications and experiments • Offers measurable data for easy comparison and analysis • Provides significant surface energy and liquid interaction data	• Surface irregularities may affect contact angle readings • Temperature and humidity impact measurements • Characterization may need different liquids since liquid choice affects contact angle • Only covers surface qualities, not mass
Membrane autopsy	• Allows tailored cleaning techniques depending on fouling conditions, boosting membrane life and performance • Assesses how operating circumstances affect membrane performance, enabling efficient operation and maintenance • Helps to identify potential issues early, preventing catastrophic failures and downtime • Provides thorough information on fouling kinds and causes (organic, inorganic, biological, particle) for effective mitigation techniques	• A complete membrane autopsy involves specialized knowledge, skills, and equipment that not all institutions have • Autopsies require many sample, processing, and testing stages, which might delay operational decisions • Membrane autopsies need expensive SEM, EDS, FTIR, and XPS analysis • Destructive membrane autopsy prevents reuse, raising replacement costs • It helps analyze and mitigate fouling after severe performance degradation or failure, not prevent it

is directed at a tiny probe that moves over a specimen's surface. As the electrons pass through the surface, photons and electrons are released due to the beam's and the material's interactions. The released particles are gathered with the right detector to provide surface data. An image results from the electron beam colliding with the sample's surface topology. The electron microprobe is a useful SEM accessory (Kumar et al. 2025a). A sample surface is exposed to an electron beam, which ionizes the material to a few micrometers. The X-rays' energies and wavelengths released during the deexcitation cycle represent the elements in the sample's upper layers. Although the electron microprobe enables the investigation of different places on the sample surface, the result is not a true surface study. In 1988, environmental scanning electron microscopy (ESEM) went into commercial use for the first time. A scanning electron microscope (SEM) intended for dual function (ESEM/SEM) may work well in both modes despite the high cost of converting a regular SEM to an ESEM. One benefit of employing an ESEM is that it may be used in "wet mode," eliminating the requirement to conduct nonconductive material. Material samples may retain their original properties for further testing or modification as they don't need to be dehydrated or coated with carbon or gold–palladium. Without rendering the sample useless after coating it to make it conductive, one may image it, alter it, and image it again, ad infinitum. Dynamic studies may also be carried out using the ESEM in wet mode. A tiny sample can be heated to as high as 1500 °C using one of the hot stages and imaged during the heating and cooling. To reject thermal electrons, bias has to be altered when a particular temperature is reached, over 1100 °C, although this process is simple. Working within 20 °C of the ambient temperature is possible with the Peltier heating/cooling stage, and achieving 100% relative humidity (RH) at the sample surface is possible when low temperature (e.g., 4 °C) and high water vapor pressure (e.g., 6.1 Torr) are combined. Samples are not dehydrated throughout the imaging procedure at 100% RH. A wet sample loses water continuously at less than 100% relative humidity (RH) due to the chamber's vacuum pumping action; this is shown as continuous sample movement in the "scope"(Ebnesajjad 2011; Hortola 2005).

4.3.2 Interferential Optical Microscopy (IFM)

The classical configuration of optical microscopy generates images of the sample under observation, with the lateral magnification and resolution contingent upon the objective employed. The axial dimensions are lost, and the contrast between the two imaged sites is solely determined by the difference in reflection from the surface at these locations rather than their axial distance. This limitation can be surmounted by more sophisticated systems, including the optical profilometer and the confocal microscope. An optical profilometer using a Mireau interference objective is employed in this investigation. The waveform generated by the reflection from a reference plane is superimposed on the surface images. The imaging system acquires interference fringes under these circumstances. The surface wave phase can be reconstructed by moving the objective along its z-axis and acquiring several successive interferential images with a white source. The measured object's tridimensional profile can be reconstructed at this point. It is always feasible to reconstruct the non-measured points by interpolating with the adjacent pixels, provided they are not excessively numerous. The images can be numerically processed following the measurement. For instance, the roughness properties of curved surfaces can be revealed by flattening them by subtracting them from a mean surface. It is also feasible to extract the roughness parameters of the sample, including membranes. This method enables the acquisition of a scan size of up to square millimeters with a vertical resolution of approximately 0.1 Å. In this manner, optical interferometry offers a more thorough examination of membrane irregularity, particularly when integrated with AFM for multi-scale analysis. The membranes' non-contact surface topography can be analyzed using this microscopy (Wyart et al. 2014, 2008).

4.3.3 *Atomic Force Microscopy (AFM)*

One of the most common techniques for seeing and working with nanoscale structures such as membranes is atomic force microscopy. AFM does have certain aberrations, however, much as other microscopy techniques. Imaging artifacts are particularly common when using tapping mode imaging, which involves an oscillating probe with a cantilevered tip that hits the surface periodically in the kHz range, as shown in Fig. 4.2. As a result, interpreting the resulting topographical maps often requires more than a cursory glance. The recorded topography may result from various causes, including the limited size of the scanning tip, the non-negligible thermal drift of the AFM scanner, adsorbed layers of liquids on the surface, and other physical characteristics of the material and AFM probe. When it comes to traits that are nanoscale in size, the significance of these contributions seems to be particularly evident. On the other hand, thorough observations and theoretical considerations enable the subtraction of hidden aspects of the structure under study while interpreting AFM data. Most of these studies concentrated on determining which parameter impacted the imaging process most. It was demonstrated that the intricate relationship between the AFM probe and sample surface can result in contrast artifacts, that the "tapping" can occasionally be "hammering" because of unforeseen high imaging forces, that the dynamics of the probe are intrinsically nonlinear, leading to bistabilities and chaotic behavior; that the accurate explanation of cantilever bending transcends the use of linear springs; and that the cantilever oscillation's resonant harmonic response has much better sensitivity for material parameters when a higher harmonic of the fundamental mode is used for excitation. Analytical imaging mapping of material characteristics was often explored based on these concepts. As a result, phase imaging patterns were understood to represent maps of energy dissipation, charge distribution, and elasticity. Additionally, surface energy has been seen to affect AFM-performed elasticity measurements. However, several other features of high-resolution imaging, such as the precision of the morphological measurements, have received little attention up to this point (Qiu et al. 2016; Mechler et al. 2005).

4.3.4 *Porometry*

A membrane's pore size distribution significantly affects its separation properties. An accurate pore size distribution assessment is required to fully understand the membrane's filtering performance. Many techniques have been documented in the literature for determining the pore size distributions of porous membranes. Most of these approaches use the pore radius to determine the critical pressure required to

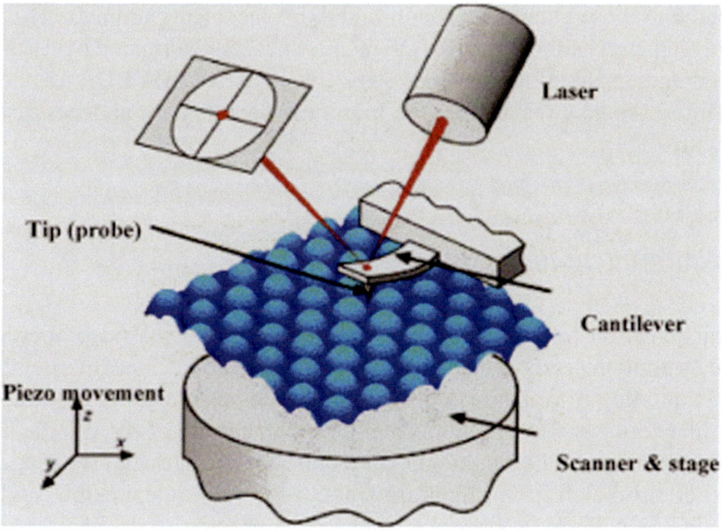

Fig. 4.2 Schematic representation of an AFM Wyart et al. (2008)

displace one fluid from another from the pore. When the transmembrane pressure is gradually raised and the flow is monitored simultaneously, the observed flow may be used to calculate the number of holes from which a fluid displacement occurred at the specified radius. A fluid resistance model such as Hagen–Poiseuille's law is required for this quantification. While measuring the flow and, hence, the permeability spectrum is helpful in and of itself, translating from the permeability domain to the pore size domain is especially useful since the pore size is often connected to a solute retained by the membrane pore. Erbe originally documented this methodology in 1933. The traditional capillary flow porometry and liquid–liquid displacement techniques are two approaches that use flow measurements to assess pore size distribution. These flow porometry procedures are well-established, and an ASTM standard exists for them (Bannwarth et al. 2017).

Synthetic membranes are employed in applications that necessitate an understanding of pore size distribution (PSD), including controlled drug release, artificial organs, clever sensors, and water treatment. Dedicated and/or costly apparatus and specialized expertise may be necessary to implement current PSD characterization methods. Evapoporometry (EP) is a novel characterization technique that utilizes a microbalance and a simple diffusion cell to determine the pore size of ultrafiltration membranes precisely. This technique is based on vapor pressure depression. A comprehensive error analysis and a straightforward characterization procedure are described. It is demonstrated that the EP characterization of the PSD of a 100 nm bilayer Anopore™ membrane, which was employed as a standard, agrees with the nominal pore diameter, FESEM analysis, and analyses conducted by previous researchers. In addition, EP is employed to characterize PSD of commercially fabricated 20 and 50 nm polyvinylidene fluoride (PVDF) membranes, and the results are compared to those obtained through liquid displacement porometry (LDP). The mean pore diameters of the 50 nm PVDF membranes, as determined by EP and LDP, are in good agreement. However, the lower values derived via LDP for the 20 nm PVDF membranes may result from the high pressures that are unnecessary for EP (Krantz et al. 2013).

4.3.5 Nanoindentation

The nanoindentation test measures materials' elastic and plastic properties on a nanoscale by applying electromagnetic stress and measuring capacitive depth. Additionally, as an add-on to nanoindentation, the nano-scratch test shifts the sample perpendicular to the scratch probe to measure the strength of the interfacial adhesion by film delamination brought on by the build-up of interfacial shear stress. The nanoindentation/scratch approach offers a quick, easy, accurate, flexible, and speedy way to evaluate small-volume samples' interfacial adhesion and mechanical strength compared to the tensile and peeling tests. Researchers have developed an effective characterization approach to quantify the interfacial adhesion of different composite membranes used for gas separation and pervaporation based on the nanoindentation/scratch methodology. When basic tensile tests are difficult to conduct, the bulging and nanoindentation tests are great instruments for analyzing mechanical behavior, especially for thin films. One of the primary uses of thin films in microtechnology is the mechanical certification of microvalves, and here is where the bulging test is outstanding. While the penetration of the indenter tip into the sample surface is continually monitored during nanoindentation, the applied load may be adjusted constantly. This test, also known as the constant-load indentation creep test, is often used to examine how crystalline materials change over time. There are two phases to the nanoindentation creep: the transient primary creep and the steady-state secondary creep (Poilane et al. 2000; Hang et al. 2015).

4.3.6 Fourier-Transform Infrared (FTIR) and Raman Spectroscopy

An infrared portion of the electromagnetic spectrum is exposed to a sample while it is analyzed using infrared spectroscopy (IR). The kind of chemical groups present in the sample determines which wavelengths it absorbs. The wave number (cm^{-1}), derived by dividing 10,000 by the wavelength in microns, defines these wavelengths (Kumar et al. 2025b). It is also possible to get infrared spectra by reflecting the IR beam onto a sample's surface. This technique is used when a transmission approach to get an IR spectrum is not feasible. Using a high refractive index crystal (such as thallium bromo-iodide) on the sample surface, numerous internal reflectances of the IR beam are used to calculate attenuated total reflectance (ATR), sometimes referred to as attenuated total internal reflectance, or ATIR. An identical spectrum to the transmission is achieved after about 25 internal reflections (5 cm) of the IR beam entering the crystal at an angle. The foundation of FTIR spectroscopy is the absorption of infrared light energy, namely the middle infrared region, which excites molecular bonds vibrationally. An infrared absorption spectrum resembling a molecular “fingerprint” for a material such as a membrane may be generated from the total vibrational spectra of its constituent substances. By clustering, this spectrum may be used for typing or identification applications. A reference spectrum library uses various substances that have been well-studied. Any unexplained isolate’s FTIR spectrum may be measured using parameters identical to those used for the reference spectra, and it can then be contrasted with various spectra from the reference spectrum collection. Identification is feasible if the library has a spectrum that is the same or very comparable. As such, the method’s effectiveness is directly correlated with the reference spectrum library’s complexity (Taha et al. 2013).

A portion of the incident light is dispersed when it hits the sample. Rayleigh scattering occurs when most scattered light has the same wavelength as the incoming light. Raman scattering is the process of scattering some light at a different wavelength. The Raman shift is the energy difference between the incident light and the Raman dispersed light. It is equivalent to the energy the molecule needs to spin or vibrate. It is common to find many Raman-shifted signals in a single sample; each signal is linked to a distinct rotational or vibrational motion of the molecules in the sample. The Raman signals detected depend on the specific molecule and its surroundings. A laser is the incident light source because the Raman effect is so weak. A Raman spectrum represents the Raman intensity against the frequency of the Raman shift. Sharp bands that indicate the functional groups of the chemicals or materials, such as membranes, are often present. It is possible to identify the substances present and establish their chemical structure by interpreting this information. Because it measures molecular vibrations using a technique different from FTIR, it is supplementary to that technique. Raman spectroscopy is a useful method for qualitative analysis and separating organic and/or inorganic chemicals in mixtures of materials. Samples as little as one millimeter may provide a Raman spectrum. The sensitivity of the individual vibrations to the Raman effect determines the band intensities in a Raman spectrum, which are proportional to concentration. Raman spectra may thus be used for quantitative and semiquantitative membrane analysis (Ebnesajjad 2011).

4.3.7 X-ray Photoelectron Spectroscopy (XPS)

A sensitive, quantitative spectroscopic method for examining a material’s surface chemistry is X-ray photoelectron spectroscopy (XPS), often referred to as electron spectroscopy for chemical analysis (ESCA). A common Al Kα or Mg Kα source irradiates a material to get an XPS spectrum. At the same time, the kinetic energy and amount of electrons that escape from atoms on the material's surface are concurrently measured. An XPS spectrum is produced by calculating the binding energy (derived from the observed kinetic energy recorded) and the intensity of the electron exiting from the surface. Because the analyzer is typically one meter distant from the X-ray irradiation surface in XPS surface analysis, a UHV system is required to acquire the maximal electron count during the spectrum capture. The elemental composition of elements with an atomic number of lithium or higher may be estimated qualitatively and quantitatively using the ultra-high vacuum (UHV) surface analysis method. It also helps to determine the kinds of chemical bonds and oxidation states present on the material surface. In a multitude of disciplines, including biology, electronics, and forensics, to mention a few, XPS has been widely used to investigate surface chemistry. However, due to the need for a UHV environment, traditional XPS can only analyze solid sample surfaces. Its major areas of interest include membranes, semiconductor surfaces, biosensor surfaces, thin-film coatings (such as metal/metal oxide/DNA layer), and dry powder medicinal materials (Korin et al. 2017).

XPS is becoming increasingly important for many kinds of analysis for several reasons. First, the true chemical makeup and condition of surfaces and interfaces, which are the primary characteristics of nanostructured materials, can be ascertained using X-ray photoelectron spectroscopy. The performance of synthetic biomaterials and biological organisms' functioning depends on surfaces. Second, although XPS is usually not thought of as a tool with "nano" resolution, the electrons associated with it can provide a great deal of information about the structure of nanometer-sized features of a sample near a surface region because they can travel only nanometer-long distances, as explained below. While many researchers may be more interested in using samples with flat surfaces to obtain nanometer-scale information, XPS spectra can yield information about coatings and layers on particles and the size of nanoparticles (sometimes in situations where other high spatial resolution measurements are not feasible) (Baer and Engelhard 2010).

4.3.8 Zeta Potential

There is no reliable method for determining the surface charge of tiny particles in liquid. It is customary to measure a particle's electric potential someplace in the diffuse layer, distant from the particle surface. The sliding or shear plane is the name given to this region, which is associated with particle movement in liquid. Zeta potential, a crucial characteristic for colloids or suspended nanoparticles, is the potential measured at this plane. It is highly crucial in determining the surface charge of membranes. Its importance is directly linked to particle surface shape and suspension stability. It can be highly beneficial in maximizing the characteristics of the membrane for its applications such as wastewater treatment. As a result, investigations on product stability and surface adsorption often use it. Zeta potential is a feature that affects not only the particles but also their surroundings, such as pH, ionic strength, and even the kind of ions in the suspension, in contrast to particle size or molecular weight. As a result, even though measuring the zeta potential of suspended particles after dilution can often yield high resolution and accurate results, these results would typically differ significantly from their true values in the original environment, making them potentially misleading or of little practical use. It may also determine a solid surface or film's zeta potential. A combination of transparent electrode technology, high-sensitivity and quick-response photo-detector, and precision optics may be used to describe molecules or nanoparticles at different concentrations (Bumb et al. 2008; Xu 2008; Delgado et al. 2007).

4.3.9 Contact Angle Measurement

The membrane material's surface-wetting qualities have drawn much interest from both an application and fundamental surface science perspective. It has a significant impact on the permeate flow, rejection, and fouling characteristics of the membrane, as well as its performance throughout the separation process. The interaction between the membrane surface and liquid macromolecules may also be influenced by pH, temperature, charge density, and surface energy. The water contact angle of a pure water drop over the surface membrane might be used to measure this. The interaction between the water molecules and the material composition of the membrane, together with the surface chemistry that results, determines the wettability of the membrane (Pal and Sen 2025c). Surface tension for a solid material may be computed by measuring the contact angle between the solid surface and a droplet of liquid on the surface. A lower contact angle value indicates that the material is hydrophilic, meaning that water molecules have a higher affinity for the substrate. The term "hydrophilic" refers to materials that readily adsorb water molecules because they contain active polar functional groups. The increased contact angle suggests that the surface is hydrophobic. Those with this property, called hydrophobicity, react to water contact the opposite way as those with hydrophilic properties. The physicochemical characteristics of the substrate influence the wettability characteristic of nonideal surfaces. Consequently, the physical characteristics of the materials have a big impact on the measurement of the contact angle, such as the four primary factors, including particle size, shape, surface roughness, and heterogeneity (Xiao et al. 2022), (Hebbar et al. 2017).

4.3.9.1 Membrane Autopsy

Most membranes will eventually experience fouling, negatively impacting membrane function and needing cleaning. Whether the deposit is organic fouling, inorganic scaling, or colloidal deposition will determine how it has to be cleaned. Regardless of the strength of the circumstantial evidence supporting the foulant's identification, a membrane autopsy is the only dependable way to determine the foulant's real identity accurately. The plant's performance is impacted differently by various deposits.

Deposits like organic fouling raise the differential pressure along the membranes, lower production, and often need greater operating pressure to sustain output. Inorganic deposits may cause an irreversible increase in salt passage. All types of membrane fouling significantly increase plant downtime and maintenance expenses. After thoroughly analyzing membrane fouling during the autopsy process, a cleaning regimen can be provided. Proposals for antiscalant treatments that prevent deposits and sometimes delay membrane fouling may be used to assist cleaning operations. Deterioration of the salt rejection layer often results in a significant increase in salt passage. This may be brought on by oxidation of the membrane, inorganic salt deposition, abrasion, or even ripping of the membrane leaves themselves. It is sometimes linked to an increase in product flow. Recurrence of the issue may often be avoided if the fouling source is found, albeit in some circumstances, salt rejection performance might not be reversible. Membrane autopsy continues to be an effective diagnostic technique that provides a valuable understanding of membrane function and optimal cleaning methods. Reduced permeate flow, increased operating pressure, and shortened membrane lifespan are all results of membrane fouling, which impairs the efficiency of the membrane filtering process. As a result, it is essential to identify the root cause of membrane contamination and implement the appropriate safety measures. Membrane autopsy provides valuable information regarding potential operating mistakes and is a dependable technique for characterizing the foulants. Membrane autopsy has been employed in several investigations to examine the predominant foulant kinds on the membrane surface. A thin film composite membrane underwent a membrane autopsy, and by analyzing the outcomes of both fractal analysis and membrane autopsy, the fouling propensity of the membrane was predicted. The fouling mechanisms of the nanofiltration (NF) membranes used in a drinking water treatment facility were examined through membrane autopsy analysis. A membrane autopsy was conducted to analyze the RO membrane fouling used in a large-scale saltwater desalination plant. They observed that an organic percentage and mostly inorganic deposits, such as magnesium silicate, iron, and aluminum, made up a heterogeneous fouling layer on the membrane surface (Unal 2022; García-Triñanes et al. 2022).

Characterizing physical morphology requires a detailed analysis of the structural and physical characteristics of membranes. SEM makes it possible to take precise images of the surfaces and cross-sections of membranes, making it easier to examine the pore architecture, surface roughness, and thickness. EDS makes detecting components in certain membrane regions possible when used with SEM. This method is invaluable for figuring out how the nanoparticles are distributed throughout the membrane. When examining the dispersion of nanoparticles within composite membranes, TEM is particularly helpful since it offers a thorough examination of the internal architecture of membranes. Porometry is a method for figuring out a membrane's pore size and configuration, which is crucial for understanding how successfully the membrane filters and chooses things. Because AFM provides a three-dimensional depiction of surface topography, it makes it possible to determine pore sizes and surface roughness precisely at the nanoscale (Wyart et al. 2008). Tensile strength and elongation tests assess membranes' mechanical robustness and pliability, ensuring sustained stability under operating loads. A method called nanoindentation is used to measure the hardness and elastic modulus of materials, which helps to better understand their mechanical strength and endurance, especially when it comes to membrane materials. Precise physical characterizations are essential for creating membranes with the required structural characteristics to meet specific filtration requirements.

Identification and analysis of the chemical components of membrane materials are necessary for the characterization of chemical composition. Functional groups, chemical bonds, and elemental composition are all factors that affect how well a membrane functions and interacts with contaminants. One method for determining the functional groups and chemical bonds present in membrane materials is FTIR spectroscopy. This examination makes a deeper comprehension of the chemical alterations or surface treatments performed on the materials possible. By providing comprehensive molecular vibrational data, Raman spectroscopy complements FTIR investigation and is particularly useful for characterizing carbon-based nanomaterials. The elemental composition and chemical states of elements present on a membrane's surface may be examined using the XPS method. Understanding the surface chemistry and functionalization of the membrane depends on this investigation. The measurement of the contact angle assesses the hydrophilicity or hydrophobicity of the surface, showing how easily the membrane may absorb water or reject fouling. The electrical charge on the surface is evaluated by zeta potential testing, which affects how charged pollutants or foulants interact with the membrane. To increase the selectivity, fouling resistance, and overall performance of membranes, it is necessary to characterize their chemical composition and modify their surfaces.

4.4 Conclusions and Future Perspectives

Membrane technologies are widely used for separation and purification in various industries, including fuel cells, gas separation, food chemistry, pharmaceuticals, water treatment, and extract concentration/separation. Without a doubt, membranes are now competitive with traditional procedures and have a broad range of commercial and scientific uses. In the context of the applications of porous membranes in separation processes, the significance of reliable characterization of pore sizes and distributions necessitates the following: the ability to define the active part of the pore structure, which contributes to the permeability, increased measurement resolution, the ability to distinguish between artifacts induced by the characterization technique itself, swelling effects, and other factors, the ability to handle pronounced asymmetric pore-size distributions, including bimodal ones, and the characterization of the three-dimensional morphology of porous membranes. Recent standard methods for characterizing membrane surfaces, cross-section, and porosity are presented and discussed within the context of the current issue of depth profile examination, nondestructive sample preparation, and wet-state characterization. Porosity and pore size distribution studies, in conjunction with factors such as tortuosity, pore shape, and connectivity, connect to recent developments that are critical to the performance of membranes in various applications.

The need for better separation technologies, such as membrane filtration, has been highlighted by the inefficiency of traditional treatment methods in removing a broad range of pollutants, including recently found contaminants, disease-causing substances, and very microscopic pollutants. These novel and efficient filtration technologies are more important than ever due to growing water scarcity issues, stringent regulations, and elevated environmental concerns. A confluence of industrial, health, and environmental concerns necessitates effective, efficient, and sustainable water treatment technologies requiring membrane filtration. Often, new contaminants, including medicines, personal care items, endocrine disruptors, and other micropollutants, are not completely removed by conventional water treatment methods. Because membrane filtering techniques like NF and RO have tiny pore diameters, they efficiently eliminate contaminants. High nitrate and phosphate agricultural runoff causes eutrophication in aquatic bodies, which leads to an overabundance of algae growth and oxygen depletion. Membrane filtration may effectively reduce nutrient concentrations and protect aquatic habitats.

Characterization approaches may be divided into static and dynamic categories. Static approaches provide information on membrane shape, structure, and chemical/physical characteristics. Dynamic approaches are critical for investigating membrane performance. Some characterization procedures are damaging to the membrane, whereas non-destructive methods are used to monitor membrane function during usage. By closely investigating the microscopic features, chemical compositions, and fouling behaviors of membranes, characterization methodologies provide information crucial for guiding membrane fabrication, increasing performance, and assessing efficacy. A thorough understanding of the topic allows scientists and engineers to precisely tailor membranes, emphasizing certain traits and qualities necessary for increased effectiveness, robustness, and selectivity. Furthermore, the constant evolution of techniques for analyzing and characterizing material characteristics propels new developments in membrane science and engineering. Future developments in membrane characterization seem promising as the need for environmentally friendly water treatment methods grows and global environmental and public health issues need to be addressed.

A significant breakthrough will be the next generation of desalination membranes, which are very selective. Improving water quality and reducing energy usage and costs associated with desalination are two important goals of enhanced membrane selectivity. One such goal is to reduce the concentration of boric acid in seawater desalination to acceptable levels for agriculture. Eliminating the need for additional separation stages can be achieved. Nevertheless, despite these enormous efforts, it cannot reliably produce even small-area membranes that outperform the most cutting-edge TFC membranes regarding water permeability and selectivity. Making thin, defect-free selective layers on top of microporous ones will be difficult for several materials. It will be much more difficult to scale up the small-area membranes to inexpensive, industrial-size modules once their principles have been proven at the laboratory scale. Since innovative materials with excellent separation performance are often incompatible with fouling resistance, fouling management will continue to be a significant concern for both present and future membranes. The most effective method for incorporating fouling resistance into next-generation isoporous membranes is likely to include surface grafting of brushes made of fouling-resistant polymers, such as zwitterionic polymers and PEO. Surface grafting cannot provide fouling resistance to the inner pore walls of "tight" UF membranes; hence, choosing novel polymer chemistries for chemical and fouling resistance while fabricating the membrane is critical. Another drawback of present TFC membranes is that the polyamide selective layer is very susceptible to oxidants like chlorine and ozone; however, this is something that next-generation desalination membranes will hopefully address. The Achilles' heel of membrane desalination, biofouling, may be better controlled this way.

Sensor-equipped intelligent membranes will provide critical data for preventive maintenance and process optimization, ensuring consistent and reliable operation. The environmental impact of membrane manufacture may be reduced using eco-friendly approaches, such as green synthesis methods that use non-toxic solvents and energy-efficient operations. This is in keeping with global campaigns to promote ecologically sustainable industrial practices. Future research will focus on creating easily recyclable and reused membranes. The sustainability of membrane filtration systems would be improved by promoting renewable resources and improving the effectiveness of regeneration procedures. New membrane designs and materials will be developed to reduce the energy required for filtering processes. Developing membranes with high permeability and low fouling tendencies may lower operating pressures and energy consumption.

References

Akhtar N, Syakir Ishak MI, Bhawani SA, Umar K (2021) Various natural and anthropogenic factors responsible for water quality degradation: a review. Water 13:2660. https://doi.org/10.3390/w13192660

Baer DR, Engelhard MH (2010) XPS analysis of nanostructured materials and biological surfaces. J Electron Spectros Relat Phenomena 178–179:415–432. https://doi.org/10.1016/j.elspec.2009.09.003

Bannwarth S, Breisig H, Houben V, Oberschelp C, Wessling M (2017) Membrane impedance porometry. J Memb Sci 542:352–366. https://doi.org/10.1016/j.memsci.2017.08.019

Boczkaj G, Fernandes A (2017) Wastewater treatment by means of advanced oxidation processes at basic pH conditions: a review. Chem Eng J 320:608–633. https://doi.org/10.1016/j.cej.2017.03.084

Bumb A, Brechbiel MW, Choyke PL, Fugger L, Eggeman A, Prabhakaran D, Hutchinson J, Dobson PJ (2008) Synthesis and characterization of ultra-small superparamagnetic iron oxide nanoparticles thinly coated with silica. Nanotechnology 19:335601. https://doi.org/10.1088/0957-4484/19/33/335601

Cai Z, Dwivedi AD, Lee W-N, Zhao X, Liu W, Sillanpää M, Zhao D, Huang C-H, Fu J (2018) Application of nanotechnologies for removing pharmaceutically active compounds from water:

development and future trends. Environ Sci Nano 5:27–47. https://doi.org/10.1039/C7EN00644F

Cevallos-Mendoza J, Amorim C, Rodríguez-Díaz J, Montenegro M (2022) Removal of contaminants from water by membrane filtration: a review. Membranes (Basel) 12:570. https://doi.org/10.3390/membranes12060570

Delgado AV, González-Caballero F, Hunter RJ, Koopal LK, Lyklema J (2007) Measurement and interpretation of electrokinetic phenomena. J Colloid Interface Sci 309:194–224. https://doi.org/10.1016/j.jcis.2006.12.075

Ebnesajjad S (2011) Surface and material characterization techniques. In: Handbook of adhesives and surface preparation. Elsevier, pp 31–48

Figueras MJ, Borrego JJ (2010) New perspectives in monitoring drinking water microbial quality. Int J Environ Res Public Health 7:4179–4202. https://doi.org/10.3390/ijerph7124179

García-Triñanes P, Chairopoulou MA, Campos LC (2022) Investigating reverse osmosis membrane fouling and scaling by membrane autopsy of a bench scale device. Environ Technol 43:3198–3211. https://doi.org/10.1080/09593330.2021.1918262

Hang Y, Liu G, Huang K, Jin W (2015) Mechanical properties and interfacial adhesion of composite membranes probed by in-situ nano-indentation/scratch technique. J Memb Sci 494:205–215. https://doi.org/10.1016/j.memsci.2015.05.076

Hebbar RS, Isloor AM, Ismail AF (2017) Contact angle measurements. In: Membrane characterization. Elsevier, pp 219–255

Hortola P (2005) SEM examination of human erythrocytes in uncoated bloodstains on stone: use of conventional as environmental-like SEM in a soft biological tissue (and hard inorganic material). J Microsc 218:94–103. https://doi.org/10.1111/j.1365-2818.2005.01477.x

Hosseini SS, Bringas E, Tan NR, Ortiz I, Ghahramani M, Alaei Shahmirzadi MA (2016) Recent progress in development of high performance polymeric membranes and materials for metal plating wastewater treatment: a review. J Water Process Eng 9:78–110. https://doi.org/10.1016/j.jwpe.2015.11.005

Hube S, Eskafi M, Hrafnkelsdóttir KF, Bjarnadóttir B, Bjarnadóttir MÁ, Axelsdóttir S, Wu B (2020) Direct membrane filtration for wastewater treatment and resource recovery: a review. Sci Total Environ 710. https://doi.org/10.1016/j.scitotenv.2019.136375

Iqbal J, Shah NS, Ali Khan J, Naushad M, Boczkaj G, Jamil F, Khan S, Li L, Murtaza B, Han C (2024) Pharmaceuticals wastewater treatment via different advanced oxidation processes: reaction mechanism, operational factors, toxicities, and cost evaluation—a review. Sep Purif Technol 347:127458. https://doi.org/10.1016/j.seppur.2024.127458

Ji C, Zhai Z, Jiang C, Hu P, Zhao S, Xue S, Yang Z, He T, Niu QJ (2021) Recent advances in high-performance TFC membranes: a review of the functional interlayers. Desalination 500:114869. https://doi.org/10.1016/j.desal.2020.114869

Korin E, Froumin N, Cohen S (2017) Surface analysis of nanocomplexes by X-ray photoelectron spectroscopy (XPS). ACS Biomater Sci Eng 3:882–889. https://doi.org/10.1021/acsbiomaterials.7b00040

Koul B, Yadav D, Singh S, Kumar M, Song M (2022) Insights into the domestic wastewater treatment (DWWT) regimes: a review. Water 14:3542. https://doi.org/10.3390/w14213542

Krantz WB, Greenberg AR, Kujundzic E, Yeo A, Hosseini SS (2013) Evapoporometry: a novel technique for determining the pore-size distribution of membranes. J Memb Sci 438:153–166. https://doi.org/10.1016/j.memsci.2013.03.045

Kumar A, Gangawane KM, Ramanjaneyulu B (2022) Ferrofluids for waste-water treatment. In: Advances in chemical, bio and environmental engineering. Springer, pp 723–744

Kumar A, Gangawane KM, Jena HM (2025a) Variations in magnetic behaviour of ferrimagnetic nanoparticles with temperature. Indian Chem Eng 1–10. https://doi.org/10.1080/00194506.2025.2471412

Kumar A, Jena HM, Gangawane KM (2025b) Iron oxide nanoparticles as efficient adsorbents for malachite green dye removal: isotherms, kinetics, and thermodynamics analysis. Environ Sci Pollut Res 1–28. https://doi.org/10.1007/s11356-025-36291-w

Kumar A, Gangawane KM (2022a) Effect of precipitating agents on the magnetic and structural properties of the synthesized ferrimagnetic nanoparticles by co-precipitation method. Powder Technol 401:117298. https://doi.org/10.1016/j.powtec.2022.117298

Kumar A, Gangawane KM (2022b) Synthesis and effect on the surface morphology & magnetic properties of ferrimagnetic nanoparticles by different wet chemical synthesis methods. Powder Technol 410:117867. https://doi.org/10.1016/j.powtec.2022.117867

Kumar A, Gangawane KM (2023a) Molecular dynamics for the membrane process. Advanced computational approaches for water treatment. CRC Press, New York, pp 13–38

Kumar A, Gangawane KM (2023b) Advanced computational approaches for water treatment. Advanced computational approaches for water treatment. CRC Press, New York, pp 1–12

Mamba FB, Mbuli BS, Ramontja J (2021) Recent advances in biopolymeric membranes towards the removal of emerging organic pollutants from water. Membranes (Basel) 11:798. https://doi.org/10.3390/membranes11110798

Mechler Á, Kopniczky J, Kokavecz J, Hoel A, Granqvist C-G, Heszler P (2005) Anomalies in nanostructure size measurements by AFM. Phys Rev B 72:125407. https://doi.org/10.1103/PhysRevB.72.125407

Mourdikoudis S, Pallares RM, Thanh NTK (2018) Characterization techniques for nanoparticles: comparison and complementarity upon studying nanoparticle properties. Nanoscale 10:12871–12934. https://doi.org/10.1039/C8NR02278J

Nagandran S, Goh PS, Ismail AF, Wong T-W, Binti Wan Dagang WRZ (2020) The recent progress in modification of polymeric membranes using organic macromolecules for water treatment. Symmetry (Basel) 12:239. https://doi.org/10.3390/sym12020239

Pal D, Sen S (2023a) Emerging petroleum pollutants and their adverse effects on the environment. In: Impact of petroleum waste on environmental pollution and its sustainable management through circular economy. Springer, pp 103–137

Pal D, Sen S (2023b) In-depth coverage of petroleum waste sources, characteristics, environmental impact, and sustainable remediation process. In: Impact of petroleum waste on environmental pollution and its sustainable management through circular economy. Springer, pp 1–38

Pal D, Sen S (2025a) Toxicity and health impacts of emerging pollutants. In: Biotechnological interventions in removal of emerging pollutants. Springer, pp 55–79

Pal D, Sen S (2025b) Circularity and sustainable economy model towards supply chain management and zero waste. In: Environmental hydrocarbon pollution and zero waste approach towards a sustainable waste management. Springer, pp 353–383

Pal D, Sen S (2025c) Sustainable technologies in the development of novel processes and products from environmental hydrocarbon pollutants. In: Environmental hydrocarbon pollution and zero waste approach towards a sustainable waste management. Springer, pp 201–240

Pal D, Sen S (2024a) Valorization of sponge iron industrial waste into iron-modified zeolite X for ciprofloxacin removal: a multi-parameter optimization study. Environ Sci Pollut Res 31:53381–53398. https://doi.org/10.1007/s11356-024-34730-8

Pal D, Sen S (2024b) Optimal synthesis of dolochar derived faujasite zeolite X for highly effective Cd(II) removal. Environ Res 240:117494. https://doi.org/10.1016/j.envres.2023.117494

Poilane C, Delobelle P, Lexcellent C, Hayashi S, Tobushi H (2000) Analysis of the mechanical behavior of shape memory polymer membranes by nanoindentation, bulging and point membrane deflection tests. Thin Solid Films 379:156–165. https://doi.org/10.1016/S0040-6090(00)01401-2

Qiu H, Gao Y, Boott CE, Gould OEC, Harniman RL, Miles MJ, Webb SED, Winnik MA, Manners I (2016) Uniform patchy and hollow rectangular platelet micelles from crystallizable polymer blends. Science (80-) 352:697–701. https://doi.org/10.1126/science.aad9521

Rezakazemi M, Khajeh A, Mesbah M (2018) Membrane filtration of wastewater from gas and oil production. Environ Chem Lett 16:367–388. https://doi.org/10.1007/s10311-017-0693-4

Siddique T, Dutta NK, Choudhury NR (2021) Mixed-matrix membrane fabrication for water treatment. Membranes (Basel) 11:557. https://doi.org/10.3390/membranes11080557

Taha M, Hassan M, Essa S, Tartor Y (2013) Use of Fourier transform infrared spectroscopy (FTIR) spectroscopy for rapid and accurate identification of Yeasts isolated from human and animals. Int J Vet Sci Med 1:15–20. https://doi.org/10.1016/j.ijvsm.2013.03.001

Tan X, Rodrigue D (2019) A review on porous polymeric membrane preparation. Part II: production techniques with polyethylene, polydimethylsiloxane, polypropylene, polyimide, and polytetrafluoroethylene. Polymers (Basel) 11:1310. https://doi.org/10.3390/polym11081310

Titus D, Samuel EJ, Roopan SM (2019) Nanoparticle characterization techniques. In: Green synthesis, characterization and applications of nanoparticles. Elsevier, pp 303–319

Unal BO (2022) Membrane autopsy study to characterize fouling type of RO membrane used in an industrial zone wastewater reuse plant. Desalination 529:115648. https://doi.org/10.1016/j.desal.2022.115648

Werber JR, Osuji CO, Elimelech M (2016) Materials for next-generation desalination and water purification membranes. Nat Rev Mater 1:16018. https://doi.org/10.1038/natrevmats.2016.18

Wyart Y, Georges G, Deumié C, Amra C, Moulin P (2008) Membrane characterization by microscopic methods: multiscale structure. J Memb Sci 315:82–92. https://doi.org/10.1016/j.memsci.2008.02.010

Wyart Y, Tamime R, Siozade L, Baudin I, Glucina K, Deumié C, Moulin P (2014) Morphological analysis of flat and hollow fiber membranes by optical and microscopic methods as a function of the fouling. J Memb Sci 472:241–250. https://doi.org/10.1016/j.memsci.2014.08.012

Xiao M, Yang F, Im S, Dlamini DS, Jassby D, Mahendra S, Honda R, Hoek EMV (2022) Characterizing surface porosity of porous membranes via contact angle measurements. J Membr Sci Lett 2:100022. https://doi.org/10.1016/j.memlet.2022.100022

Xu R (2008) Progress in nanoparticles characterization: sizing and zeta potential measurement. Particuology 6:112–115. https://doi.org/10.1016/j.partic.2007.12.002

Yin J, Deng B (2015) Polymer-matrix nanocomposite membranes for water treatment. J Memb Sci 479:256–275. https://doi.org/10.1016/j.memsci.2014.11.019

Ying Y, Ying W, Li Q, Meng D, Ren G, Yan R, Peng X (2017) Recent advances of nanomaterial-based membrane for water purification. Appl Mater Today 7:144–158. https://doi.org/10.1016/j.apmt.2017.02.010

Chapter 5
Membrane Solutions for Next-Gen Drinking Water Purification

Shubham More, Ephrem Alemu, and Ashootosh Mandpe

Abstract Membrane-based filtration has emerged as a cornerstone technology in next-generation drinking water treatment, offering precise, energy-efficient, and scalable solutions for removing a broad-spectrum of contaminants. This chapter provides a comprehensive overview of membrane fundamentals, including material types, separation mechanisms, and process classifications ranging from microfiltration to reverse osmosis. The discussion highlights how advances in polymeric, ceramic, metallic, and thin-film composite membranes have improved selectivity, durability, and fouling resistance. Applications in pathogen removal, turbidity reduction, natural organic matter control, and desalination are examined, demonstrating the superior and consistent water quality achievable through membrane systems. The chapter also analyzes operational parameters and challenges such as fouling, energy demand, and concentrate management, emphasizing the need for optimized pretreatment and system design. Recent innovations—including nanocomposite materials, hybrid treatment configurations, and smart monitoring technologies—illustrate the ongoing evolution of membrane science toward more sustainable and resilient water treatment solutions. By synthesizing principles, applications, and emerging trends, this chapter underscores the critical role of membrane filtration in addressing global water scarcity and ensuring safe drinking water in an era of increasing environmental pressures.

Keywords Desalination · Drinking water treatment · Membrane filtration · Nanocomposite membranes · Sustainable water treatment · Thin-film composite (TFC) membranes

S. More · E. Alemu · A. Mandpe (✉)
Department of Civil Engineering, Indian Institute of Technology Indore, Indore, Madhya Pradesh, India
e-mail: as_mandpe@iiti.ac.in

E. Alemu
Faculty of Civil and Environmental Engineering, Jimma Institute of Technology, Jimma University, Jimma, Ethiopia

Department of Civil Engineering, College of Engineering and Technology, Wachemo University, Hossana, Ethiopia

A. Mandpe et al. (eds.), *Membrane Filtration and Membrane Bioreactors for Wastewater Treatment*, Advances in Wastewater Research,
https://doi.org/10.1007/978-981-95-6706-5_5

5.1 Introduction

5.1.1 *Objectives of the Chapter*

The objective of this chapter is to provide a comprehensive overview of membrane-based filtration technologies for safe drinking water treatment. The chapter aims to explain the fundamental principles, operational mechanisms, and applications of membrane filtration processes in removing impurities from water sources. The chapter explores advancements in material science and process optimization to enhance performance and scalability regarding membrane filtration. The chapter also evaluates the advantages of membrane systems over conventional treatment methods. Additionally, it addresses challenges related to membrane filtration such as membrane fouling, energy consumption, and cost barriers. Generally, the chapter pursues to inform researchers, engineers, and policymakers about the critical role of membrane filtration in addressing global water scarcity through synthesizing technical insights.

5.1.2 *Background of Membrane Filtration*

Membrane filtration is a separation technology which uses semi-permeable barriers to selectively separate components based on size, charge, or molecular affinity from a fluid (Khery et al. 2022). The technology originated in the eighteenth century with the study of natural membranes. The concept of membrane filtration developed significantly in the mid-twentieth century with the development of synthetic polymeric membranes (Ajizah 2023). These synthetic polymers enabled precise control over pore sizes and separation mechanisms (Ajizah 2023). Early applications of membrane filtration majorly focused on laboratory-scale practices, but due to advancements in materials science during the 1960s and 1970s expanded its use into industrial sectors such as water treatment and pharmaceuticals (Wang et al. 2021). Due to its ability to operate without thermal input makes, it energy efficient and suitable for heat-sensitive substances. The membrane technology gained importance, specifically due to its energy efficiency. Currently, inventions like modular designs and asymmetric membranes have improved efficiency, scalability, and cost-effectiveness. Specifically, contaminant removal efficiency makes membrane filtration the cornerstone of modern separation technology (Bressloff 2022).

The membrane technique encompasses various treatment methods such as microfiltration, ultrafiltration, nanofiltration, and reverse osmosis. The techniques are majorly differentiated by pore size and separation capacities (Kerkulah et al. 2023). Microfiltration and ultrafiltration target suspended solids and macromolecules. Nanofiltration and reverse osmosis remove dissolved ions and organic contaminants.

Nanofiltration and reverse osmosis have critical roles in desalination and wastewater reuse (Yue et al. 2022). The demand for high-purity products further accelerated its adoption, particularly in addressing global water scarcity. Current trends emphasize driving research into bio-based membranes, fouling-resistant materials, and hybrid systems integrating renewable energy. Today, membrane filtration is integral to advancing circular economies and achieving sustainable development goals. There are ongoing innovations on membrane that are expanding its applications in biotechnology, energy production, and resource recovery (Kaufman et al. 2023).

5.1.3 Overview of Drinking Water Treatment

Drinking water treatment is a critical process that is designed to remove all impurities, pollutants, contaminants, and pathogens from raw water sources. Drinking water treatments ensure safety to provide potable water for human consumption (Yan et al. 2022). A conventional treatment method typically involves multiple stages such as coagulation, flocculation, sedimentation, and filtration to remove impurities (Yan et al. 2022). Disinfection through chlorination, ultraviolet light or ozonation is applied after filtration to eliminate harmful microorganisms (Li et al. 2022). In recent decades, advanced technologies like membrane filtration are enhancing the removal efficiency of pollutants at molecular levels (Zhang et al. 2020). Regulatory standards and emerging contaminants are driving endless innovation in treatment processes to meet safety and quality standards. Increasing concerns over water scarcity and pollution have also spurred the adoption of hybrid treatment systems (Franke et al. 2021).

5.2 Fundamentals of Membrane Filtration

5.2.1 Definition and Principles of Membrane Filtration

Membrane filtration is a separation process that selectively removes particles, microorganisms, or dissolved substances from a fluid based on their size, charge, or chemical affinity using semi-permeable membranes. The fundamental principle of membrane filtration depends on applying a driving force such as pressure, concentration gradient, and electric potential to push the fluid through the membrane. Pollutants, pathogens, and contaminants larger than the pore size membrane or incompatible with its chemical properties are retained (Bressloff 2022). Membrane filtration process is governed by principles of sieving, adsorption, and diffusion through ensuring precise selectivity and efficiency (Bressloff 2022). Membrane filtration operates at ambient temperatures unlike thermal methods with preserving heat-sensitive components (Abdel-Hai et al. 2024). The modular design of membranes

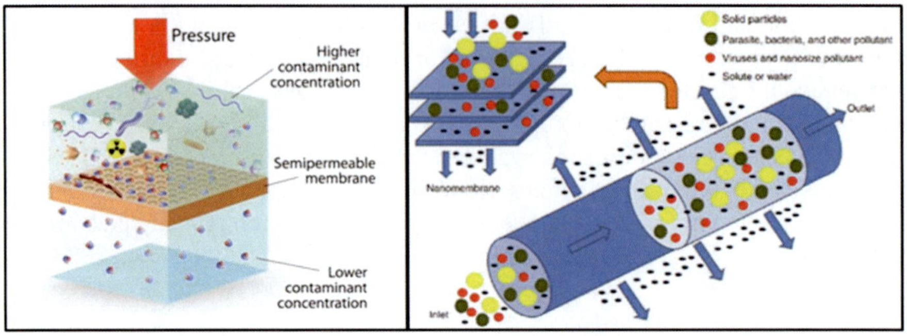

Fig. 5.1 Fundamental principles of membrane filtration. *Source* Researchgate.com; uetwater.com

allows scalability and adaptability across industries, behind applications from water purification to biotechnology. Currently ongoing advancements are focusing on enhancing durability, reducing fouling, and improving sustainability (Razmgar and Nasiraee 2021) (Fig. 5.1).

5.2.2 *Membrane Materials and Their Properties*

I. **Polymeric membranes**

Due to their versatility, adaptability, flexibility, and ease of fabrication, polymeric membranes are among the most widely utilized in filtration. A common material of polymeric membrane includes polyamide, polysulfone, polyethersulfone, and polyvinylidene fluoride. These membranes are engineered with tailored pore sizes and surface chemistries. They are designed to achieve selective separation for applications such as water purification, gas separation, and dialysis (Jiang and Ladewig 2020). Their flexibility and lightweight nature make them ideal for spiral-wound or hollow fiber module designs. Conversely, polymeric membranes can be prone to fouling, chemical degradation, and thermal instability. Their failure is limiting their use in extreme pH and high-temperature environments (Sutrisna et al. 2020). New innovations like thin-film composite membranes and hydrophilic coatings have enhanced filtration performance in reverse osmosis and ultrafiltration processes (Wu et al. 2021) (Fig. 5.2).

II. **Ceramic membranes**

Ceramic membranes are made up from Alumina, Zirconia, or Titania (Zhang et al. 2023). Ceramic membranes offer exceptional thermal, chemical, and mechanical stability. Due to their rigid porous structure, they are able to operate in harsh conditions. They can operate at high temperatures greater than 100 °C. Ceramic membranes can also handle aggressive solvents and abrasive fluids (Wu et al. 2023). All these properties make them suitable for industrial wastewater treatment, food processing,

Fig. 5.2 Polymeric membranes. *Source* Adwynwater.com; waterlandtech.com

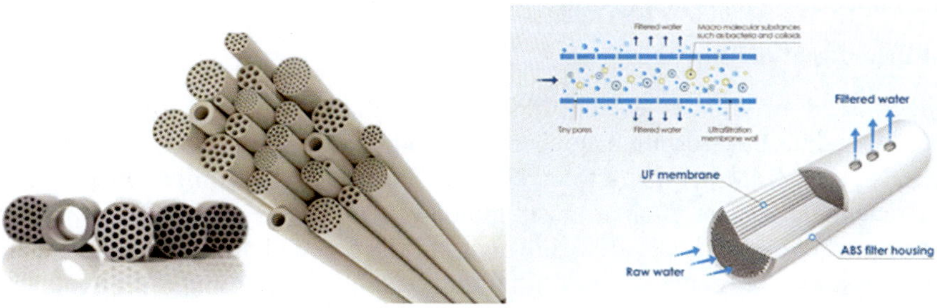

Fig. 5.3 Ceramic membranes. *Source* Newater.com

and petrochemical industries (Chai et al. 2024). Ceramic membranes exhibit excellent resistance to fouling (Wu et al. 2023). They can be cleaned with strong acids or alkalis, ensuring long lifespans. One of their drawbacks is high manufacturing costs and brittleness. Recent advances in nanostructured ceramics and hybrid organic–inorganic coatings aim to reduce costs. Recent advances are also improving selectivity and permeability for various industrial applications like hydrogen separation and catalytic reactors (Chai et al. 2024) (Fig. 5.3).

III. **Metallic membranes**

Metallic membranes are made from materials like stainless steel, nickel, or palladium alloys. Metallic membrane materials are highly valued for their robustness and conductivity. These dense and porous membranes excel in high-pressure and temperature environments. Metallic membranes materials are particularly used in gas separation such as hydrogen purification and catalytic processes. For example, palladium-based metallic membranes are highly selective for hydrogen due to their ability to adsorb and diffuse hydrogen molecules. Metallic membranes materials have superior mechanical strength and thermal resilience. Their major limitations are their high cost, vulnerability to corrosion, and their heavy weight (Zhang et al. 2020). These limitations restricted their use to specialized industrial applications. Currently researchers are focusing on the development of composite metallic membranes and nanostructured surfaces to enhance efficiency.

IV. **Biological membranes**

Biological membranes are inspired by natural systems like cell membranes such as lipid bilayers or protein-based structures. These biological membranes are designed for high biocompatibility and molecular recognition. Their biocompatibility and molecular recognition make them ideal for medical applications. They are used for drug delivery, artificial organs, and biosensors. Materials for biological membranes include phospholipids, aquaporin proteins, and engineered peptides that simulate natural transport mechanisms. Biological membranes offer unparalleled selectivity for specific ions or biomolecules. Their major drawback is that mechanical fragility and sensitivity to environmental conditions pose challenges for industrial use. Innovations in synthetic biology and hybrid membranes combining biological components with synthetic polymers aim to bridge this gap (Wild et al. 2022).

V. **Fiber-based membranes**

Fiber-based membranes leverage high surface-area-to-volume ratios for efficient filtration, such as electrospun nanofibers and hollow fibers. Fiber-based membrane materials range from polymers to carbon nanotubes and glass fibers. Fiber-based membrane materials are often arranged in nonwoven mats or woven fabrics. These types of membrane materials are lightweight, flexible, and highly permeable. Fiber-based membranes are suitable for air filtration, wearable sensors, and water treatment modules. Interconnected structures in fiber-based membranes enable rapid flux while trapping fine particulates or pathogens due to their porous nature. Challenges related to fiber-based membrane materials include fiber brittleness in dry conditions and clogging in humid environments (Zhou et al. 2022). Inventions of fibers like functionalized nanofibers with antimicrobial coatings and embedded nanoparticles are enhancing their performance. Their current innovations are driving applications in environmental remediation and smart textiles. Tunable properties of each fiber type allow using for specific separation needs which leads to balancing efficiency and practicality (Table 5.1).

5.3 Types and separation mechanisms of membranes

This section discusses types and separation mechanisms of membrane filtration technology in water treatment. Membranes can be categorized based on their material composition and filtration process. Based on their material composition, membranes can be primarily classified as polymeric, ceramic, and metallic membranes. Also, based on filtration processes they facilitate, membranes can be classified as microfiltration, ultrafiltration, nanofiltration, and reverse osmosis. Each type of membrane and process has unique advantages and limitations. Understanding the classifications of membranes and separation mechanisms is important for selecting the appropriate filtration method for specific water treatment requirements.

Table 5.1 Membrane filtration materials properties

Membrane type	Materials	Key properties	Advantages	Limitations
Polymeric membranes	• Polyamide, • PVDF, PES, • Polysulfone	• Moderate thermal stability • Tunable pore size • Hydrophobic • Hydrophilic	• Low cost • Flexible and • Lightweight Scalable	• Prone to fouling • Limited chemical/ thermal resistance • Degrades under harsh conditions
Ceramic membranes	• Alumina, • Zirconia, • Titania	• High thermal stability • Chemical stability • Rigid structure • Inert to solvents	• Durable in harsh environments • Resistant to fouling • Long lifespan	• High cost • Brittle • Heavy
Metallic membranes	• Stainless steel, • Palladium alloys	• High mechanical strength • Conductivity • Thermal resilience	• Suitable for high pressure and temperature • Selective for gases	• Expensive • Susceptible to corrosion • Heavy
Biological membranes	• Lipid bilayers • Aquaporin proteins	• Biocompatible • Molecular recognition • Selective ion transport	• Mimics natural systems • High specificity for biomolecules	• Fragile • Sensitive to pH/ temperature • Difficult to scale
Fiber-based membranes	• Electrospun polymers, • Carbon nanotubes	• High surface area • Flexible and porous • Tunable fiber density	• Lightweight • High permeability • Customizable for specific needs	• Prone to clogging • Limited strength in dry conditions

5.3.1 Classification of Membranes Based on Materials

Based on the materials they are composed of, membranes used in water treatment are classified. The most commonly used membrane materials in water treatment are classified as polymeric membranes, ceramic membranes, metallic membranes, and thin-film composite (TFC) membranes. The choice of membrane material depends on many factors such as contaminant type, operating conditions, and desired lifespan. In this subsection, classification of material based on material is discussed in detail.

I. **Polymeric membranes**

Polymeric membranes are made from polyamide, polysulfone, or cellulose acetate. Polymeric membranes are the most widely used type of membranes due to their flexibility and chemical resistance. Specifically such type of membranes is ideal for processes like RO (reverse osmosis) and UF (ultrafiltration).

Polymeric membranes materials are extensively used in drinking water treatment due to their extraordinary advantages. This membrane types are cost-effective to produce and install compared to other types of membrane materials. Their cheapness during production and installation make them accessible for large-scale applications. Their flexibility and chemical resistance make them to effectively remove diverse contaminants (Hansen et al. 2020). Through processes like RO (reverse osmosis), UF (ultrafiltration), and NF (nanofiltration), polymeric membranes effectively remove pathogens, organic matter, and dissolved ions. Polymeric membranes those made from polyamide and polysulfone exhibit high selectivity for specific pollutants. This high selectivity for specific pollutant ensures compliance with water quality standards. Polymeric membranes modular design enables easy integration into existing water treatment systems and scalability due to their energy-efficient operation in low-pressure processes like UF further enhances their demand for drinking water purification.

Polymeric membranes face some limitations. One of the major limitations of polymeric membranes is fouling. Fouling is caused by the accumulation of particles, organic matter and microbes on the membrane surface. Fouling is a persistent challenge that reduces permeability and increases maintenance costs. Repeated chemical cleaning to mitigate fouling can degrade the membrane material and shorten its lifespan (Yuan et al. 2020). They are also sensitive to extreme pH levels, high temperatures, and chlorine exposure. For example, polyamide RO membranes are damaged when exposed to chlorine. To operate in harsh conditions, they need rigorous pretreatment. Polymeric pore sizes may struggle with very small dissolved ions and emerging pollutants. For removing dissolved ions and microplastics, they require hybrid systems. Since polymeric membranes are not easily biodegradable, environmental concerns also arise from the disposal after use. There are ongoing advancements to enhance fouling resistance, durability and sustainability in drinking water treatment.

II. **Ceramic membranes**

A ceramic membrane has advantages in drinking water treatment. Due to their exceptional thermal and chemical stability they can operate in harsh conditions. They can operate without degradation in high temperatures or exposure to corrosive disinfectants. Ceramic membranes are made up from materials like alumina and zirconia. Ceramic membranes exhibit mechanical strength and longevity. They have lower replacement frequency compared to polymeric membranes. Due to their smooth and inert surfaces, ceramic membranes are highly resistant to fouling. They also tolerate aggressive cleaning methods such as steam and acids. Also, their precise pore sizes effectively remove pathogens, turbidity, and some viruses.

Ceramic membranes have some drawbacks such as their energy-intensive production process. Their high initial cost is due to their expensive raw materials and energy-intensive production process. Ceramic membranes require robust infrastructure since they are brittle, heavy and have complicated installation. Even though ceramic membranes are fouling resistant, they are not immune to biofouling. Due to their rigid structure, they have limited adaptability for variable water quality and flow demands. Moreover, their limited pore-size variety restricts finer separations such as dissolved salts. Their limited pore-size varieties need hybrid systems with nanofiltration and reverse osmosis.

III. **Metallic membranes**

Metallic membranes are constructed from materials such as stainless steel, titanium, and nickel alloys. They have excellent mechanical strength and durability that withstand high pressures, temperatures, and abrasive conditions. Particularly in industrial and high-stress environments, metallic membranes are advantageous in drinking water treatment. Due to their inherent resistance to chemical corrosion, they are suitable for treating water with aggressive disinfectants such as chlorine, acidic, and alkaline streams. Metallic membranes tolerate rigorous cleaning methods such as steam sterilization and strong solvents. Their long-term replacement costs are lower when compared to other types of membranes. Their precise and uniform pore structures allow reliable microfiltration for removing suspended solids, bacteria, and particulates. Metallic membranes smooth surfaces minimize fouling compared to other membrane materials. Properties of metallic membranes are suitable for specialized applications in pretreatment of desalination.

Metallic membranes face significant limitations. They are not economical due to their expensive raw materials and complex manufacturing. Their weight and rigidity complicate modular installation. Even though resistant to fouling, metallic membranes are not immune to scaling from minerals like calcium and silica. Calcium and silica can clog pores and reduce efficiency. Operational parameters can be expensive since their energy demands may rise in high-pressure systems. In addition, their limited pore size range restricts finer separations such as dissolved salts and viruses.

IV. **TFC (thin-film composite) membranes**

Thin-film composite membranes are emerging materials, which combine layers of polymers and nanomaterials. Their combined layers enhance selectivity and fouling resistance in desalination and nanofiltration. Due to their high selectivity and permeability, thin-film composite membranes are widely preferred in drinking water treatment. They are ideal particularly for desalination and nanofiltration. They are made up of a thin polyamide active layer at the top and a porous polymer support. They are excellent in rejecting salts, heavy metals, and micropollutants while allowing efficient water flux. They have reduced energy demands compared to other traditional membranes. Due to their chemical stability, TFC perform treatment in diverse pH conditions. TFC membranes are also compact and modular, making them critical for production of potable water from salty and seawater sources.

The polyamide layer in TFC is highly vulnerable to chlorine degradation. They require extensive pretreatment such as dechlorination to avoid irreversible damage to the polyamide layer. Fouling from organic matter remains a persistent problem that reduces the efficiency of TFC membrane. Due to fouling there is frequent chemical cleaning, which in turn weakens the membrane. TFC membranes have high production costs and the complexity of manufacturing. Advancements in coating technologies such as graphene oxide layers and antifouling surface modifications

Table 5.2 Membrane material and process compatibility summary

Membrane material	Compatible processes	Key features
Polymeric	MF, UF	Flexible, low-cost, chemical-resistant
Ceramic	MF, UF	Thermal/chemical stability, long lifespan
Metallic	MF	Robust, reusable, handles abrasive feeds
TFC	NF, RO	High selectivity, dense structure for fine separation

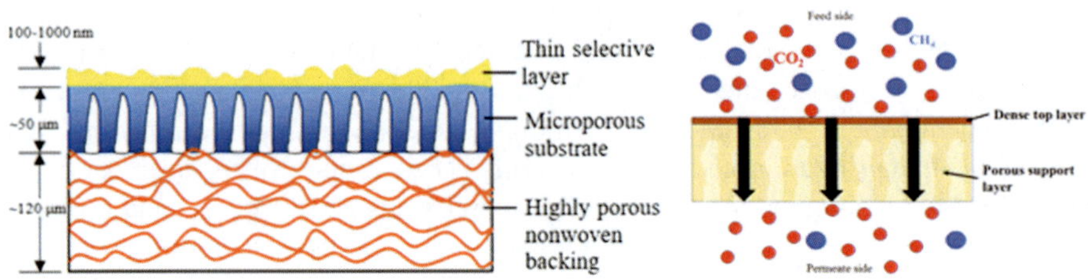

Fig. 5.4 Thin-film composite membranes. *Source* Researchgate.com

aim to improve durability of TFC membranes. Improvements in TFC membrane layers can make them a cornerstone of modern desalination and advanced water purification methods (Table 5.2; Fig. 5.4).

5.3.2 Classification of Membranes Based on Process

Membranes in water treatment are classified based on their separation processes. Their type can determine their applications and efficiency in water treatment. There are four types of membranes based on their separation processes. They are MF (microfiltration), UF (ultrafiltration), NF (nanofiltration), and RO (reverse osmosis). Pressure-driven methods include microfiltration and ultrafiltration. Pressure-driven methods use porous membranes to physically filter particles, bacteria, and macromolecules. Nanofiltration and reverse osmosis use dense membranes and higher pressures to remove dissolved ions, organic molecules, and salts. This subsection discusses in detail about each type of membrane based on their process.

I. **Microfiltration (MF)**

Microfiltration types of membranes are widely used in water treatment to remove suspended solids, bacteria and large pathogens. They have pore sizes ranging from 0.1 to 10 µm. They are commonly made up of polymers like polypropylene and polyvinylidene fluoride. These types of membranes operate at relatively low pressures which make them energy-efficient. They are used for applications such as surface water clarification, wastewater pretreatment. They are also used to safeguard downstream processes such as reverse osmosis. Microfiltration membranes are effective for turbidity reduction and microbial removal. They do not remove dissolved contaminants like salts and viruses. Even though they are susceptible to fouling, their modular design and cost-effectiveness make them ideal for industrial water treatment systems.

II. **Ultrafiltration (UF)**

Ultrafiltration membranes are used in water treatment for removing fine particles, colloids, viruses, and macromolecules. Their pore sizes range between 0.01 and 0.1 μm. Ultrafiltration membranes are made up of polymers like polysulfone and polyethersulfone. These types of membranes operate at low to moderate pressures. Since they can operate in low pressures, they are energy efficient. Ultrafiltration membranes remove pathogens and turbidity while retaining important minerals. They are widely used in drinking water purification, wastewater recycling, and as a pretreatment for reverse osmosis. An ultrafiltration membrane effectively reduces microbial risks and organic loads. They are prone to fouling from organic matter and biofilms. Regular backwashing and chemical cleaning can maintain their performance. They are reliable for high-quality water production.

III. **Nanofiltration (NF)**

Nanofiltration membranes are made from thin-film composites and charged polymers. Nanofiltration membranes are a type of membrane that bridges the gap between ultrafiltration and reverse osmosis. They have pore sizes between 0.001 and 0.01 μm. Nanofiltration membranes selectively remove divalent ions such as calcium and magnesium. In addition, they filter organic molecules, pesticides, and micropollutants. Nanofiltration membranes do not remove smaller monovalent ions like sodium. Nanofiltration membranes operate at moderate pressures. They operate at lower energy than reverse osmosis. Nanofiltration membranes are excellent in water softening, color removal, and reducing hardness. They remove heavy metals such as arsenic in drinking water and retain beneficial minerals. Nanofiltration membranes are susceptible to fouling from organic matter. They require pretreatment. They are ideal for treating surface water, industrial wastewater, and brackish sources. They are less effective for full desalination treatments.

IV. **Reverse osmosis (RO)**

Reverse osmosis membranes are fundamental in advanced water treatment. They are made up of dense and semi-permeable membranes having pore size less than 1 nm. They are designed to remove dissolved salts, ions, organic contaminants, and pathogens through high-pressure-driven separation technique. They are commonly structured as thin-film composites with polyamide active layers.

Reverse osmosis is the best method in desalinating seawater and brackish water. Reverse osmosis produce ultrapure water for industrial and municipal use. Reverse osmosis has the ability to remove 99% of dissolved solids. This extra removal capacity makes it crucial in addressing drinking water scarcity. Reverse osmosis demands significant energy for pressurization. They also face fouling risks requiring robust

Table 5.3 Integrating membrane processes and materials for drinking water treatment

Process	Materials	Pore size	Key applications (removal)	Advantages	Limitations
MF	• Polymeric • Ceramic • Metallic	0.1–10 μm	• Bacteria • Suspended solids	• Low energy, • Durable • Cost-effective	• Limited to large particles • Fouling prone
UF	• Polymeric • Ceramic	0.01–0.1 μm	• Virus • Colloid • Macromolecule	• Better pathogen control • Modular design	• Requires frequent cleaning • Moderate cost
NF	TFC	~ 1 nm	• Softening • Organic • Divalent ion	• Selective for ions • Retains some minerals	• Chlorine-sensitive • Scaling risk
RO	TFC	< 1 nm	• Desalination • Dissolved salt	• High salt rejection (>99%) • Pure water	• Energy-intensive • Generates brine waste

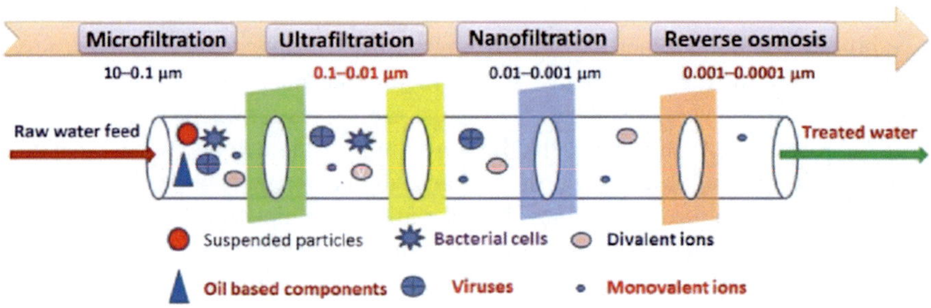

Fig. 5.5 Types of membranes based on process. *Source* Researchgate.com

pretreatment such as ultrafiltration. Reverse osmosis generates concentrated brine waste. It has drawbacks like important mineral removal that affect taste of treated water. Reverse osmosis remains the best and efficient process for safe drinking water treatment (Table 5.3; Fig. 5.5).

5.3.3 Separation Mechanisms of Membranes

Membranes in drinking water treatment depend on different physical and chemical processes to separate contaminants and pollutants. The primary mechanism of separation through membrane is size exclusion or sieving. During sieving, membranes act as physical barriers with specific pore sizes to block particles larger than membrane pore

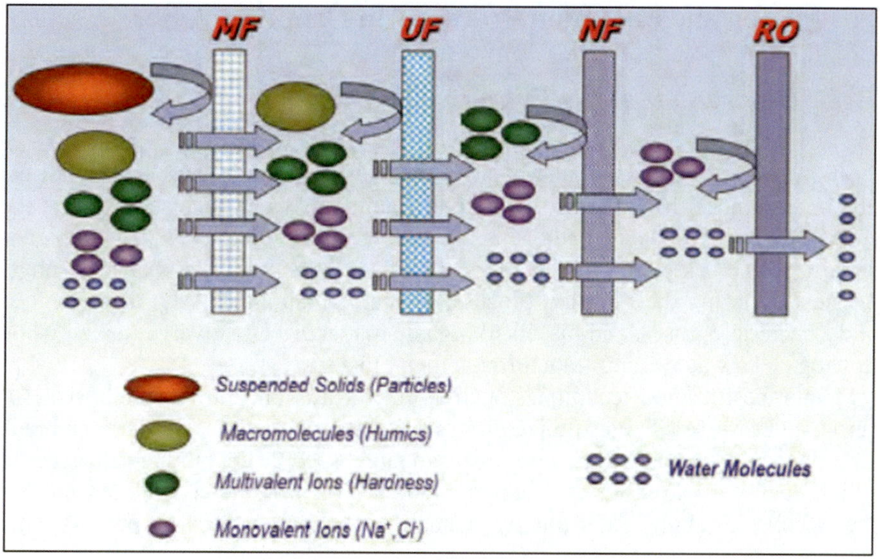

Fig. 5.6 Component separation during pressure-driven membrane process (Source: researchgate.com)

size. For instance, microfiltration membranes have a pore size range of 0.1 to 10 μm and can sieve suspended solids, bacteria, and protozoa. Ultrafiltration membranes have a pore size of 0.01 to 0.1 μm and can remove viruses and colloids. This sieving effect is straightforward and effective for particulate matter. Size exclusion or sieving does not remove dissolved ions or smaller organic molecules (Fig. 5.6).

In denser membranes like nanofiltration and reverse osmosis, separation of solution–diffusion and electrostatic interactions are executed. In these separation mechanisms, water molecules dissolve into membrane material and diffuse through it, whereas dissolved ions are rejected due to differences in solubility and diffusivity. Nanofiltration membranes additionally leverage electrostatic repulsion called Donnan exclusion to reject divalent ions such as $Ca2^+$, $Mg2^+$, and charged organics. These make nanofiltration membrane ideal for water softening. Reverse osmosis membranes having ultra-dense structures with pore sizes less than 1 nm can remove nearly all dissolved salts and uncharged contaminants. Reverse osmosis can remove nearly all pollutants through a combination of steric hindrance and diffusion resistance that enable desalination.

5.4 Membrane Filtration Processes in Drinking Water

5.4.1 Description of the Filtration Process

The filtration process in drinking water treatment involves the use of selective barriers called membranes. Membranes separate contaminants from water and ensure the removal of impurities and pathogens. During the filtration process, raw water is pressurized and passed through membranes with specific pore sizes that act as physical sieves. During the filtration process, several contaminants such as suspended solids, bacteria, viruses, and dissolved organic matter are effectively retained while allowing clean water to infiltrate through membranes.

Membrane filtration techniques such as microfiltration, ultrafiltration, nanofiltration, and reverse osmosis utilize different membrane materials. Each membrane filtration technique has different pore sizes to target specific contaminants that result in high-quality drinking water. The membrane filtration process is efficient and can meet specific drinking water quality standards. Membrane filtration process is an essential technology in modern drinking water treatment facilities.

5.4.2 Flow Configurations

Flow configurations in membrane filtration mainly include dead-end and cross-flow setups. In dead-end filtration setup, feed water flows perpendicular to the membrane surface. All fluid is forced through the membrane perpendicularly in dead-end setup. This setup is ideal for low suspended solid applications such as municipal pretreatment due to its simplicity and low energy demand. There is a high fouling risk due to accumulated particles forming a cake layer.

Cross-flow filtration setup feeds water tangentially across the membrane surface. This setup generates shear forces that wash away debris and reduce fouling. Cross-flow configuration is comfortable in high turbidity and industrial feeds. Hybrid systems like semi-dead-end with periodic backwashing can balance fouling control and energy efficiency. Flow configuration choice depends on water quality, operational costs, and scalability. Flow configuration can impact system life-span. Additionally, configuration setup can affect water treatment consistency in applications ranging from ultrafiltration pretreatment to reverse osmosis desalination processes (Fig. 5.7).

5.4.3 Operating Conditions and Parameters

Operating conditions and parameters of membrane filtration in drinking water treatment majorly influence the performance and durability. Key parameters include

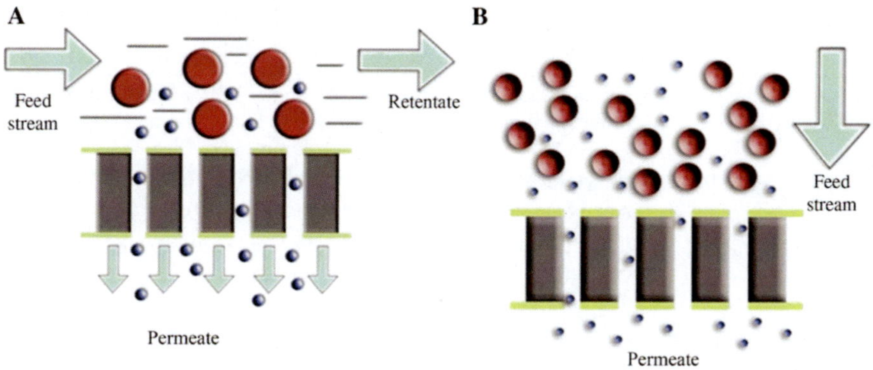

Fig. 5.7 Flow configurations A. Dead-end setup B. Cross-flow setup (Source: researchgate.com)

TMP (transmembrane pressure) that drives water flux through the membrane. Cross-flow velocity that mitigates fouling by sweeping away accumulated particles is also another important parameter. Specifically in dense membranes such as reverse osmosis, higher TMP increases permeate flux but risks compaction or irreversible fouling.

One of the important operational parameters is temperature and affects viscosity and diffusion rates. Warmer water increases permeability but degrades polymeric membranes. Feed water quality that has turbidity, organic content, and ionic strength needs pretreatment, such as coagulation and pH adjustment. Pretreatment prevents scaling, biofouling, and chemical degradation. For instance, reverse osmosis requires antiscalants to prevent mineral precipitation. In addition, ultrafiltration and microfiltration systems use backwashing cycles to maintain their flux.

Optimization of operating parameters comprises balancing energy efficiency with fouling and scaling control. Low-pressure processes like ultrafiltration membranes prioritize frequent backwashing and moderate cross-flow velocities to limit their energy use. High-pressure reverse osmosis methods focus on accurate TMP (transmembrane pressure) control and pretreatment to minimize scaling. pH adjustments can improve contaminant removal and protect chlorine-sensitive membranes (Table 5.4).

Table 5.4 Operating conditions and parameters for membrane filtration

Parameter	Description	Key considerations
Transmembrane pressure/TMP	Driving force for water flux through the membrane	Higher TMP increases flow but risks fouling or membrane compaction
Cross-flow velocity	Tangential flow rate across the membrane surface	Higher velocity reduces fouling but increases energy consumption
Feed water quality	Includes turbidity, organic content, ionic strength, pH, and microbial load	Dictates pretreatment to prevent fouling
Temperature	Influences water viscosity and membrane stability	Higher temps improve flux but may degrade polymeric membranes
pH	Affects membrane charge, and chemical compatibility	Critical for NF/RO ion removal; Adjust to minimize scaling ($CaCO_3$)
Flux rate	Permeate flow per unit membrane area	Operate below "critical flux" to avoid rapid fouling
Recovery rate	Ratio of permeate to feed water volume	High recovery concentrates contaminants, Increasing scaling and fouling risks
Backwashing frequency	Regular cycles to remove fouling layers	Frequent cleaning restores flux but may shorten membrane lifespan
Fouling control	Use of antiscalants, coagulants and oxidants	Requires compatibility with membrane material
Membrane material compatibility	Resistance to chemicals (chlorine, oxidants) and pH extremes	Ceramic membranes tolerate harsh conditions, polymers may degrade
Operational mode	Dead-end and cross-flow configuration	Dead-end: low energy, high fouling; Cross-flow: higher energy, low fouling
Critical flux management	Threshold flux beyond which fouling accelerates	Monitoring ensures sustainable operation and minimizes cleaning frequency
Pressure drop (ΔP)	Pressure loss across the membrane module	Rising ΔP indicates fouling or clogging; Requires maintenance
Energy consumption	Linked to TMP, cross-flow velocity	Optimize for cost-effectiveness and sustainability (low-pressure UF and high-pressure RO)
Pretreatment requirements	Steps like filtration, coagulation and disinfection	Protects membranes from clogging, scaling and chemical damage

5.5 Applications of Membranes in Drinking Water Treatment

5.5.1 Removal of Pathogens and Other Contaminants

Membrane filtration provides a physical barrier to pathogens—including bacteria, viruses, and protozoan parasites—and can effectively block many harmful contaminants. MF and UF work principally by size exclusion: MF membranes (pore ~ 0.1 μm) can retain most bacteria and protozoa (e.g., *Giardia* cysts, *Cryptosporidium* oocysts) (Franke et al. 2021), while UF (pore ~ 0.01 μm) also retains smaller microbes like viruses to a greater extent. For instance, UF processes have demonstrated 0.5–5.9 log removal values (LRV) for pathogens, compared to 0.7–4.6 LRV for MF. In practice, properly operated UF can achieve > 3–4 log removal of viruses and > 5 log for larger organisms, meeting stringent drinking water standards. NF and RO membranes offer an even more robust barrier their nanometer-scale pores and dense polymer matrices remove > 99.9% of all pathogens, effectively sterilizing the water (Shukla et al. 2023).

Recent Case Studies: In Europe, several drinking water plants have adopted UF as a barrier against *Cryptosporidium* after conventional filtration, given UF's reliable removal of oocysts below detectable levels. In the United States, concern over per- and polyfluoroalkyl substances (PFAS) has driven the use of high-pressure membranes: NF has been shown to reject over 90–99% of PFAS contaminants in affected water supplies, and full-scale systems integrating NF or RO are being implemented to meet new PFAS regulations (Al-Maas et al. 2021).

Membranes also target other contaminants through physical sieving and chemical interactions. NF and RO, in particular, are effective for removing many micropollutants—such as pesticides, pharmaceutical residues, and heavy metals—that are dissolved in water. While MF/UF primarily remove particulate-bound contaminants, NF/RO can reject dissolved solutes via size/charge exclusion. For example, NF membranes (200–300 Da cutoff) significantly reduce heavy metal ions and organic micropollutants; one study reported > 99% retention of certain phthalate plasticizers using an NF membrane. RO, being the tightest membrane, can eliminate virtually all ionic and organic contaminants, including nitrate, arsenic, pesticides, and emerging pollutants, often achieving > 95–99% rejection for these compounds. This broad-spectrum removal capability of RO is leveraged in advanced treatment schemes (e.g., indirect potable reuse). A notable example is Singapore's NEWater facilities, which use RO (after MF/UF pretreatment) to polish treated wastewater—achieving 99% removal of dissolved organic carbon and salts—thereby producing ultra-pure reclaimed water for potable use. NEWater now supplies about 40% of Singapore's total water demand, illustrating the effectiveness of membrane barriers against a wide range of contaminants (Qin et al. 2023). A key challenge in sustained membrane performance is membrane fouling, especially biofouling by microbial growth on the membrane surface. Pathogens and natural microbes can form biofilms on membranes, causing flux decline and reduced removal efficacy. Biofouling can account for up to ~

40% of fouling in RO systems. To mitigate this, water utilities employ pretreatments and novel membrane designs. Pretreatment disinfection (e.g., chlorination, ozone, UV) is widely used to inactivate microbes and prevent biofilm growth on membranes. For instance, low-dose chlorination upstream of an MF/UF unit can significantly curb bacterial fouling (though any residual must be removed to protect downstream NF/RO). Coagulation–flocculation pretreatment is another strategy to remove colloids and biopolymers that could otherwise form a cake layer conducive to biofilm growth. Periodic backwashing and chemical cleaning (CIP) of membranes are standard practices to remove accumulated biofilms and recover permeability. Additionally, antimicrobial coatings are being developed: Researchers have impregnated RO membranes with silver or zinc oxide nanoparticles to inhibit microbial attachment, resulting in enhanced biofouling resistance and sustained flux. Such integrated approaches—pretreatment, operational controls, and advanced materials—help maintain the high pathogen and contaminant removal efficiency of membranes over the long term (Abusultan et al. 2024).

5.5.2 *Reduction of Turbidity and Organic Matter*

Membrane filtration excels at producing water with very low turbidity and substantially reducing natural organic matter, which are essential for both esthetic quality and health (by controlling disinfection by-products). Turbidity removal: Conventional filters can typically produce water with turbidity below 0.1–0.3 NTU under good conditions. Membranes, however, act as an absolute barrier for particulates, consistently yielding turbidity often < 0.1 NTU. For example, an ultrafiltration plant might guarantee permeate turbidity < 0.1 NTU 95% of the time (and always < 0.15 NTU) as a performance specification. Even if the source water turbidity spikes into the dozens of NTU (during a storm event on a surface water source), membranes can still produce extremely low turbidity effluent, albeit with adjustments like more frequent backwashing or reduced flux (Tang et al. 2020). Because membranes physically remove even sub-micron colloids, the clarity of the water (low turbidity) is typically better than what can be achieved by granular media. Low turbidity is not just a regulatory goal but also improves downstream disinfection. By delivering water of < 0.1 NTU, UF/MF membranes provide an excellent safety margin for disinfection (pathogens are less likely to be shielded by particles). This is one reason membranes are often installed as a polishing step in treatment. In some cases, membranes have replaced multi-stage coagulation and filtration processes—for instance, a UF system can take a moderately turbid river water and directly produce < 0.1 NTU water, whereas a conventional plant would require optimized coagulation and filter performance to approach that level (Langton 2021).

Natural organic matter (NOM) removal: Dissolved organic matter (humic and fulvic acids from decaying vegetation, for example) is a precursor to disinfection by-products (DBPs) like trihalomethanes. Reducing NOM is important for meeting DBP regulations. MF and UF membranes by themselves have limited effectiveness for NOM, because many NOM molecules are < 0.01 μm and pass through UF pores. They might remove the higher molecular weight fraction or NOM that is attached to colloids, but typically an UF yields only 10–30% reduction in total organic carbon (TOC) unless enhanced by pretreatment (like adding a coagulant that binds NOM into removable flocs). Hence, hybrid processes such as coagulation UF are used: a coagulant (alum, ferric, etc.) is added to the feed, which agglomerates NOM into larger flocs that the UF can then reject. This approach can boost NOM removal significantly (often to 50% or more) while the UF membrane handles the floc filtration. In contrast, nanofiltration and low-pressure RO are very effective at NOM removal due to their small pore size and charge repulsion mechanisms. NF membranes are often used for color reduction in water with high humic content—they can remove 60–90% of NOM, yielding water low in DBP precursors. For example, an NF plant treating a colored surface water can reduce dissolved organic carbon from, say, 5 mg/L down to < 1 mg/L, simultaneously softening the water by removing multivalent ions. Reverse osmosis is even more aggressive, typically removing > 95% of NOM and any micropollutants present. RO is used in advanced treatment (like indirect potable reuse) specifically to remove not only NOM but also trace organic contaminants such as pharmaceuticals. In a drinking water context, RO might be applied to groundwater high in organics or in specialized cases like algal toxin removal (Hsieh et al. 2020).

In summary, for turbidity, membranes provide a near-total physical barrier—producing sparkling clear water and protecting against particulate breakthroughs. For organics, MF/UF alone are complementary to conventional methods (often needing coagulation aid), while NF/RO directly target dissolved organics and are chosen when high removal of NOM or specific organic chemicals is required. The combined removal of turbidity and organics improves disinfectant efficiency and stability (lower NOM means less chlorine demand and fewer DBPs). Many membrane plants report that their distribution system water remains biologically stable with lower chlorine residuals because the water is low in both turbidity and assimilable organic carbon after membrane treatment. This demonstrates how membranes contribute to multiple treatment objectives at once.

5.5.3 Desalination of Water

Membrane technology plays a central role in desalination, which is the process of removing salts to produce fresh drinking water from brackish or seawater sources. Reverse osmosis is the workhorse of modern desalination: It accounts for roughly 70% of installed desalination capacity globally, largely because of its efficiency and steadily decreasing costs. In RO desalination, saline water is pressurized against semi-permeable membranes that allow water to pass but reject the dissolved salts (primarily NaCl). The result is a low-salinity permeate that can be used as drinking water, and a concentrated brine that is discharged or further treated. Membrane desalination is used at various scales, from small solar-powered RO units supplying remote communities to mega-scale plants treating hundreds of millions of liters per day for major cities. For example, in the Middle East and North Africa, RO plants produce a significant portion of municipal water supply. One of the largest RO desalination plants is in Saudi Arabia (the Ras Al Khair plant, combined with other processes) and new plants like the UAE's Taweelah RO (909,000 m^3/day) are coming online, reflecting the confidence in RO for large-scale production (Shukla et al. 2023).

The RO membranes used for desalination are typically thin-film composite polyamide, configured in spiral-wound modules. Seawater RO operates at pressures around 55–70 bar to overcome the high natural osmotic pressure of seawater (~35,000 mg/L TDS). Brackish water RO uses lower pressures (10–20 bar) as the salinity is lower. Nanofiltration is sometimes used for partial desalination or selective removal of hardness from brackish water where complete demineralization is not needed—NF operates at lower pressure and lets some monovalent ions through. An example is using NF to treat slightly brackish well water to remove sulfate and reduce total dissolved solids to an acceptable level, while keeping some minerals for taste.

Membrane desalination's advantages include modularity (plants can be scaled by adding more membrane modules), and much lower energy consumption than thermal distillation processes. Energy recovery devices and improved membrane permeability have driven down the energy cost of RO desalination to as low as ~ 3 kWh/m^3 for seawater in state-of-the-art facilities. This makes RO competitive even in regions without abundant energy resources. The quality of RO permeate is very high—essentially distilled quality—which often requires post-treatment (remineralization by adding calcium/magnesium or blending with raw water) to make the water less corrosive and more palatable (Sahu et al. 2023).

Aside from seawater desalination, RO membranes are used for treating brackish groundwater in inland areas (to provide potable water from aquifers that are slightly saline) and for advanced reuse (treating reclaimed wastewater to potable standards by removing not just salts but also contaminants of emerging concern). In reuse applications like the well-known NEWater in Singapore or Orange County's Groundwater Replenishment System in California, RO is employed after MF/UF to produce ultrapure water. This showcases RO's ability to remove virtually all contaminants: salts, microbes, organics, etc., making it indispensable for the highest purity requirements (Nidzhom Zainol Abidin et al. 2022).

In summary, membrane desalination (primarily RO) has transformed the availability of drinking water in arid coastal regions by providing a reliable source from the sea. It offers flexibility to also tackle brackish water and even treat challenging sources like arsenic-laden groundwater or contaminated surface water, since RO will strip out a broad range of pollutants. The main limitations—high-pressure energy cost and concentrate disposal—have been mitigated through technological improvements and innovative brine management. For instance, energy recovery turbines greatly reduce net energy use, and efforts are underway to use brine in salt gradient power or mineral extraction to lessen environmental discharge concerns. As a result, membrane desalination is poised to play an even greater role in augmenting drinking water supplies under climate change and growing demand, supported by continued improvements in membrane durability and efficiency.

5.6 Advantages and Limitations of Membrane Filtration

5.6.1 Membrane Filtration Benefits in Drinking Water Treatment

Membrane filtration offers several compelling advantages over conventional water treatment methods. A primary benefit is the consistent and high quality of treated water. Because membranes are absolute barriers, they can produce water with virtually no turbidity and extremely low particle counts regardless of fluctuations in raw water quality. This consistency improves downstream disinfection and overall safety. Membrane systems also achieve superior pathogen removal without reliance on chemicals—UF/MF provide a physical 4-log (or greater) pathogen reduction on their own, which is a significant health safeguard. Another advantage is the small footprint of membrane plants. A membrane skid can replace large sedimentation basins and rapid filters, saving space—an important factor in urban areas or retrofitting old plants. Membrane plants are often modular, so capacity expansion is as simple as adding more modules within a compact area. Additionally, membranes reduce the use of treatment chemicals: traditional plants consume coagulants, flocculants, filter aids, etc., whereas a direct membrane plant might use only a minor dose of chlorine or none at all during filtration. Less chemical usage means fewer issues with sludge disposal and potentially lower costs and environmental impacts (Gohil et al. 2023).

Membrane systems are also highly automated and flexible. They can easily ramp flow up or down to follow demand by adjusting pumps and valves, and they handle intermittent operation well (e.g., small community systems that shut down at night). The modular nature allows tailoring the capacity and adds resilience—if one membrane train is offline for maintenance, others can continue treating water. In contrast, conventional filters when taken out for backwashing or maintenance may temporarily reduce plant capacity more significantly. Membranes enable integrated multi-barrier strategies: For example, a UF + RO combination addresses virtually all classes of contaminants (particles, microbes, dissolved solids, organics) within one treatment train, which would otherwise require a series of different treatment steps.

From an operational standpoint, membranes often provide faster start-up and shutdown. There is no need to wait for filter media to ripen or for flocculation to stabilize—a membrane plant can start producing specification water within minutes of start-up. They are thus convenient for variable source waters or emergency use. Safety and handling is another advantage: Operators are not directly handling as many hazardous chemicals (though membrane cleaning chemicals like acids and caustics are used in contained systems). There is also no risk of media leakage into water which can happen if a sand filter underdrain fails; with membranes, any breach is quickly detected by loss of integrity and can be addressed (Contreras-Martínez et al. 2021).

Economically, membranes have become more competitive. The costs of membranes have dropped over the decades, and their longevity has improved many UF membranes last 7–10 + years; RO membranes are similar. The reduced labor due to automation and smaller footprint can offset some of the higher equipment costs. Case studies have shown that for small systems in particular, membranes can be more cost-effective than building large clarifiers and filters. Moreover, membranes can handle water quality challenges (algal blooms, high turbidity spikes, pathogen surges) more robustly, potentially avoiding costly plant shutdowns or boil-water advisories (Demirel et al. 2021).

In summary, the benefits of membrane filtration in drinking water treatment include enhanced removal of contaminants (especially microbes and fine particles), consistent output water quality that often exceeds regulatory requirements, compact and modular plant design, reduced chemical usage, and flexible, automated operation. These strengths make membranes an attractive option, particularly as water sources become more compromised and standards for treated water tighten (lower DBP limits, higher pathogen log removal requirements, etc.). Membranes effectively add a robust safety barrier in water treatment, contributing to the reliability and security of drinking water supplies.

5.6.2 Challenges and Limitations of Membrane Filtration

Despite their advantages, membrane technologies come with several limitations and challenges that must be managed. The foremost challenge is membrane fouling. As membranes filter out contaminants, those materials accumulate on or in the membrane, leading to clogging of pores or formation of a cake layer. Fouling increases the resistance to flow, causing flux decline or rising pressure requirements over time. All types of foulants can affect membranes: Suspended solids form cakes, organic matter can cause gel layers and pore blocking, microbes can grow biofilms, and dissolved minerals can precipitate as scale on RO membranes. Fouling is inevitable and requires regular cleaning, which adds complexity and cost. Frequent chemical cleaning can also degrade the membrane material over time. Membrane operations thus need careful pretreatment (to remove as much fouling load as possible) and well-designed cleaning cycles (backwashing, chemical cleans) to control fouling. In systems with inadequate pretreatment, fouling can lead to rapid flux decline and the need to replace membranes much sooner than their nominal life (Park et al. 2023).

Another limitation is cost and energy consumption. Membrane systems, especially high-pressure ones, can be energy-intensive. RO for desalination requires significant pumping energy to generate the high pressures; although energy recovery devices mitigate this, energy use is still a large portion of RO operating cost. Low-pressure UF/MF have lower energy usage but still require pumping and air for backwash. The capital cost of membranes and associated equipment (high-pressure pumps, automated controls) can be higher than conventional plants of the same capacity. While membrane costs have decreased, they remain a substantial investment, and periodic replacement (every few years to decade) is needed, whereas a concrete settling tank or sand filter bed can last for decades. For small systems, the cost of skilled operation and maintenance of membranes can be a barrier (though automation helps, it still requires technical know-how to troubleshoot). In developing countries, the higher capital and operational cost of membrane technology compared to conventional systems has been a key issue slowing adoption (Al-Maas et al. 2021).

Concentrate disposal is a limitation particularly for RO/NF processes. The concentrate (brine) contains the removed salts and other contaminants at higher concentrations. Coastal RO plants can often discharge brine to the ocean (with environmental considerations), but inland plants must find alternatives like deep well injection, evaporation ponds, or costly volume reduction. This waste stream management adds to the complexity and cost of membrane desalination. MF/UF have less concentrate (only backwash water, a few percent of throughput), but if the feed water is high in certain contaminants (e.g., arsenic), the backwash waste may require handling as well.

Membrane integrity and failure is another concern. Membranes are not fail-safe; they can develop flaws such as pinholes or broken fibers that could allow contaminants to pass. Ensuring integrity requires continuous or regular monitoring (such as particle counts, pressure decay tests, or marker tests), which adds operational burden. A single pinhole in an RO element might allow a small amount of salt to leak, which is not critical, but a broken UF fiber could bypass protozoa. Redundancy and prompt repair protocols are needed to maintain protection. By contrast, a sand filter has more forgiving failure modes (though breakthrough can happen, it's usually gradual).

Operationally, membranes can be sensitive to upset conditions: for instance, if feed water is contaminated with oil/grease, many membranes will foul severely and require intensive cleaning or replacement. They also generally have upper limits on tolerance for chlorine or other oxidants (especially polyamide RO/NF which can be damaged by chlorine). So upstream process upsets (like an overdosing of oxidant) can harm membranes. Temperature sensitivity is another limitation—cold water greatly reduces membrane permeability (flux declines), meaning a plant sized for summer flow might struggle in winter unless more membrane area is added or flux is reduced seasonally (di Vincenzo et al. 2021).

From a technical skill perspective, membrane plants often require well-trained operators and maintenance of complex instruments (transmitters, PLCs, automated valves). In some settings, this need for technical capacity is a limitation compared to simpler conventional systems. However, as membranes become more common, this gap is closing with improved training and support.

Finally, one must consider lifecycle and waste: used membranes (at end-of-life) are a waste product that must be disposed of or hopefully recycled; there is ongoing work on recycling RO elements into UF modules or other second-life uses, but this is not yet widespread. The environmental footprint of manufacturing membranes (polymers, solvents) is also a factor, though not usually a deciding limitation compared to the environmental benefits of providing safe water.

In summary, the main limitations of membrane filtration are fouling propensity, requiring careful pretreatment and cleaning; higher costs and energy use than some conventional methods (especially for RO); the challenge of concentrate disposal for RO; the need to maintain membrane integrity and handle potential failures; and operational sensitivities (chemical tolerance, temperature, etc.). These challenges are actively being addressed by research and engineering: for example, development of more fouling-resistant and easily cleanable membranes, lower-pressure membranes to cut energy, smarter monitoring systems to detect and isolate leaks, and hybrid processes to reduce fouling and waste. In the next section, we explore recent advances aiming to overcome many of these limitations.

5.7 Advances in Membrane Technology

5.7.1 Innovations in Membrane Materials

In the past 10 years, there have been significant innovations in membrane materials designed to improve performance and durability. One major area is the development of nanotechnology-enhanced membranes. Researchers have incorporated nanomaterials (like metal or metal oxide nanoparticles, carbon nanotubes, graphene oxide sheets, etc.) into membrane matrices to create nanocomposite membranes. These materials can impart beneficial properties: for instance, adding silver or copper nanoparticles can give the membrane surface antimicrobial properties to inhibit biofilm growth. Titanium dioxide (TiO_2) nanoparticles can make membranes more hydrophilic and even photocatalytic (self-cleaning under light). Graphene oxide, with its ultra-thin structure, has been used to create nanolayered membranes that offer high flux and tunable selectivity. Such 2D-material membranes (including graphene and molybdenum disulfide) can achieve nearly frictionless water transport in lab studies, pointing toward future ultra-permeable RO membranes. Another innovation is bio-inspired membranes, such as those incorporating aquaporin proteins—nature's extremely efficient water channels. Aquaporin Inside™ technology has commercialized a thin-film composite membrane with aquaporin proteins embedded, aiming for high water flux at lower pressures (Patel et al. 2021).

Material innovations also target fouling resistance. Surface modifications like grafting zwitterionic or PEG (polyethylene glycol) chains onto membrane surfaces have been developed to create ultra-hydrophilic, non-stick surfaces. These resist foulant attachment (because a hydrated hydrophilic surface is harder for hydrophobic organics or bacteria to adhere to). Stimuli-responsive coatings have also been explored—for example, membranes coated with a temperature-responsive polymer that can alternate between hydrophilic and hydrophobic states with a small temperature change, aiding in fouling release (these belong to the "smart membrane" category, discussed in 7.3). Some new RO membranes incorporate tailored surface charge to repel scaling ions or organic foulants. For instance, modifying the RO membrane surface to be more electronegative can reduce organic fouling and biofouling, as many problematic foulants (humics, bacteria surfaces) are negatively charged and will be electrostatically repelled (Cescon and Jiang 2020).

The conventional polyamide RO membrane has also seen innovation to break the permeability-selectivity tradeoff. By synthesizing membranes with ultrathin polyamide layers (on the order of 10–20 nm) or by incorporating nanoporous fillers (like porous silica or metal–organic frameworks) into the polyamide, researchers have achieved significantly higher water permeability without sacrificing salt rejection. These are known as thin-film nanocomposite (TFN) RO membranes. Commercial products now exist that offer 50–100% higher flux at the same pressure compared to older-generation membranes, directly reducing energy consumption.

Ceramic membranes too have advanced new fabrication methods produce ceramic filters with finer pore control and even multi-layer ceramics that can achieve NF-like selectivity. There's also work on graphene oxide laminates that act as molecular sieves, potentially replacing polyamide in RO for improved chlorine resistance and antifouling (Yang et al. 2022).

In UF/MF, manufacturers have introduced reinforced hollow fibers (with braided cores) to allow higher operating pressures and backwash frequencies, improving robustness. PVC and PVDF blends have improved to yield stronger fibers with less breakage. Another novel concept is electrically conductive membranes, where carbon nanotubes or conductive polymers are integrated so that an electric potential can be applied on the membrane surface to repel charged foulants or even generate local oxidants to kill biofilm. This can be seen as a merging of material science with electrochemistry to yield self-cleaning features (Alibeigi-Beni et al. n.d.).

Overall, these material innovations aim to make membranes more selective, permeable, and fouling-resistant. For example, a smart composite UF membrane might self-adjust its surface properties in response to the water quality, maintaining higher flux. Many of these innovations are moving from the lab to the market, with some already in pilot or full-scale use. The trend is toward membranes that require less energy, less chemical cleaning, and have longer lifespans—directly tackling the traditional limitations. While cost of new materials can be a factor, improved performance can justify it in many cases. The next few years will likely see wider adoption of nanocomposite and surface-tailored membranes as water utilities seek more resilient filtration solutions.

5.7.2 *Hybrid Water Treatment Systems with Membranes*

Hybrid systems combine membrane processes with other treatment techniques to capitalize on their synergies. In recent years, there has been a surge in developing integrated membrane systems that address multiple treatment goals or mitigate membrane drawbacks like fouling. A classic example is coagulation MF/UF, which is widely used and has been further refined. Instead of coagulation settling, sand filtration, many plants now dose a coagulant like alum or polyaluminum chloride directly ahead of UF membranes. The coagulant causes colloids and dissolved organics to form larger flocs that UF can easily filter, dramatically improving organics removal and reducing membrane fouling. This hybrid approach has been shown to increase virus removal as well—studies found that adding coagulation and a sedimentation step before UF improved virus LRV by 1.8–4.2 logs beyond UF alone. The flocculation provides an additional barrier or adsorptive removal for viruses and then the UF filters out the floc-bound viruses, yielding higher overall removal (Abdelrady et al. 2020).

Another common hybrid is adsorption + membrane. For instance, powdered activated carbon (PAC), UF: PAC is added to water to adsorb taste and odor compounds, algal toxins, or synthetic organic chemicals, and then a UF membrane filters out the carbon particles along with contaminants. This combination can address micropollutants that the membrane pore size alone would not reject. Research in the past few years on emerging contaminants like pesticides, pharmaceuticals, and algal metabolites often points to PAC-UF or carbon nanotube enhanced membranes as effective solutions. Similarly, ion exchange + membrane can remove specific ions (ion exchange grabs them, membrane ensures no resin fines or byproducts go into finished water).

Membrane bioreactors (MBRs), while primarily used in wastewater treatment, represent a biological + membrane hybrid. In potable reuse, an MBR (which integrates biological degradation of contaminants with UF membrane separation) can serve as an advanced pretreatment before RO, reducing biofouling potential and breaking down complex organics. Recent advances in MBR technology and understanding have allowed some direct potable reuse schemes to consider MBR as a key component, leveraging both biological and physical removal in one step.

Advanced oxidation membranes is another emerging hybrid. For instance, applying UV/H_2O_2 or ozonation upstream of NF/RO can break down organic foulants or micropollutants, so the membrane has less fouling and primarily rejects any remaining oxidized byproducts. Some setups use ozone to effectively "soften" the foulant layer on UF membranes, making backwash more effective (often called ozone-assisted membrane filtration). Conversely, membranes can be used downstream of oxidation to remove any generated particles like after manganese greensand ozonation, a UF could remove fines (Gidstedt et al. 2022).

Electrically enhanced membranes are a novel hybrid category: for RO brine minimization, researchers have combined RO with electrodialysis in sequential or hybrid arrangements to achieve higher overall recovery than either alone, thereby tackling concentrate disposal issues. Also, coupling capacitive deionization with membranes has been explored to predesalt water slightly and reduce RO energy.

In desalination, a prevalent hybrid is UF or MF as RO pretreatment. Almost all new seawater RO plants use UF/MF instead of conventional sand filters for pretreatment, which has drastically improved RO reliability by providing consistently low-silt feed water. Another hybrid in desal is RO + energy recovery + brine concentrators, but focusing on membrane-centric: Dual-pass RO (two RO stages in series) is used in reuse to meet ultra-low dissolved solids targets, and RO + NF combinations can be used where NF softens water that then goes to RO, allowing higher RO recovery with less scaling.

The concept of integrated membrane systems extends to combining multiple membrane types in one flowsheet to tackle different contaminants. One example: UF followed by RO followed by UV disinfection is the scheme in many potable reuse plants, providing a sequential removal of particles, then salts and organics, then final disinfection. Each is a membrane or advanced step in a chain—while not "hybrid" in the sense of combined processes in one unit, it's a hybrid treatment train (Junker et al. 2021).

Recent work has also looked at membrane + photocatalyst hybrids. Embedding photocatalysts like TiO_2 on membrane surfaces (creating a photoactive membrane) means that shining UV light can induce oxidation on the membrane to self-clean or degrade pollutants that are caught. These so-called photocatalytic membranes have shown promise in lab studies for treating organic-laden waters without intensive fouling.

In summary, the last ten years have seen increasing implementation of hybrid approaches to maximize performance: By combining processes, one can overcome the limitations of a single standalone membrane. Coagulation/adsorption address dissolved matter that membranes alone can't remove; bio-processes degrade what membranes can't physically reject; oxidation prevents or breaks down fouling; and membranes themselves ensure particulate-free and pathogen-free effluent in each case. Such hybrid systems thus achieve a "better together" result—for example, a coagulation UF system can produce water with < 0.5 NTU and high NOM removal even on a highly variable raw water, where either coagulation–sedimentation alone or UF alone would struggle to do so. As regulatory demands tighten and source waters get more complex, these integrated membrane systems are becoming the norm in new advanced treatment plants.

5.7.3 Smart Membranes and Their Monitoring Technologies

"Smart membranes" refer to next-generation membranes that can adapt or be controlled to optimize performance, as well as the use of smart monitoring technologies to manage membrane systems in real time. A key development in recent years is the creation of stimuli-responsive membranes, whose properties (pore size, hydrophilicity, charge) can change in response to external triggers like pH, temperature, light, or electric/magnetic fields. These smart membranes are made by integrating stimuli-responsive polymers or materials into the membrane structure. For example, a pH-responsive membrane might have polymer gates that remain closed at certain pH and open at another pH level, offering tunable selectivity. Temperature-responsive membranes using poly (NIPAM) can toggle between hydrophilic and hydrophobic states around a critical temperature. The appeal of smart membranes is their tunability and reversibility—they can self-adjust to optimize flux or selectivity as conditions change. For instance, in low contaminant loads they might maximize permeability, but if a certain contaminant is detected, they change to a mode that increases rejection of that contaminant. Research has shown such membranes can achieve controllable molecular sieving, and the concept has been demonstrated in lab-scale water treatment for controllable ion removal. While many smart membranes are still in experimental stages, the progress suggests a "bright future" for these materials to provide high selectivity without sacrificing permeability. One challenge remains scaling up these often complex materials and ensuring long-term stability over many cycles of switching states.

On the monitoring side, membrane plants are becoming smarter through advanced sensors and data analytics. Real-time fouling monitoring has seen progress using techniques like optical coherence tomography (OCT) to visualize fouling layers in situ and electrical impedance spectroscopy to sense deposition on membranes. A notable example is the use of the 3-omega thermal sensing method, where an insulated heater wire is embedded in a membrane to detect the growth of a fouling layer by its thermal response. This provides an in situ, non-invasive way to monitor fouling buildup in real time, allowing operators to optimize cleaning timing. There's also development of fiber optic sensors for pressure or strain that can be integrated in modules to detect any integrity breach or abnormal pressure drops at specific locations (Kamranvand et al. 2020).

Automation and control algorithms are getting more sophisticated including the use of artificial intelligence (AI) and machine learning for predictive maintenance. For example, AI models can analyze historical operation data (pressures, flows, water quality) to predict when fouling will reach a critical point or when a membrane clean or replacement will be needed. This moves membrane operation from reactive to proactive. Some large membrane plants now have "digital twins"—virtual models of the plant that simulate performance—assisting in scenario testing and optimal setpoint determination.

Another smart aspect is integrity monitoring: Instead of a once-daily manual test, continuous systems like particle counting or turbidity monitoring on each individual membrane train are used. If a sudden spike in particle count is seen in permeate from one train, that train can be automatically shut down for an integrity test, isolating a potential break immediately. This minimizes risk of pathogen passage.

In RO desalination, advanced process control systems monitor differential pressures and salt rejection in each stage and can pinpoint a suspect module indicating possible damage. Likewise, smart cleaning protocols have emerged—e.g., algorithms that decide the optimal cleaning chemical and duration based on the type of fouling inferred from sensor data like normalized flux vs. pressure trends, which can indicate whether fouling is more likely biological, organic, or scaling (Kamranvand et al. 2020).

Moreover, some modern membranes integrate a sensor function—for instance, fluorescent tags in the membrane that get quenched or emit signal when certain foulants accumulate, effectively making the membrane self-indicating of its condition.

These developments in smart membranes and monitoring aim to reduce downtime, extend membrane life, and ensure water safety. By having membranes that self-optimize and control systems that can detect and adapt to changing conditions, the reliability of membrane plants is greatly enhanced.

5.8 Conclusion

Membrane filtration has established itself as a key technology in drinking water treatment, offering an effective barrier against a broad-spectrum of contaminants. In this chapter, we reviewed the fundamentals of membrane processes (MF, UF, NF, RO, and others) and their operation principles, and saw how materials and design influence performance. We discussed the major applications: from removing pathogenic microbes and turbidity to eliminating organic pollutants and desalinating salty water. Membranes provide significant advantages—they consistently produce high-quality, low-turbidity water and achieve exceptional pathogen log removals, all within compact treatment units. They enable access to non-traditional water sources (like seawater or wastewater reuse) by providing robust treatment barriers. We also examined limitations, primarily fouling, cost, and concentrate disposal issues, which require careful management and engineering solutions. Recent advances are actively mitigating these drawbacks: Smarter membrane materials resist fouling and offer higher flux, and hybrid process designs combine the strength of membranes with complementary treatments to improve efficiency. Innovations such as stimuli-responsive "smart" membranes and advanced sensor-based monitoring promise to further enhance reliability and automate control of membrane systems.

In the future, membrane filtration is expected to play an even more prominent role in securing safe drinking water. Climate change and population growth are intensifying stress on freshwater supplies, making alternatives like seawater desalination and water reuse increasingly important—both domains where membrane processes (RO, UF) are indispensable. We anticipate more utilities upgrading conventional plants with membrane add-ons (e.g., integrating UF for pathogen safety or NF for targeted contaminant removal) to meet stricter regulations on pathogens, turbidity, and emerging contaminants. Membranes will likely be the central component of direct potable reuse schemes, providing the necessary multi-barrier treatment to convert wastewater to drinking water. Moreover, continuing research and development could yield membranes with significantly lower energy requirements and longer lifetimes, improving sustainability. The integration of renewable energy with membrane systems (such as solar-powered RO) and resource recovery from brines (mining minerals) could address economic and environmental concerns of membrane desalination.

Further research is needed in several areas to fully realize the potential of membranes. One area is fouling control: Despite progress, fouling remains inevitable, so research into more fouling-resistant materials, better pretreatment methods (like pre-ozonation, novel adsorbents), and anti-fouling surface coatings will remain critical. Membrane longevity and reuse: Developing membranes that can last longer or be easily rejuvenated/recycled will improve lifecycle costs—this includes studying the long-term degradation mechanisms of membranes and finding ways to repair or recycle used modules. Concentrate management for RO/NF is another research priority—innovative approaches like zero-liquid discharge systems, co-locating RO with beneficial reuse of brine (e.g., in salt production or algae bioreactors), or new membrane processes like forward osmosis or membrane distillation to treat brine could alleviate disposal issues. Emerging contaminants: as new pollutants (PFAS, pharmaceutical residues, algal toxins, etc.) are recognized in water supplies, research should continue on how membranes remove these and how to optimize membranes or membrane-integrated processes to target them (for instance, tailoring membrane surface chemistry to adsorb specific micropollutants).

On the operational side, continued development of smart monitoring and control algorithms—possibly AI that can manage a membrane plant proactively—is a fruitful area to ensure membranes operate at peak efficiency and safety. This includes real-time integrity monitoring technologies that could allow membranes to be used as the sole treatment barrier with absolute confidence (for small systems), because any failure would be instantly detected and contained.

References

Abdel-Hai S, El-Tonsy M, El-Henawey M (2024) Investigations for the heat treatment effects on permeability of some semi-permeable membranes. Phys Scr 99(9):095901. https://doi.org/10.1088/1402-4896/ad669a

Abdelrady A, Bachwenkizi J, Sharma S, Sefelnasr A, Kennedy M (2020) The fate of heavy metals during bank filtration: Effect of dissolved organic matter. J Water Process Eng 38. https://doi.org/10.1016/j.jwpe.2020.101563

Nidzhom Zainol Abidin M, Mahmoud Nasef M, Matsuura T, Teknologi Malaysia U, Sultan Yahya Petra J, Lumpur K (2022) Fouling prevention in polymeric membranes by radiation induced graft copolymerization. Citation: Zainol Abidin, M.N. https://doi.org/10.3390/polym

Abusultan AAM, Wood JA, Sainio T, Kemperman AJB, van der Meer WGJ (2024) A hybrid process combining ion exchange resin and bipolar membrane electrodialysis for reverse osmosis remineralization. Desalination 573. https://doi.org/10.1016/j.desal.2023.117209

Ajizah F (2023) Characteristics of PVA/GO composite membranes prepared using solution casting technique for reducing methylene blue concentration. Jurnal Ilmu Dan Inovasi Fisika 7(1):20–29. https://doi.org/10.24198/jiif.v7i1.40650

Alibeigi-Beni S, Masoud, Zare H, Chenar MP, Sadeghi M, Shirazian S (n.d.) Design and optimization of a hybrid process based on hollow-fiber membrane/coagulation for wastewater treatment. https://doi.org/10.1007/s11356-020-11037-y/Published

Al-Maas M, Hussain A, Minier Matar J, Ponnamma D, Hassan MK, al Ali Al-Maadeed M, Alamgir K, Adham S (2021) Validation and application of a membrane filtration evaluation protocol for oil-water separation. J Water Process Eng 43. https://doi.org/10.1016/j.jwpe.2021.102185

Bressloff P (2022) A probabilistic model of diffusion through a semi-permeable barrier, Proc R Soc A Math, Phys Eng Sci 478(2268). https://doi.org/10.1098/rspa.2022.0615

Cescon A, Jiang JQ (2020) Filtration process and alternative filter media material in water treatment. Water (Switzerland) 12(12). https://doi.org/10.3390/w12123377

Chai N, Yue Y, Chen X, Zeng Z, Li S, Wang X (2024) Isotropic sintering shrinkage of 3D glass-ceramic nanolattices: backbone preforming and mechanical enhancement. Int J Extreme Manuf 6(2):025003. https://doi.org/10.1088/2631-7990/ad1857

Contreras-Martínez J, García-Payo C, Khayet M (2021) Electrospun nanostructured membrane engineering using reverse osmosis recycled modules: Membrane distillation application. Nanomaterials 11(6). https://doi.org/10.3390/nano11061601

Demirel SE, Li J, Hasan MMF (2021) Membrane separation process design and intensification. Ind Eng Chem Res 60(19):7197–7217. https://doi.org/10.1021/acs.iecr.0c05072

di Vincenzo M, Tiraferri A, Musteata VE, Chisca S, Sougrat R, Huang LB, Nunes SP, Barboiu M (2021) Biomimetic artificial water channel membranes for enhanced desalination. Nat Nanotechnol 16(2):190–196. https://doi.org/10.1038/s41565-020-00796-x

Franke V, Ullberg M, McCleaf P, Wålinder M, Köhler SJ, Ahrens L (2021) The price of really clean water: combining nanofiltration with granular activated carbon and anion exchange resins for the removal of per- and polyfluoralkyl substances (PFASs) in drinking water production. ACS ES T Water 1(4):782–795. https://doi.org/10.1021/acsestwater.0c00141

Gidstedt S, Betsholtz A, Falås P, Cimbritz M, Davidsson Å, Micolucci F, Svahn O (2022) A comparison of adsorption of organic micropollutants onto activated carbon following chemically enhanced primary treatment with microsieving, direct membrane filtration and tertiary treatment of municipal wastewater. Sci Total Environ 811. https://doi.org/10.1016/j.scitotenv.2021.152225

Gohil PP, Desai H, Kumar A, Kumar R (2023) Current status and advancement in thermal and membrane-based hybrid seawater desalination technologies. In: Water (Switzerland), vol 15, issue 12. MDPI. https://doi.org/10.3390/w15122274

Hansen S, Nierychlo M, Christensen M, Nielsen P, Jørgensen M (2020) Fouling of membranes in membrane bioreactors for wastewater treatment: planktonic bacteria can have a significant contribution. Water Environ Res 93(2):207–216. https://doi.org/10.1002/wer.1392

Hsieh YZ, Lin SS, Luo YC, Jeng YL, Tan SW, Chen CR, Chiang PY (2020) ARCS-assisted teaching robots based on anticipatory computing and emotional Big Data for improving sustainable learning efficiency and motivation. Sustainability (Switzerland) 12(14). https://doi.org/10.3390/su12145605

Jiang S, Ladewig B (2020) Green synthesis of polymeric membranes: recent advances and future prospects. Curr Opin Green Sustain Chem 21:1–8. https://doi.org/10.1016/j.cogsc.2019.07.002

Junker MA, de Vos WM, Lammertink RGH, de Grooth J (2021) Bridging the gap between lab-scale and commercial dimensions of hollow fiber nanofiltration membranes. J Membr Sci 624. https://doi.org/10.1016/j.memsci.2021.119100

Kamranvand F, Davey CJ, Williams L, Parker A, Jiang Y, Tyrrel S, McAdam EJ (2020) Ultrafiltration pretreatment enhances membrane distillation flux, resilience and permeate quality during water recovery from concentrated blackwater (urine/faeces). Sep Purif Technol 253.https://doi.org/10.1016/j.seppur.2020.117547

Kaufman Y, Hunt K, Hale G, McClure M, Latulippe D, Sivan M et al (2023) Fouling of virus filtration membranes by monoclonal antibody feeds with low aggregate content. Biotechnol Bioeng 121(8):2400–2408. https://doi.org/10.1002/bit.28420

Kerkulah EN, Jin G, Huang J, Huang X, Liu Y, Ran CE, Peng H (2023) A mini review on treatment of wastewater with membrane technology. J Basic Appl Sci 19:86–96. https://doi.org/10.29169/1927-5129.2023.19.08

Khery Y, Daniar S, Nawi N, Bilad M, Wibisono Y, Nufida B et al (2022) Ultra-low-pressure membrane filtration for simultaneous recovery of detergent and water from laundry wastewater. Membranes 12(6):591. https://doi.org/10.3390/membranes12060591

Langton MJ (2021) Engineering of stimuli-responsive lipid-bilayer membranes using supramolecular systems. In: Nature reviews chemistry, vol 5, issue 1. Nature Research, pp 46–61. https://doi.org/10.1038/s41570-020-00233-6

Li J, Feng C, Jin J, Yang W, Wang Z (2022) Current understanding on antibacterial mechanisms and research progress of tea polyphenols as a supplementary disinfectant for drinking water. J Water Health 20(11):1611–1628. https://doi.org/10.2166/wh.2022.062

Park K, Mudgal A, Mudgal V, Sagi M, Standing D, Davies PA (2023) Desalination, water re-use, and halophyte cultivation in salinized regions: a highly productive groundwater treatment system. Environ Sci Technol 57(32):11863–11875. https://doi.org/10.1021/acs.est.3c02881

Patel SK, Biesheuvel PM, Elimelech M (2021) Energy consumption of brackish water desalination: identifying the sweet spots for electrodialysis and reverse osmosis. ACS ES and T Engineering 1(5):851–864. https://doi.org/10.1021/acsestengg.0c00192

Qin J, Ziemann E, Bar-Zeev E, Bone SE, Liang Y, Mauter MS, Herzberg M, Bernstein R (2023) Microporous polyethersulfone membranes grafted with zwitterionic polymer brushes showing microfiltration permeance and ultrafiltration bacteriophage removal. ACS Appl Mater Interfaces 15(14):18343–18353. https://doi.org/10.1021/acsami.3c01495

Razmgar K, Nasiraee M (2021) Polyvinyl alcohol-based membranes for filtration of aqueous solutions: a comprehensive review. Poly Eng Sci 62(1):25–43. https://doi.org/10.1002/pen.25846

Sahu A, Dosi R, Kwiatkowski C, Schmal S, Poler JC (2023) Advanced polymeric nanocomposite membranes for water and wastewater treatment: a comprehensive review. In: Polymers, vol 15, issue 3. MDPI. https://doi.org/10.3390/polym15030540

Shukla AK, Alam J, Mishra U, Alhoshan M (2023) A sustainable approach for the removal of pharmaceutical contaminants from effluent using polyamide thin-film composite membranes integrated with Zn-based metal organic frameworks. Environ Sci Pollut Res 30(51):110104–110118. https://doi.org/10.1007/s11356-023-30056-z

Sutrisna P, Hadi R, Siswanto J, Prabowo G (2020) The CO_2/CH_4 separation potential of zif-8/polysulfone mixed matrix membranes at elevated particle loading for biogas upgradation process. Int J Renew Energy Dev 10(2):213–219. https://doi.org/10.14710/ijred.2021.33118

Tang N, Li Y, Ge J, Si Y, Yu J, Yin X, Ding B (2020) Ultrathin cellulose voronoi-nanonet membranes enable high-flux and energy-saving water purification. ACS Appl Mater Interfaces 12(28):31852–31862. https://doi.org/10.1021/acsami.0c08504

Wang D, Sant S, Ferrell N (2021) A biomimetic in vitro model of the kidney filtration barrier using tissue-derived glomerular basement membrane. Adv Healthc Mater 10(16). https://doi.org/10.1002/adhm.202002275

Wild B, White C, Bourg I (2022) Molecular dynamics simulations of reverse osmosis in silica nanopores. J Phys Chem C 126(21):9161–9172. https://doi.org/10.1021/acs.jpcc.2c01815

Wu X, Kohl T, Tsanaktsidis J, Xie Z (2021) Improved performance and mitigated internal concentration polarization of thin-film composite forward osmosis membrane with polysulfone/polyaniline substrate. ACS Appl Poly Mater 3(11):5758–5766. https://doi.org/10.1021/acsapm.1c00964

Wu Z, Li S, Fan X, Vogel F, Mao J, Tu X (2023) In-situ preparation of robust self-lubricating composite coating from thermally sprayed ceramic template. J Adv Ceram 12(2):357–372. https://doi.org/10.26599/jac.2023.9220689

Yan R, Zhu J, Xi F, Chen A (2022) Development and application of a hydrogeochemical model for the groundwater treatment process in waterworks. Water 14(13):2103. https://doi.org/10.3390/w14132103

Yang Y, Bar-Zeev E, Oron G, Herzberg M, Bernstein R (2022) Biofilm formation and biofouling development on different ultrafiltration membranes by natural anaerobes from an anaerobic membrane bioreactor. Environ Sci Technol 56(14):10339–10348. https://doi.org/10.1021/acs.est.2c02007

Yuan B, Zhao S, Hu P, Cui J, Niu Q (2020) Asymmetric polyamide nanofilms with highly ordered nanovoids for water purification. Nat Commun 11(1). https://doi.org/10.1038/s41467-020-19809-3

Yue Y, Hou K, Chen J, Cheng W, Wu Q, Han J et al (2022) Ag/agbr/agvo3 photocatalyst-embedded polyacrylonitrile/polyamide/chitosan nanofiltration membrane for integrated filtration and degradation of RHB. ACS Appl Mater Interf 14(21):24708–24719. https://doi.org/10.1021/acsami.2c04988

Zhang Z, Zhang W, Hu X, Li K, Luo P, Li X et al (2020) Evaluating the efficacy of point-of-use water treatment systems using the water quality index in rural southwest china. Water 12(3):867. https://doi.org/10.3390/w12030867

Zhang Z, Zhou W, Wang T, Gu Z, Zhu Y, Liu Z et al (2023) Ion–conducting ceramic membrane reactors for the conversion of chemicals. Membranes 13(7):621. https://doi.org/10.3390/membranes13070621

Zhou W, Zhang X, Gong X, Ding M, Yu J, Zhang S et al (2022) Environmentally friendly polyamide nanofiber membranes with interconnective amphiphobic channels for seawater desalination. ACS Appl Mater Interf 14(30):35287–35296. https://doi.org/10.1021/acsami.2c12061

Chapter 6
Membrane Filtration for Seawater Desalination

Veena N Bhajantri, Shreya Dixit, and Ashootosh Mandpe

Abstract Seawater desalination is a critical technology for addressing global freshwater shortages, particularly in water-scarce regions. Among various methods, membrane-based techniques—especially reverse osmosis (RO)—are widely adopted due to their high salt rejection, scalability, and operational flexibility. This review explores key aspects of membrane desalination, including membrane types, operating principles and system configurations. RO is examined in detail, highlighting its mechanism, pre/post-treatment needs, and common challenges like fouling and scaling. Emerging alternatives such as nanofiltration (NF), forward osmosis (FO) and hybrid systems offer the potential for energy savings and specialized applications. Advances in membrane materials—such as polymeric, ceramic, and composites with antifouling coatings—enhance durability and performance. System design factors, including module setup, energy use, and monitoring, are crucial for cost-effective operation. Lifecycle assessments (LCA) provide insights into environmental and economic impacts, emphasizing energy consumption, carbon emissions, and water footprint. Notably, Energy Recovery Devices (ERDs) significantly reduce energy demands in large-scale RO plants, improving overall sustainability. The integration of advanced membrane materials, process optimization and sustainable energy strategies is essential to improving the efficiency and environmental compatibility of seawater desalination. This highlights the importance of innovation in desalination a sustainable component of global water security frameworks.

Keywords Desalination · Energy recovery devices · Fouling · Membrane filtration · Reverse osmosis · Sustainability · Water scarcity

V. N Bhajantri · S. Dixit · A. Mandpe (✉)
Department of Civil Engineering, Indian Institute of Technology Indore, Indore, Madhya Pradesh, India
e-mail: as_mandpe@iiti.ac.in

A. Mandpe et al. (eds.), *Membrane Filtration and Membrane Bioreactors for Wastewater Treatment*, Advances in Wastewater Research,
https://doi.org/10.1007/978-981-95-6706-5_6

6.1 Membrane Filtration for Seawater

6.1.1 Introduction

Seawater is one of the important factors to reducing the global water scarcity problem. Due to increasing global population and growing agricultural and industrial water demand, diminishing freshwater availability, making desalination an important solution (Aende et al. 2020) Desalination technology makes it possible to convert a huge amount of seawater into fresh water. This process involves removing salt and minerals from seawater, so it meets the standards of water quality required.

6.1.2 Importance of Desalination in Water Scarcity Management

A United Nations report reveals that more than 800 million people are facing severe water scarcity. It is assumed that by the year 2025, this number is assumed to extend significantly to around 2.8 billion people worldwide (Aende et al. 2020).

Especially in drought-affected areas with rising demand due to growing population and industries, desalination of seawater is the best long-term alternative because traditional sources of freshwater are not enough to fulfil the requirement (Ezugbe and Rathilal 2020)

Desalination plays an important role in maintaining water security by providing regular potable water even at the time of drought and climate change by maintaining proper access of water to the society and helping in economic growth (DeNicola et al. 2015)

6.1.3 Purpose of Membrane Filtration in Desalination

For desalination, membrane filtration technology is one of the most important processes. Out of various desalination processes, membrane-based methods, especially reverse osmosis, are mostly used due to their high efficiency and low-energy consumption instead of another distillation process such as thermal-based processes like multistage flash distillation and multi-effect desalination (Aladwani et al. 2021)

There are several benefits to using membrane filtration, such as easy modular setup, simple operation and flexibility to handle various levels of water salinity. The use of a permeable membrane system to filter the dissolved salt and impurities, which results in clean and high-quality portable water (Ezugbe and Rathilal 2020). Its high performance, efficiency, and flexibility make it more adaptable for desalination methods in the modern era.

6.2 Principle of Membrane Filtration

Membrane technology is used to filter different substances within the liquid or gases, which works as a selective filter to separate the specific molecules to pass through that filter while barriers to others according to their physical and chemical properties (permogorv16). The force that drives separation in the membrane method is different based on their specific type such as pressure, concentration difference, and electrical potential (Aliyu et al. 2018). The membrane technology mainly depends on different factors such as membrane material, size of pore and operating conditions.

6.2.1 Classifications of Membrane for Desalination

There are many types of filtration membrane used in the desalination process, which have specific characterization and applicability

Reverse Osmosis (RO): Reverse osmosis is one of the most important membrane technologies used in seawater desalination, which accounts for about 80% of world desalination. In this process, water goes through a semi-permeable membrane with high pressure, and the membrane retains the salts and other dissolved solids (Aliyu et al. 2018). Reverse osmosis effectively treats brackish water and seawater with high salinity levels (Fig. 6.1).

Nanofiltration (NF): Nanofiltration is a process in which membrane properties are in between ultrafiltration and reverse osmosis (Abdel-Fatah 2018). This method typically used for desalination, treating brackish water, and removing specific oil and organic substance (Abdel-Fatah 2018). Nanofiltration has a lower pressure than

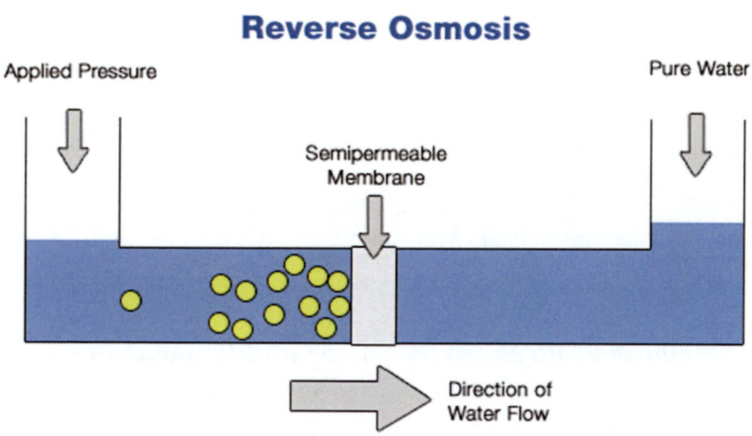

Fig. 6.1 Process of reverse osmosis (https://www.avidityscience.com/en_gb/news/water-purification/understanding-water-purification)

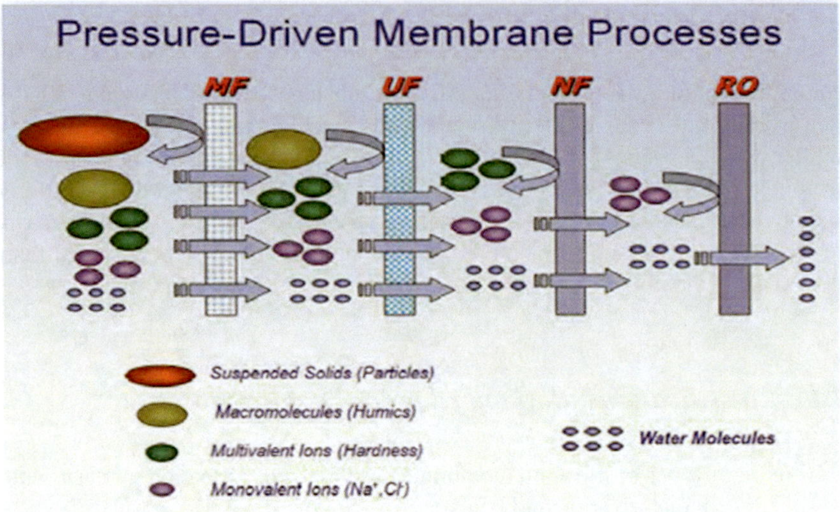

Fig. 6.2 Technique of pressure-driven membrane processes (www.Wateronline.com)

Table 6.1 Membrane filtration classification according to pore size and target contamination

Type of filtration	Pore size	Target contaminants
Reverse osmosis (RO)	< 0.001 μm	Heavy metals, dissolved salts and all dissolved solids
Nanofiltration (NF)	0.001–0.01 μm	Dissolved salts, organic compounds
Ultrafiltration (UF)	0.01–1 μm	Viruses, proteins and macromolecules
Microfiltration (MF)	1–1 μm	Suspended solids, bacteria

reverse osmosis, and it is more suitable for the application of partial desalination (Fig. 6.2).

Forward Osmosis (FO): In this process, osmotic pressure gradient is used in membrane method for driving water through a semi-permeable membrane (Aende et al. 2020) It provided benefits like minimum consumption of energy and decreased fouling tendency unlike reverse osmosis (Skuse et al. 2021). Forward osmosis uses the dry solution to establish the osmotic gradient while recovering the dry solute continues to be a significant challenge (Table 6.1).

6.2.2 Membrane Filtration: Selectivity and Permeability

There are two main factor that is selectivity and permeability that determine how the membrane filtration process works.

Selectivity: The process in which the membrane can select and allow the component to pass through the membrane while rejecting others (Werber et al. 2016). It requires high selectivity for performing effective separation and producing high-quality parameters. For this process, the membrane depends on the size of pores, membrane material and interaction between membrane and component which are going to be separate.

Permeability: Permeability is the process by which it measures the membranes' ability to allow the liquid to pass through it. (Werber, 16). For achieving the high production rate and reducing the consumption, it is desirable to have high permeability. However, permeability is inversely proportional to selectivity; thus, decrease in selectivity may lead to increase in permeability.

6.3 Rivers Osmosis for Seawater Desalination

6.3.1 Reverse Osmosis Mechanism

Mechanism of reverse osmosis works by pressure applied on saline water, which drives water through a semi-permeable membrane from an area with higher dissolved solids to one with a lower concentration (Aliyu et al. 2018). Unlike other natural osmosis that occurs randomly, reverse osmosis needs external pressure to push water molecules through a semi-permeable membrane against the concentration gradient; thus, it is called reverse osmosis (6). For this process, pressure must be the original pressure to overcome the natural flow of water to ensure water purification.

6.3.2 System Components and Configuration

For efficient work in a reverse osmosis system, it depends on different necessary components that function collectively to enable effective desalination. These are the following elements of the intake system: in this system, seawater is taken from the source, and by using the bar screen, the coarse materials and suspended solids are filtered before further treatment (Park et al. 2020).

Pretreatment: Pretreatment is crucial to prevent the membrane fouling and scaling. In this process, coagulation, flocculation, sedimentation and fine-filtration process are used to remove the collider solids, organic contaminants, and microbial contents by ensuring the membrane durability (Laqbaqbi et al. 2017)

High-Pressure Pump: In this process, pump exceeds the pressure of pre-treated seawater over its natural osmotic pressure (Chun et al. 2017).

Reverse Osmosis Membrane Module: This is the main part of the system, the semi-permeable membrane, which is made up of polyamide with a thin film that allows the water molecule to pass through while significantly blocking the salts and other impurities.

Pressure Vessels: They are designed to hold the high operational pressure; these vessels contain the reverse-osmosis membrane elements for providing mechanical stability (Aladwani et al. 2021).

Posttreatment System: After the desalination process, water undergoes conditioning to meet the potable standards. In this process, adjustment like pH correction and disinfection realization in potable water is ensured.

Concentrate disposal system: The concentrated brine by-product requires careful disposal to minimize the ecological harm (Permogorov 2016)

6.3.3 Pretreatment and Post Treatment Requirements

Pretreatment is essential for ensuring the sustainable performance and reliability of reverse osmosis membrane. Advantage of this is to eliminate scale-forming material and foulants that diminished membrane flux and enhance energy utilization (Laqbaqbi et al. 2017). Common pretreatment methods include:

Coagulation and Flocculation: In this process, collided particles and suspended solids are eliminated by destabilizing them and forming large aggregates that can be easily filtered (Park et al. 2020).

Sedimentation: In this process, a large aggregate formed during the coagulation and flocculation is to be settled and removed from the water.

Filtration: The federation eliminates remaining suspended solids and particulate matter by using various filter types like sand filter, cartridge filter and membrane filter (Krzeminski et al. 2017).

Additional Agents for Anti-scaling: Antiscalants are the chemicals which are used to prevent the buildup of mineral scale on membrane surfaces.

Disinfection: In this process, microorganisms are inactivated or killed to remove the biofouling of the membrane (Krzeminski et al. 2017).

Post-treatment is the most important process to stabilize the desalination of water to make it as freshwater.

Post-treatment process involves:

pH Adjustment: Rising pH of desalinated water to a slightly alkaline level helps to prevent corrosion in the distribution system.

Disinfection: It ensures that the desalinated water is free from harmful microorganisms. (Krzeminski et al. 2017).

Remineralization: This process involves additions of essential minerals such as calcium and magnesium to the desalinated water to improve its taste and make it more suitable consumption.

6.3.4 Issues and Remedies in Reverse Osmosis Membrane Systems

Reverse osmosis is an important desalination process despite that reverse osmosis has many issues that are required to be overcome to enhance performance and reliability.

Membrane fouling is a significant problem in reverse osmosis desalination due to the presence of organic matter, inorganic scales and microorganisms build up on the surface (Laqbaqbi et al. 2017). Fouling must be managed because it reduces membrane flux, increases consumption and decreases membrane lifetime. Pretreatment, cleaning of membranes and design of fouling-resistant membranes are some of the solutions to membrane fouling (Laqbaqbi et al. 2017).

Mineral Scaling: Mineral scaling takes place when sparingly soluble salts like calcium sulphate, calcium carbonate and silica deposit on the membrane surface area, causing membrane decreasing. Particularly, solutions to minerals scaling range from antiscalant addition to pH correction and membrane revision (Laqbaqbi et al. 2017).

Energy Consumption: Reverse osmosis, a key part of the desalination process, contributes to high operation costs and greenhouse emissions. The energy reduction solutions would comprise Energy Recovery Devices, renewable energy supplies and umproved system design (Shahid et al. 2023).

Brine Disposal: This is the salt-rich solution that was obtained during the process of reverse osmosis (RO) desalination. Disposing of brine harms ocean ecosystems and infects water resources. Brine disposal effectively managed through deep-well injection, diluted surface-water discharge, and by involvement of advanced brine-concentration processes.

6.4 Alternative Membrane Filtration Methods

6.4.1 Nanofiltration (NF) for Brackish Water Desalination

In this process, nanofiltration (NF) is a membrane separation process located in the range of pore sizes and operating pressure between reverse osmosis (RO) and ultrafiltration (UF) (Abdel-Fatah 2018). NF membranes usually have pore sizes from 1 to 10 nanometres and are operated at lower pressures compared to reverse osmosis membranes. NF is especially applicable to the desalination of brackish water when the salinity is lower than that of seawater (Abdel-Fatah 2018).

NF has numerous benefits for desalination of brackish water:

Lower Energy Consumption: NF, compared to RO, works at low pressures and thus has low-energy utilization, which leads to decreased operational costs (Abdel-Fatah 2018).

Partial Water Desalination: NF can specifically reject divalent ions like calcium and magnesium that contribute to the hardness of water while permitting monovalent ions like sodium and chloride to permeate (Abdel-Fatah 2018). This degree of desalination may be well-suited for certain applications that do not require total desalination.

Organic Matter Rejection: The NF membranes are also capable of rejecting both natural organic matter (NOM) as well as low-MW synthetic organic compounds, thus improving the quality of the effluent water.

6.4.2 Advantages of Forward Osmosis (FO)

In this process, new technology of membrane that employs an osmotic pressure gradient to propel water through a semi-permeable membrane (Aende et al. 2020) In FO, osmotic pressure differential across the membrane is formed by using a highly concentrated draw solution, allowing water to be pulled from the feed solution (e.g. seawater) into the draw solution. Finally, the diluted draw solution is treated to recover the water and regenerate the draw solution to its initial concentration for reuse (Aende et al. 2020).

Forward osmosis has several potential benefits when compared to traditional pressurization-driven membrane processes such as RO:

Reduced Energy Utilization: FO functions on less or no hydraulic head and hence consumes less energy than RO (Skuse et al. 2021).

Lower Tendency towards Fouling: Compared to RO membranes, FO membranes have higher permeated product quality because of less hydraulic pressure and no pressure gradients application, resulting in a lower tendency towards fouling in FO (Chun et al. 2017).

Greater Water Recovery: Forward osmosis can achieve higher water recovery than RO, which increases concentration volume and reduces the volume of waste for disposal (Skuse et al. 2021).

Wide Applications: FO can be applied in various sectors, such as desalination, wastewater treatment and food processing (Chun et al. 2017).

However, FO has its own challenges that need to be overcome before it can become widely adopted:

Feed Water: The diluted draw solution must be purged, and the draw solute must be recovered. This draw solute recovery step can be of critical importance in the FO process and is typically energy-intensive (Aende et al. 2020).

Concentration Polarization: It refers to the phenomenon that the solutes concentration increases near the membrane surface, causing the osmotic driving force to decrease and the water flux to limit (Chun et al. 2017).

Membrane Development: Advancing high-performance FO membranes having higher water selectivity and permeability is a key to improving the effectiveness of FO processes.

6.4.3 Hybrid Membrane System for Enhanced Efficiency

The hybrid membrane system is the combination of two or more membrane processes for lifting their individual strength, efficiency performance and reliability and diminishing their limitations, which also enhances the working process. This process includes different membrane technology, such as reverse osmosis, forward osmosis, ultrafiltration, nanofiltration and microfiltration in one set-up system.

There are some customized designs such as ultrafiltration and reverse osmosis, which are commonly used for seawater desalination; microfiltration, forward osmosis and reverse osmosis (MF + FO + RO) can be used in high-strength seawater treatment; and nanofiltration and reverse osmosis allow for selective removal of divalent salt before complete desalination.

RO+FO Hybrid Systems: Reverse osmosis and forward osmosis hybrid process are integrated to take advantage of both reverse osmosis and forward osmosis processes. It is a typical system in which forward osmosis is used as a pretreatment while RO reduces fouling and improves the quality of feed water to the river osmosis membrane (Gohil et al. 2023). The diluted draw solution from the FO process is then fed to the RO system for desalination.

Nano Filtration and Reverse Osmosis Hybrid System: In this process, nanofiltration and reverse osmosis system combine to optimize the desalination performance. Nanofiltration is used to remove divalent ions and organic matter while reducing the fouling potential by reverse osmosis membrane (Gohil et al. 2023). The NF permeate is then fed to the RO system for further desalination.

- **Membrane Distillation (MD)-RO Hybrid Systems**: The water vapour from saline solution can be separated using a hydrophobic membrane in a thermal membrane process. MD-RO hybrid systems combine RO and MD processes to reduce energy consumption and improve desalination efficiency (Al-Obaidi et al. 2024). The overall energy footprint can be reduced by using the waste heat from the RO system to drive the MD process (Avlonitis et al. 2003). The benefits of a hybrid system are given in Table 6.2.

Table 6.2 Components and uses of the hybrid system

Hybrid system	Uses	Components
Thermal-RO hybrid	RO performance is improved by utilizing thermal plant coolant, increasing flux by 1.5–3% per °C	RO + MED/MSF
Ceramic coated membranes	At (2–4 bar) low pressures, > 80% of pharmaceuticals and phosphorus are removed while operating.	polyelectrolyte coatings
FO- (MD)	Effectively treats high-TDS wastewater (>60,000 ppm) and yields distilled-quality water	Waste heat utilizing MD and FO
NF integrated with RO	The lifespan of a membrane and RO efficiency improved as well as removed scaling ions (e.g. Mg^{2+}, Ca^{2+}).	RO + NF

6.5 Fabrication and Membrane Materials

6.5.1 Classification of Membrane Materials

The choice of membrane material is important in essential for both efficiency and lifespan of membrane filtration. For desalination, a variety of membrane materials are used, each offering a unique characteristic which is suitable for a specific purpose and operational situations.

- **Ceramic Membranes**: It is constructed from inorganic compounds such as zirconia titania and alumina (Eray et al. 2021). It provides outstanding thermal, mechanical and chemical stability, making them suitable for harsh operating conditions (Meulenberg et al. 2019). Ceramic membranes are less prone to fouling and can be easily cleaned (Eray et al. 2021). However, ceramic membranes are generally more expensive than polymeric membranes and may have lower permeability (Meulenberg et al. 2019).
- **Polymeric Membranes**: Among all membrane materials used in desalination, polymeric membranes are the most employed, thanks to their affordability, ease of manufacturing and adaptability (Qadir et al. 2017). Frequently used polymers include polyamide (PA), polysulfone (PS), polyethersulfone (PES) and cellulose acetate (CA). These membranes strike a good balance between selectivity and permeability; however, they can be prone to fouling and may degrade under certain chemical exposures (Yadav et al. 2021).
- **Composite Membranes**: Composite membranes combine the advantages of different materials to achieve enhanced performance. Thin-film composite (TFC) membranes, consisting of a thin selective layer on top of a porous support layer, are

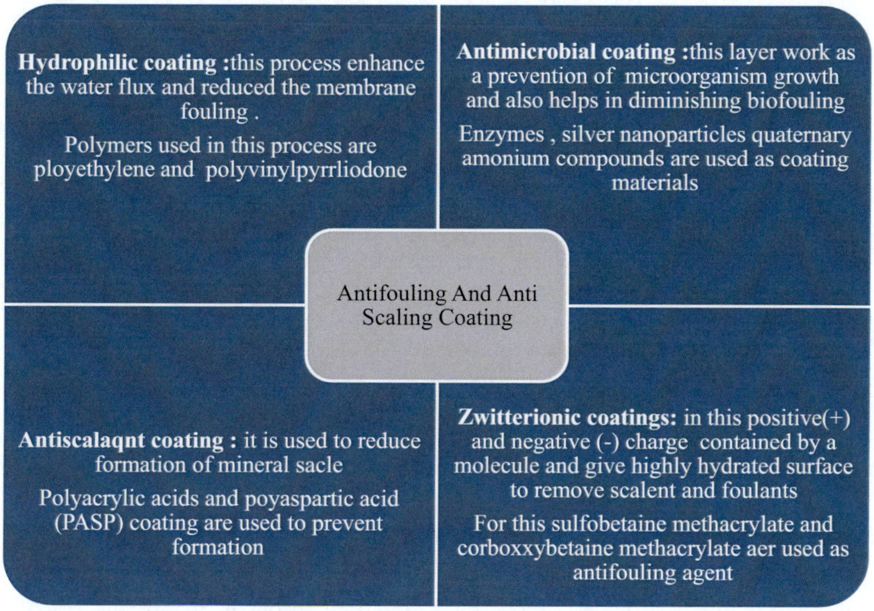

Fig. 6.3 Antifouling and antiscaling coating

widely used in RO desalination (Werber et al. 2016). The selective layer provides high selectivity, while the porous support layer provides mechanical strength and permeability (Werber et al. 2016). Mixed matrix membranes (MMMs) incorporate inorganic fillers into a polymeric matrix to improve membrane properties such as permeability, selectivity, and fouling resistance (Qadir et al. 2017).

6.5.2 Recent Developments in Membrane Surface

See Fig. 6.3.

6.6 Operational Considerations and Design

6.6.1 Configuration of Membrane Modules

Membrane modules are the main component of membrane filtration which is mainly used for providing the path for flow to feed and permeate streams. There are different types of membrane modules such as tabular modules, hollow fibre modules, spiral modules and plate and frame module.

Type of modules	Structure	Flow pattern	Suitable for
Tabular modules	Membrane are employed on inner surface of tube	Feed flow through tube and permeate passes outwards	Waste water and high suspended solids
Hollow fibre	Bundle of small dia hollow fibres are tightly packed	Feed flow outside and inside fibre	High solid feed and low compact system
Spiral wound	Flat sheet membrane warp around central perforated tube	Feed flow axially through spiral layer	Reverse osmosis desalination
Plate and fame	Flat sheet membrane between plates and forming system	Feeds and permeate flow through alternate channels	Viscous flow stream

6.6.2 Advantage of Membrane Modules

Membrane modules provide compact and modular design for membrane. Tabular modules are strong and effective for water and wastewater treatment having high suspended solids, hollow fibres module provides high surface area in small volume, spiral-wound module used in reverse osmosis desalination having compact design, cost-effectiveness, and easy handling and plate and frame module is easy to access and prevent fouling.

6.6.3 Energy Consumption and Optimization Strategies

Energy usage plays an important role in desalination, representing a significant portion of the operating costs. Various strategies can be employed to optimize energy consumption in membrane filtration systems.

- **Energy Recovery Devices**: Energy Recovery Devices (ERDs) capture the energy contained in the high-pressure concentrate stream and redirect it to the incoming feed stream, improving overall system efficiency [30]. Technologies like pressure exchangers and energy recovery turbines play a key role in lowering the energy demands of reverse osmosis desalination (Yadav et al. 2021).
- **Optimized System Design**: Enhancing system design—through careful selection of membranes, strategic module configuration, and fine-tuning of operating parameters—can lead to significant reductions in energy usage (Chun et al. 2017). Computer modelling and simulation tools can be used to optimize system design and predict performance (Pichardo-Romero et al. 2020).

- **Variable Frequency Drives**: Variable frequency drives (VFDs) can be used to control the speed of pumps and motors, reducing energy consumption during periods of low demand (Aladwani et al. 2021).
- **Renewable Energy Sources**: Integrating renewable energy sources such as solar, wind, and geothermal energy can reduce the reliance on fossil fuels and lower the carbon footprint of desalination (Shahid et al. 2023).
- **Two-Pass RO Systems**: Implementing a two-pass RO system, where the permeate from the first pass is further treated in a second RO pass, can improve water quality and reduce the need for post-treatment, potentially saving energy (Skuse et al. 2021).

6.6.4 Process Monitoring and Control Systems

Process monitoring and control systems are essential for ensuring the stable and efficient operation of membrane filtration systems. These systems continuously monitor key parameters such as pressure, flow rate, temperature and water quality, providing real-time feedback to operators (Aladwani et al. 2021).

- **Instrumentation**: Instrumentation includes sensors and transmitters that measure process parameters and transmit data to the control system (Aladwani et al. 2021)
- **Programmable Logic Controllers**: Programmable logic controllers (PLCs) are used to automate and control the operation of the membrane filtration system (Aladwani et al. 2021). PLCs can adjust operating conditions, control pumps and valves and trigger alarms in response to changing process conditions (Aladwani et al. 2021).
- **Supervisory Control and Data Acquisition Systems**: Supervisory control and data acquisition (SCADA) systems provide a centralized interface for monitoring and controlling the entire membrane filtration system (Aladwani et al. 2021). SCADA systems collect data from various instruments and PLCs, display real-time information to operators, and allow operators to remotely control the system (Aladwani et al. 2021).
- **Data Analysis and Optimization**: Data collected from the process monitoring system can be analysed to identify trends, optimize operating conditions and predict membrane performance (Laqbaqbi et al. 2017). Advanced data analytics techniques such as machine learning can be used to improve process control and optimize energy consumption (DeNicola et al. 2015).

6.7 Fouling, Scaling and Mitigation

The membrane technology usage made seawater desalination one of the economically feasible approaches, as the durable RO technology reduces the life-cycle operational costs by extending the membrane life by providing high-quality performance.

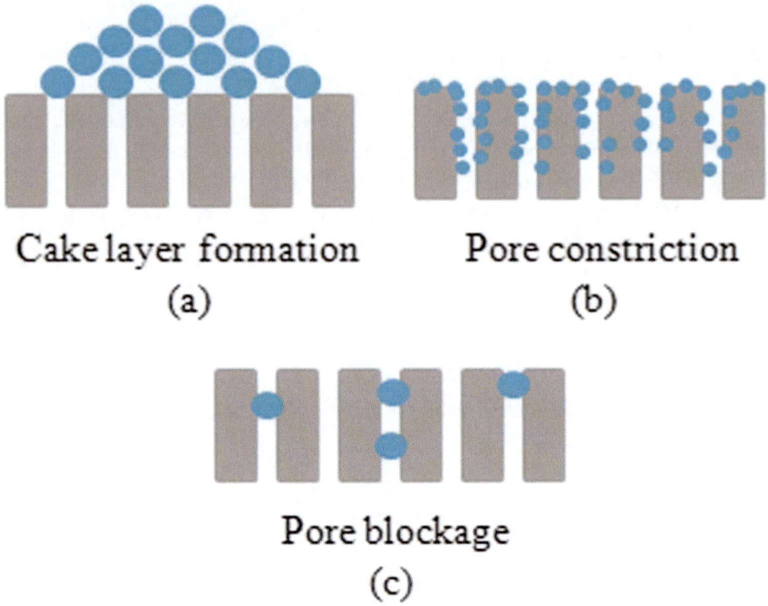

Fig. 6.4 Fouling mechanisms in membrane technology. *Source* Cynthia Hallé (2009)

Despite the advantages of separating minerals and salts from water, it reduces the efficiency of ion-exchange membranes by membrane fouling. Fouling is the deposition and adsorption of foulants and accumulation (substances causing fouling) on the membrane surface and inside its pores. These results increased operational inputs, voltage or pressures (Wenten et al. 2024). Membrane fouling is based on feed characteristics, membrane properties, and operational conditions. This is usually classified by pore constriction, blockage and cake formation and is shown in Fig. 6.4. The particles form a separate layer on top of the membrane called cake layer formation, in which the membrane pores are completely closed causing the blockage of water that passes through it. Pore constriction minimizes the available area for filtration while the particulates adsorb within the membrane pores. In pore blockage, membrane pores are sealed by the dissolved particles. (Cynthia Hallé 2009).

6.7.1 Types of Fouling in Membrane Technology

The fouling types can be classified based on foulant type. Membrane fouling classification is divided into particulate, organic, scaling, chemical and biological (El Batouti et al. 2021). Figure 6.5 shows the various types of fouling in membrane technology of seawater desalination which are discussed.

1. **Particulate Fouling**:

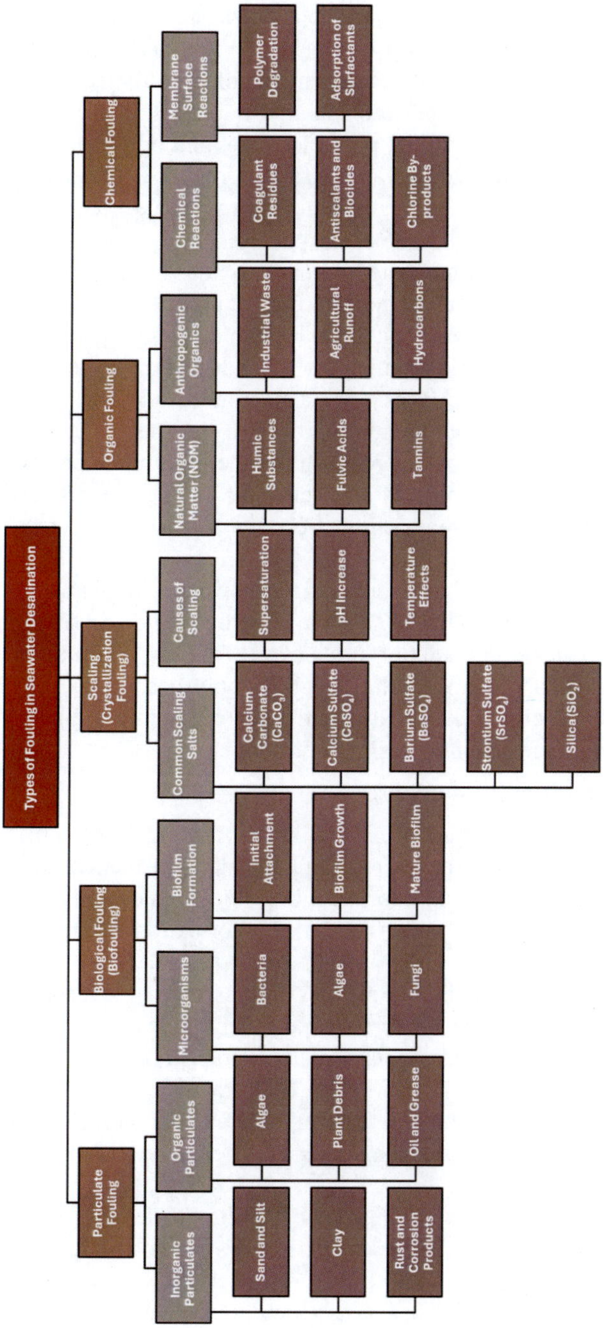

Fig. 6.5 Classification of fouling based on the type of foulants

The particulate fouling appears when the colloidal particles in feed water deposit on the membrane surface (Henry et al. 2012). The particulate fouling is called colloidal fouling. The major foulants causing colloidal fouling are humic substances, inorganic, microbially generated foulants and microbes. The fouling mechanisms involved are cake layer formation, pore blockage, and pore constriction (Xia et al. 2019). The deposited particle size in this fouling ranges between 0.0002 and 1 μm (Stewart 2021). These particles can be removed from molecular diffusion and shear-induced diffusion based on the size of the particles. The commonly found inorganic colloidal foulants in membrane systems are iron oxides/hydroxides, silica, elemental sulphur and aluminium silicate minerals (Tang et al. 2011). These can be classified into rigid inorganic groups and organic macromolecule groups. The rigid inorganic groups include aluminium silicate minerals, silica and iron hydroxide. These particles are rigid and compact. Whereas, the organic macromolecule groups are classified into rigid biopolymers, fulvic compounds and flexible polymers. The colloidal particles in solution and their effect on the propensity of the foul membrane depends on their size and shape, the charge of particles, and their interaction with feed water ions. Table 6.3 summarizes different colloidal classes and their properties (Abdulrahim et al. 2017).

The molecular weight of humic acid ranges between several kDa to a few hundred kDa depending on their sources. These are highly polydisperse as they comprise heterogeneous organic degradation products. Its acidity is around 8 meq/g (Tang et al. 2011). Whereas, the molecular weight of polysaccharides ranges from a few hundred to a few thousand kDa. The multivalent metal ions are responsible for the extensive three-dimensional cross-linked structures.

2. **Biological fouling (Biofouling)**

Biological fouling is also called biofouling. Biofouling is one of the fouling mechanisms caused due to bacteria. After the attachment to the membrane surface eventually, they multiply and grow and form the biofilm which decreases the membrane's performance (Matin et al. 2021). The biofilm formation mechanism is shown in Fig. 6.6 and is summarized.

If bacteria are present during the process of transferring the feed water through the membrane, they will attach to the surface of the membrane, where reversible adhesion occurs. Subsequently, this is associated with the biosynthesis of extracellular polymeric substances (EPS) (Najid et al. 2022). Subsequently, they grow and multiply by utilizing the feed water nutrients. This forms the complex biofilm. Biofilm formation involves biological functions among cells in different levels of the biofilm (Najid et al. 2022).

Microscopic examination is considered the direct method used to identify the biofilm. The indirect method is the analysis of system performance by determining the solute rejection and permeate flux. Conversely, the surface of the membrane degrades, and the formed biofilm acts as a membrane. The indices used to determine biofouling are the silt density index (SDI) and modified fouling index (MFI). These impacts decrease in the membrane flux, an increase in differential

3. **Scaling**

Table 6.3 List of various colloids with their characteristics Abdulrahim et al. (2017)

Colloidal type	Particles charge	Shape and size	Examples
Particulate	At pH = 7, negatively charged	Angular	Aluminium silicate minerals
	'+ve' charged. pHpzc for goethite ~ 9	Alters depending on crystalline or amorphous	Hydroxide/ferric oxides
	(pHpzc= 3) '-ve' charged	Round	Silica
Polysaccharides	Neutral	400–500 kDa, rigid rod-like	Schizophyllan
	Negatively charged	100–2500 kDa, linear	Xanthan and gellan
	Negatively charged, ~ 3 meq/g (up to 6 meq/g)	200–2000 kDa, extended random coil	Alginate
Humic acid	Negative charged (pHpzc < 3), typical total acidity: 5–10 meq/g	Linear under high pH, low ionic strength, and low concentration (Mw of a few kDa to a few hundred kDa. Globular molecule)	IHSS-International Humic Substance Society standards
Proteins	$pH_{IEP} = 4.7$ (total acidity 1.5 meq/g including both carboxylic and amine groups)	67 kDa	Bovine serum albumin
	$pH_{IEP} = 6.6$	155 kDa	Bovine immunoglobulin G
	$pH_{IEP} = 7.1$	68 kDa	Bovine haemoglobin
	$pH_{IEP} = 7.8$	13.7 kDa	Bovine pancreas ribonuclease A
	$pH_{IEP} = 11.0$	14.4 kDa	Lysozyme

When the inorganic salt concentration in the seawater becomes supersaturated (exceeds the solubility product equilibrium), this forms a scale on the membrane surface. The main stages involved in the mineral scale formation are nucleation, crystal growth, and deposition. The mineral salts responsible for scaling potential increase are $CaCO_3$, $BaSO_4$ and $CaSO_4$. Scaling also precipitation kinetics. The major fouling salts such as $CaCO_3$, $BaSO_4$, $Ca_3(PO_4)_2$, $SrSO_4$, $CaHPO_4$ and $CaSO_4$ have solubility products such as 2.8×10^{-9}, 1.1×10^{-10}, 2.07×10^{-29}, 3.8×10^{-7}, 1.0×10^{-7} and 4.93×10^{-5} respectively (Matin et al. 2019). Among them, $CaCO_3$ is the primary source in the scale formation process. There are three types of scaling elements based on their characteristics and are namely alkaline, non-alkaline and silica-based. Calcium carbonate is the most widespread alkaline salt.

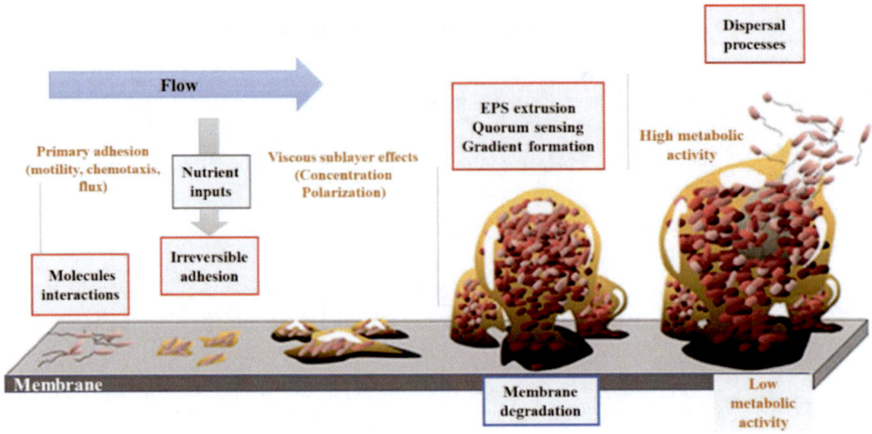

Fig. 6.6 Pictorial representation of biofilm formation mechanism (Najid et al. 2022)

It is available in a hydrated and anhydrous crystalline polymorph state. The calcite, aragonite and vaterite are the three possible phases of anhydrous state. Whereas, amorphous, monohydrate and hexahydrate are the phases of the hydrous state (Matin et al. 2019). Calcium sulphate is the most common non-alkaline salt. The scaling is mainly due to surface and bulk crystallization of calcium sulphate. Depending on the hydration state, it is available in three different forms: hemihydrate, anhydride, and gypsum. Compared to alkaline and non-alkaline, silica fouling is considered the major and complex process. It is available in three different forms such as silicic acid in monomeric form, colloidal pressure, biodegradation of membrane, reduction in energy efficiency and water quality and lowered salt rejection efficiency (Nejati et al. 2019).

Silica in polymeric form, and particulate silica at pH-7, the solubility rate of silica is 100–150 mg/L. If it exceeds, results in the scale formation on the membrane surface. Fluid velocity, pH, temperature and salt concentrations are the main factors that affect the scaling (Najid et al. 2022).

4. **Organic fouling**

Among all the fouling mechanisms, organic fouling remains the main. The primary contributor of organic fouling is natural organic matter (NOM). There are three parts in the categorization of organic fouling characterization technology. These parts include structural and morphological characterization, chemical composition characterization, and non-invasive techniques. Various scientific instruments are used to study and analyse the characterization of organic fouling and are described as follows (Ding et al. 2025).

i. Scanning electron microscopy (SEM) is used to determine the morphological characteristics of the samples.

ii. An atomic scope microscope (AFM) is used to study membrane fouling and membrane surface.
iii. The energy-dispersive X-ray spectrometer is used to analyse the chemical composition of the membrane surface.
iv. Attenuated total reflection Fourier-transform infrared (ART-FTIR) is used to identify and characterize the deposition of foulants on the membrane surface.
v. Liquid chromatography-organic carbon detection (LC-OCD) is used to perform dissolved organic carbon (DOC) analysis.
vi. To analyse the organic matter in the samples, the excitation-emission matrix fluorescence spectroscopy (EEM) is used.
vii. Two-dimensional gas chromatography is used to examine organic mixture components.
viii. Electrochemical impedance spectroscopy (EIS) is used to characterize the sample coating.
ix. Ultrasonic time domain reflectometry (UTDR) is used to detect the reflected wave, which will characterize the membrane-wetting process.
x. Optical coherence tomography (OCT) is used for the real-time monitoring of the membrane.

Organic fouling forms a distinct gel layer on the membrane's surface and is responsible for pore blocking. This type of fouling is considered harmful. Organic fouling can lead to biofouling, as the metabolism and growth of microorganisms can occur due to the presence of dissolved organic compounds (Najid et al. 2022).

6.7.2 Prevention and Cleaning Protocols

1. **Pretreatment Systems**

Pretreatment removes the foulants and transparent exopolymer particles (TEP) through which the membrane fouling can be minimized. The widely used pretreatment processes are adsorption and coagulation. The coagulating agents such as aluminium sulphate and ferric sulphate are used with optimum dosage and pH. Effective coagulation and flocculation require 1 min rapid mixing with 300 rpm speed and 15 min slow mixing. The activated carbon is used to verify the carbon adsorption (Baek and Chang 2009). The pretreatment method selection depends on the characteristics of feed water. A suitable pretreatment method will enhance the quality of the water produced. The various pretreatment methods other than coagulation and adsorption are electrocoagulation, advanced oxidation and membrane-based filtration (microfiltration and ultrafiltration). The advantages of pretreatment include reduced total water cost and reduced footprint. In practical applications, integrating conventional processes into membrane-based pretreatment systems proves beneficial for effectively removing a wide range of foulants. Despite notable progress in the development of advanced membrane materials and pretreatment strategies aimed at minimizing fouling, these approaches alone do not fully eliminate all contaminants present in the feed water. As a result, fouling continues to occur over time, making periodic chemical cleaning essential to restore membrane performance and maintain operational efficiency. To ensure the reliable operation of desalination plants, membrane cleaning—whether through clean-in-place or ex-situ methods—must be incorporated as a core component of the treatment process. (El Batouti et al. 2021; Goh et al. 2018) explained a variety of cleaning methods are employed, including physical techniques such as backwashing, pulsing, forward flushing with air and sonication, along with chemical cleaning. Alkaline and acidic solutions are typically used in chemical cleaning: alkaline agents help remove organic fouling and biofilms, while acidic agents are effective against scaling.

Membrane modification has emerged as a critical strategy for mitigating fouling in membrane-based water treatment systems by tailoring membrane surfaces or internal structures to better suit specific operational environments. Fouling, especially organic fouling, significantly compromises membrane efficiency and longevity, primarily due to hydrophobic interactions between membrane surfaces and organic solutes such as natural organic matter (NOM), proteins and polysaccharides. To counter this, modifying membrane surfaces to enhance hydrophilicity has proven highly effective. One of the most promising approaches involves the grafting or incorporation of hydrophilic functional groups—such as hydroxyl (–OH), carboxyl (–COOH), and sulfonic ($-SO_3H$) groups—onto the membrane surface. These groups facilitate the formation of a stable hydration layer composed of structured water molecules that physically and energetically repel hydrophobic foulants. This hydration barrier not only reduces foulant adhesion but also minimizes its penetration into membrane pores, thereby preserving flux and selectivity over prolonged operation. Advanced membrane modification techniques have been extensively investigated to improve antifouling characteristics. These include surface coating with ultrathin hydrophilic layers containing zwitterions, polyethylene glycol (PEG), or other functional nanomaterials such as graphene oxide and titanium dioxide (TiO_2), which provide both hydrophilicity and additional functionalities like photocatalytic self-cleaning. Bulk modification methods, such as blending antifouling agents into polymer dope solutions during membrane fabrication, enable uniform distribution of functional components throughout the matrix. Furthermore, post-fabrication treatments like plasma-induced polymerization, UV-assisted grafting, or oxidative functionalization (e.g. using ozone or permanganate) alter surface energy and wettability, resulting in decreased contact angles and improved resistance to foulants.

Such modifications are particularly beneficial in high-fouling scenarios, including membrane distillation, tertiary wastewater treatment and industrial effluent processing, where conventional membranes often suffer from rapid performance decline. Overall, hydrophilic surface engineering not only mitigates fouling but also enhances membrane durability, cleaning efficiency, and operational sustainability in challenging water treatment applications (Ding et al. 2025).

6.7.3 Chemical Cleaning

Chemical cleaning is widely used to restore membrane performance, with alkaline cleaning generally more effective than acid cleaning for removing organic fouling. This is due to the ability of alkaline solutions to hydrolyze, solubilize and saponify organic matter. Hydroxide ions also help break down polysaccharides and proteins into smaller fragments. Sodium hydroxide (NaOH) flushing has shown greater effectiveness than deionized water or NaCl, though its efficiency varies with foulant type. In some cases, acid cleaning results in higher dissolved organic carbon (DOC) due to the release of organic matter bound to inorganic scale. However, repeated chemical cleaning can negatively affect membrane properties, particularly by reducing hydrophobicity and liquid entry pressure, leading to irreversible wetting and performance decline. Therefore, the cleaning frequency and intensity must be carefully managed to minimize long-term damage (Madaeni and Samieirad 2010).

6.7.4 Physical Cleaning

Osmotic backwashing (OBW) is a physical cleaning technique that has shown promise in removing organic fouling from membrane surfaces. This method involves introducing a highly saline draw solution to the feed side of the membrane system without applying external pressure. The resulting osmotic pressure difference between the feed side (with high salinity) and the permeate side (typically reverse osmosis permeate) generates a reverse water flux, which flows from the permeate to the feed channel. This reverse water movement can help dislodge compact and dense fouling layers or weaken the adhesion of accumulated foulants, thereby improving membrane performance. OBW has been observed to enhance water permeability by effectively reducing fouling. In comparative evaluations, OBW has demonstrated higher cleaning efficiency in nanofiltration systems than other physical methods like salt cleaning. Furthermore, the reverse water flux generated during OBW has been linked to better flux recovery, even in cases of irreversible fouling, sometimes outperforming conventional chemical cleaning approaches. Although OBW has gained attention for its cleaning potential, the underlying mechanisms responsible for foulant removal and its synergistic effects in membrane cleaning are still not fully understood and require further investigation (Park et al. 2022).

6.7.5 Surface Modifications

Surface modification of membranes can be categorized into surface coating (physical) and surface grafting (chemical). Plasma treatment is a widely used surface coating method because it is solvent-free and eco-friendly. In one approach, hydrophilic

PAMAM was sprayed onto the membrane surface post-interfacial polymerization, resulting in a 20–25% increase in water permeability (Goh et al. 2018). Surface grafting enables the creation of tailored membrane surfaces with specific functionalities. This can be achieved using single or mixed monomers, typically initiated by UV or plasma treatment. Plasma polymerization techniques have been employed to enhance hydrophilicity; for instance, incorporating PEG-like triglyme reduced the contact angle from 32° to 7°, indicating improved surface wettability (Goh et al. 2018).

Various membrane modifications using polymer additives are given in Table 6.4.

6.7.6 Innovations in Fouling-Resistant Membranes

The advancement of RO technology is vital in addressing global water scarcity, and future research must strategically target three key areas: innovative cleaning methods, fouling-resistant membranes and sustainable system practices.

1. **Innovation in Cleaning Techniques**

Membrane fouling remains a major operational bottleneck, reducing RO efficiency and increasing maintenance. Conventional chemical cleaning, though effective, often degrades membranes and generates hazardous waste. Emerging non-chemical methods such as ultrasonic agitation, hydrodynamic shear and electrically induced cleaning offer promising alternatives by physically dislodging foulants without damaging the membrane surface. Additionally, optimizing pH control and exploring biodegradable, low-toxicity agents can enhance cleaning efficacy while minimizing environmental impact. Research should also emphasize pretreatment optimization to match feed water characteristics and lower fouling risk upstream (Ding et al. 2025).

2. **Development of Fouling-Resistant Membranes**

Surface modification strategies like polymer grafting, plasma treatment and nanocomposite layering (e.g. with graphene oxide, and silver nanoparticles) are enabling membranes to resist organic and microbial fouling while retaining high flux. Advanced membranes incorporating hydrophilic or zwitterionic coatings can reduce biofilm formation. Furthermore, novel materials such as thin-film nanocomposites (TFNs) and biopolymer hybrids show potential for long-term performance with reduced cleaning frequency, contributing to lower life-cycle costs (Tayeh et al. 2025).

3. **Sustainable RO Applications**

Sustainability in RO systems involves reducing energy consumption, waste generation and membrane disposal impacts. Research should prioritize integrating renewable energy sources (solar, wind) and Energy Recovery Devices (ERDs) to lower the carbon footprint. Meanwhile, pretreatment innovations like biofiltration and ceramic

Table 6.4 Various membrane modifiers with method and performance enhancement Goh et al. (2018)

Modifier type	Key materials/ approach	Modification method	Performance enhancements
Poly (ethylene glycol) (PEG)	PEG, Triazine ring, Melamine, MPD, PIP	In situ incorporation, interfacial polymerization (IFP)	Enhanced antifouling
			Reduced bacterial adhesion
			Improved hydrophilicity
			Lower surface roughness
Hyperbranched polymers	PAMAM dendrimers, PAMAM-PEG, Hyperbranched polyesteramide	Surface coating, chemical coupling	High hydrophilicity
			Low protein adsorption
			High chemical stability
			Maintained salt rejection
Zwitterionic polymers	Sulfobetaine, Phosphorylcholine, L-cysteine	Graft polymerization (ATRP, free radical), Chemical coupling	Superior antifouling
			Hydration layer formation
			Stable in high salinity
			Smoother morphology
Polydopamine (PDA)	Dopamine solution	Bio-inspired surface coating	Improved hydrophilicity
			Reduced surface roughness
			Enhanced fouling resistance (depending on pH and time)
Nanoparticle additives	Ag, Cu, GO, CNTs, Clays	Incorporated into PA layer	Antibacterial properties
			Enhanced antifouling and mechanical stability

membranes can preemptively mitigate fouling, thus reducing chemical usage. End-of-life strategies such as membrane recycling into nanofiltration or ultrafiltration membranes, and nutrient or metal recovery from concentrate streams, support circular economy goals and resource optimization (Tayeh et al. 2025).

In essence, the future of RO lies in the intersection of advanced materials science, green engineering and systems sustainability, ensuring high efficiency, environmental compliance, and long-term operational resilience.

6.7.7 Environmental and Economic Impacts of the Membrane Desalination Process

Membrane desalination processes pose several environmental impacts (EIs), primarily associated with five key components: (1) feed water intake, (2) pretreatment, (3) desalination operations, (4) brine discharge and (5) energy use. Among these, brine discharge and energy consumption represent the most significant environmental concerns. Brine is the main waste stream, containing concentrated salts, pretreatment chemicals, corrosion by-products and backwash residues (Elsaid et al. 2020).

Brine generated from membrane desalination, especially seawater reverse osmosis (SWRO), typically maintains a brine-to-product water ratio of 1:1–2:1, reflecting recovery rates between 35 and 50%. Its discharge into marine environments affects salinity, temperature and pH, and introduces heavy metals and residual chemicals—posing ecological risks to marine organisms. Impact assessments often focus on brine behaviour at discharge points and mixing zones (Elsaid et al. 2020).

Membrane desalination is highly energy-intensive, often reliant on fossil fuels, making it a significant contributor to greenhouse gas (GHG) emissions and climate change. While reducing energy consumption can lower emissions, only the integration of renewable energy can simultaneously reduce both fossil fuel dependency and GHG output. Desalination processes release substantial amounts of CO_2, NO_x, SO_x and particulate matter (PM), correlating directly with their energy demands. Emission levels vary based on the specific energy consumption of SWRO plants and the energy generation source (Kim et al. 2019). Mezher et al. (2011) reported CO_2 emissions of 2.79, 2.13 and 1.75 kg/m^3 for steam cycle, internal combustion engine and combined cycle power generation using natural gas, respectively.

1. **Energy recovery devices**

Seawater desalination is a highly energy-intensive process, with substantial energy losses occurring through the high-pressure brine discharge. To address this, Energy Recovery Devices (ERDs) are employed to transfer pressure from the brine stream to the incoming feedwater, significantly lowering specific energy consumption. Early ERDs, like Pelton wheels, were limited by mechanical inefficiencies. Modern isobaric ERDs—particularly rotary and positive displacement types—have since become the standard due to their higher-pressure transfer efficiency and reduced mechanical complexity. Positive displacement ERDs, such as DWEER™, utilize chambers and pistons for pressure exchange, while rotary displacement devices like the PX™ pressure exchanger use a rotor-driven design without valves or pistons, offering greater efficiency and reliability. These technologies have been widely adopted in large-scale desalination plants, where their performance and economics are favourable. However, their application in small-scale systems such as Wave-Driven Desalination Systems (WDDS) is limited due to cost, complexity and energy requirements. For such systems, pressure intensifiers like the axial piston pump/

motor by Danfoss and the Clark Pump by Spectra Watermakers offer a viable alternative. These mechanical systems are suited to variable input conditions and do not require electrical energy or booster pumps. Among them, the Clark Pump is especially suited for WDDS applications due to its full mechanical operation, adjustability via piston size, and compatibility with intermittent energy sources. Unlike conventional ERDs, it operates efficiently in low-capacity, renewable-powered systems, making it a preferred choice for decentralized desalination solutions (Das et al. 2024).

2. **Cost Analysis and Lifecycle Assessment**

The evaluation by Kurokawa et al. (2018) focused on the environmental and economic performance of domestic seawater desalination, specifically examining CO_2 emissions, water consumption, and life cycle costs. It has been observed that actual operational data showed higher environmental burdens and costs per cubic metre compared to planned projections. CO_2 emissions were predominantly linked to electricity use, contributing up to 99% of the total, while indirect water consumption—mainly for chemical production used in membrane cleaning—accounted for up to 95%. In Okinawa, emissions were based on coal-fired electricity, whereas in Fukuoka, natural-gas-powered generation was the basis. The life cycle cost in Okinawa was approximately 5.5 times higher than expected, attributed to a low operational rate of around 8% (Kurokawa et al. 2018). Sensitivity analysis and literature comparisons have confirmed that transitioning to cleaner electricity sources can significantly reduce CO_2 emissions, aligning with previous findings.

6.8 Conclusion

Seawater desalination has emerged as a critical solution to global water scarcity, particularly in arid and coastal regions. Among the various technologies available, membrane filtration has proven to be the most efficient and scalable, with reverse osmosis (RO) leading due to its high salt rejection capabilities and relative energy efficiency. The principles of membrane filtration revolve around selectivity and permeability, with RO, nanofiltration (NF) and forward osmosis (FO) offering distinct advantages based on feed water characteristics and operational goals. While RO remains dominant for seawater desalination, NF is suited for brackish water, and FO presents potential in low-energy systems. The success of membrane-based systems depends heavily on the material science underpinning membrane design. Polymeric, ceramic and composite membranes have been developed with surface modifications and coatings to enhance fouling resistance and durability. Addressing challenges such as biofouling, scaling and membrane degradation is central to ensuring long-term performance. Advanced membrane configurations, such as spiral-wound and hollow fibre modules, are integrated with energy optimization strategies, including the use of Energy Recovery Devices (ERDs), which significantly reduce operational energy requirements.

Despite technological advances, seawater desalination remains energy-intensive and poses environmental concerns, particularly from CO_2 emissions and brine disposal. Life-cycle cost assessments have shown variability based on plant operation rates and energy sources, with cleaner electricity markedly reducing carbon footprints. Integration of real-time monitoring and adaptive control systems further improves operational resilience and sustainability. In conclusion, membrane-based seawater desalination offers a promising pathway to secure freshwater supplies. Continued innovation in membrane materials, system design, and energy integration, coupled with environmental stewardship, is vital for making desalination a sustainable and economically viable component of future water resource management.

References

Abdel-Fatah MA (2018) Nanofiltration systems and applications in wastewater treatment: review article. In: Ain shams engineering journal, vol 9, issue 4. Ain Shams University, pp 3077–3092. https://doi.org/10.1016/j.asej.2018.08.001

Abdulrahim HK, Al-Rasheed AM, Hassan AS, Mabrouk ANA, Shomar B, Darwish MA (2017) Reverse osmosis desalination system and algal blooms part III: SWRO pretreatment. Desalin Water Treat 60:11–38. https://doi.org/10.5004/DWT.2017.0266

Aende A, Gardy J, Hassanpour A (2020) Seawater desalination: a review of forward osmosis technique, its challenges, and future prospects. In: Processes, vol 8, issue 8. MDPI AG. https://doi.org/10.3390/PR8080901

Aladwani SH, Al-Obaidi MA, Mujtaba IM (2021) Performance of reverse osmosis based desalination process using spiral wound membrane: Sensitivity study of operating parameters under variable seawater conditions. Clean Eng Technol5. https://doi.org/10.1016/j.clet.2021.100284

Aliyu UM, Rathilal S, Isa YM (2018) Membrane desalination technologies in water treatment: a review. In: Water practice and technology, vol 13, issue 4. IWA Publishing, pp 738–752. https://doi.org/10.2166/wpt.2018.084

Al-Obaidi M, Alsarayreh AA, Rashid FL, Sowgath MT, Alsadaie S, Ruiz-García A, Khayet M, Ghaffour N, Mujtaba IM (2024) Hybrid membrane and thermal seawater desalination processes powered by fossil fuels: a comprehensive review, future challenges and prospects. Desalination 583:117694. https://doi.org/10.1016/J.DESAL.2024.117694

Avlonitis SA, Kouroumbas K, Vlachakis N (2003) Energy consumption and membrane replacement cost for seawater RO desalination plants. Desalination 157(1–3):151–158. https://doi.org/10.1016/S0011-9164(03)00395-3

Baek SO, Chang IS (2009) Pretreatments to control membrane fouling in membrane filtration of secondary effluents. Desalination 244(1–3):153–163. https://doi.org/10.1016/J.DESAL.2008.04.043

El Batouti M, Alharby NF, Elewa MM (2021) Review of new approaches for fouling mitigation in membrane separation processes in water treatment applications. Separations 2022 9(1):1. https://doi.org/10.3390/SEPARATIONS9010001

Chun Y, Mulcahy D, Zou L, Kim IS (2017) A short review of membrane fouling in forward osmosis processes. In: Membranes, vol 7, issue 2. MDPI AG. https://doi.org/10.3390/membranes7020030

Das TK, Folley M, Frost C, Brewster P (2024) Application of an energy recovery device with RO membrane for wave powered desalination. Desalination 592:118064. https://doi.org/10.1016/J.DESAL.2024.118064

DeNicola E, Aburizaiza OS, Siddique A, Khwaja H, Carpenter DO, Arabia S (2015) Climate change and water scarcity: the case of Saudi Arabia

Ding M, Zhang S, Zhou H, Xu H, Tao H, Lin T, Gao L (2025) Critical review on organic fouling in membrane distillation for desalination and water purification. Desalination 606:118718. https://doi.org/10.1016/J.DESAL.2025.118718

Elsaid K, Kamil M, Sayed ET, Abdelkareem MA, Wilberforce T, Olabi A (2020) Environmental impact of desalination technologies: a review. Sci Total Environ 748:141528. https://doi.org/10.1016/J.SCITOTENV.2020.141528

Eray E, Candelario VM, Boffa V, Safafar H, Østedgaard-Munck DN, Zahrtmann N, Kadrispahic H, Jørgensen MK (2021) A roadmap for the development and applications of silicon carbide membranes for liquid filtration: Recent advancements, challenges, and perspectives. In: Chemical engineering journal, vol 414. Elsevier B.V. https://doi.org/10.1016/j.cej.2021.128826

Ezugbe EO, Rathilal S (2020) Membrane technologies in wastewater treatment: a review. In: Membranes, vol 10, issue 5. MDPI AG. https://doi.org/10.3390/membranes10050089

Goh PS, Lau WJ, Othman MHD, Ismail AF (2018) Membrane fouling in desalination and its mitigation strategies. Desalination 425:130–155. https://doi.org/10.1016/J.DESAL.2017.10.018

Gohil PP, Desai H, Kumar A, Kumar R (2023) Current status and advancement in thermal and membrane-based hybrid seawater desalination technologies. In: Water (Switzerland), vol 15, issue 12. MDPI. https://doi.org/10.3390/w15122274

Cynthia Hallé (2009) Biofiltration in drinking water treatment: reduction of membrane fouling and biodegradation of organic trace contaminants [University of Waterloo]. https://www.researchgate.net/publication/267682502_Biofiltration_in_Drinking_Water_Treatment_Reduction_of_Membrane_Fouling_and_Biodegradation_of_Organic_Trace_Contaminants#full-text

Henry C, Minier JP, Lefèvre G (2012) Towards a description of particulate fouling: from single particle deposition to clogging. Adv Coll Interface Sci 185–186:34–76. https://doi.org/10.1016/J.CIS.2012.10.001

Kim J, Park K, Yang DR, Hong S (2019) A comprehensive review of energy consumption of seawater reverse osmosis desalination plants. Appl Energy 254:113652. https://doi.org/10.1016/J.APENERGY.2019.113652

Krzeminski P, Schwermer C, Wennberg A, Langford K, Vogelsang C (2017) Occurrence of UV filters, fragrances and organophosphate flame retardants in municipal WWTP effluents and their removal during membrane post-treatment. J Hazard Mater 323:166–176. https://doi.org/10.1016/j.jhazmat.2016.08.001

Kurokawa K, Ono Y, Koseki Y, Sugimoto K, Ohkuma N, Itsubo N (2018) Life cycle assessment and life cycle costing of seawater desalination plants using reverse osmosis: comparison analysis of evaluation results between activities plan stage and actual observations stage. J Life Cycle Assess, Jpn 14(1):46–54. https://doi.org/10.3370/LCA.14.46

Laqbaqbi M, Sanmartino JA, Khayet M, García-Payo C, Chaouch M (2017) Fouling in membrane distillation, osmotic distillation and osmotic membrane distillation. In: Applied sciences (Switzerland), vol 7, issue 4. MDPI AG. https://doi.org/10.3390/app7040334

Madaeni SS, Samieirad S (2010) Chemical cleaning of reverse osmosis membrane fouled by wastewater. Desalination 257(1–3):80–86. https://doi.org/10.1016/J.DESAL.2010.03.002

Matin A, Rahman F, Shafi HZ, Zubair SM (2019) Scaling of reverse osmosis membranes used in water desalination: phenomena, impact, and control; future directions. Desalination 455:135–157. https://doi.org/10.1016/J.DESAL.2018.12.009

Matin A, Laoui T, Falath W, Farooque M (2021) Fouling control in reverse osmosis for water desalination & reuse: current practices & emerging environment-friendly technologies. Sci Total Environ 765:142721. https://doi.org/10.1016/J.SCITOTENV.2020.142721

Meulenberg WA, Schulze-Küppers F, Deibert W, Van Gestel T, Baumann S (2019) Ceramic membranes: materials—components—potential applications. In: ChemBioEng Reviews, vol 6, issue 6. Wiley-Blackwell, pp 198–208. https://doi.org/10.1002/cben.201900022

Mezher T, Fath H, Abbas Z, Khaled A (2011) Techno-economic assessment and environmental impacts of desalination technologies. Desalination 266(1–3):263–273. https://doi.org/10.1016/J.DESAL.2010.08.035

Najid N, Hakizimana JN, Kouzbour S, Gourich B, Ruiz-García A, Vial C, Stiriba Y, Semiat R (2022) Fouling control and modeling in reverse osmosis for seawater desalination: a review. Comput Chem Eng 162:107794. https://doi.org/10.1016/J.COMPCHEMENG.2022.107794

Nejati S, Mirbagheri SA, Warsinger DM, Fazeli M (2019) Biofouling in seawater reverse osmosis (SWRO): impact of module geometry and mitigation with ultrafiltration. J Water Process Eng 29:100782. https://doi.org/10.1016/J.JWPE.2019.100782

Park S, Son M, Shim J, Jeong K, Cho KH (2022) Physically-assisted removal of organic fouling by osmotic backwashing coupled with chemical cleaning. J Clean Prod 378:134490. https://doi.org/10.1016/J.JCLEPRO.2022.134490

Park YM, Yeon KM, Park CH (2020) Silica treatment technologies in reverse osmosis for industrial desalination: a review. In: Environmental engineering research, vol 25, issue 6. Korean Society of Environmental Engineers, pp 819–829. https://doi.org/10.4491/eer.2019.353

Permogorov N (2016) Membrane technologies for water treatment: removal of toxic trace elements with emphasis on arsenic, fluoride and uranium. Johns Matthey Technol Rev 60(4):273–276. https://doi.org/10.1595/205651316x693283

Pichardo-Romero D, Garcia-Arce ZP, Zavala-Ramírez A, Castro-Muñoz R (2020) Current advances in biofouling mitigation in membranes for water treatment: an overview. Processes 8(2). https://doi.org/10.3390/pr8020182

Qadir D, Mukhtar H, Keong LK (2017) Mixed matrix membranes for water purification applications. In: Separation and purification reviews, vol 46, issue 1. Taylor and Francis Inc., pp 62–80 https://doi.org/10.1080/15422119.2016.1196460

Shahid MK, Mainali B, Rout PR, Lim JW, Aslam M, Al-Rawajfeh AE, Choi Y (2023) A review of membrane-based desalination systems powered by renewable energy sources. In: Water (Switzerland), vol 15, issue 3. MDPI. https://doi.org/10.3390/w15030534

Skuse C, Gallego-Schmid A, Azapagic A, Gorgojo P (2021) Can emerging membrane-based desalination technologies replace reverse osmosis? In: Desalination, vol 500. Elsevier B.V. https://doi.org/10.1016/j.desal.2020.114844

Stewart M (2021) Heat transfer theory. Surf Prod Oper 361–430. https://doi.org/10.1016/B978-0-12-803722-5.00009-4

Tang CY, Chong TH, Fane AG (2011) Colloidal interactions and fouling of NF and RO membranes: a review. Adv Coll Interface Sci 164(1–2):126–143. https://doi.org/10.1016/J.CIS.2010.10.007

Tayeh YA, Alazaiza MYD, Alzghoul TM, Bashir MJ (2025) A comprehensive review of RO membrane fouling: mechanisms, categories, cleaning methods and pretreatment technologies. J Hazard Mater Adv 18:100684. https://doi.org/10.1016/J.HAZADV.2025.100684

Wenten IG, Bazant MZ, Khoiruddin K (2024) Mitigating electrodialysis membrane fouling in seawater desalination. Sep Purif Technol 345:127228. https://doi.org/10.1016/J.SEPPUR.2024.127228

Werber JR, Deshmukh A, Elimelech M (2016) The critical need for increased selectivity, not increased water permeability, for desalination membranes. In: Environmental science and technology letters, vol 3, issue 4. American Chemical Society, pp 112–120. https://doi.org/10.1021/acs.estlett.6b00050

Xia L, Vemuri B, Saptoka S, Shrestha N, Chilkoor G, Kilduff J, Gadhamshetty V (2019) Antifouling membranes for bioelectrochemistry applications. Biomass, biofuels, biochemicals: microbial electrochemical technology: sustainable platform for fuels, chemicals and remediation, pp 195–224. https://doi.org/10.1016/B978-0-444-64052-9.00008-X

Yadav P, Ismail N, Essalhi M, Tysklind M, Athanassiadis D, Tavajohi N (2021) Assessment of the environmental impact of polymeric membrane production. J Membr Sci 622. https://doi.org/10.1016/j.memsci.2020.118987

Chapter 7
Membrane Filtration for Special Applications

Udhayarasu Balamurugan, Sanjay Raj Natarajan, and Shobana Sugumar

Abstract Membrane filtration has emerged as one of the most versatile and indispensable separation technologies across biotechnology, food processing, chemical manufacturing, environmental management, and healthcare. Its ability to selectively retain molecules, operate under mild conditions, and integrate seamlessly into continuous processes has positioned membranes at the core of modern industrial practice. In the pharmaceutical and biotechnology sectors, membrane systems enable critical operations such as cell harvesting, microbial removal, protein fractionation, and nucleic-acid purification. Their precision and reliability support both upstream and downstream bioprocessing, ensuring product purity while reducing processing time and energy consumption. In food and beverage industries, membranes contribute significantly to the quality and safety of dairy products, fruit juices, sugar solutions, and wine, offering efficient clarification, concentration, and pathogen reduction without compromising nutritional or sensory attributes. Within the oil, gas, and chemical industries, advanced membrane modules—including air-gap configurations, hydrophobic contactors, aquaporin-based membranes, and organic solvent nanofiltration—offer selective separation under demanding conditions. These innovations enhance hydrocarbon treatment, desalination, metal recovery, and catalytic conversions, aligning industrial operations with sustainability and resource-efficiency goals. Semiconductor manufacturing relies on ultraclean water and defect-free processing environments, where ceramic nanofiltration, integrated adsorption–membrane systems, and hybrid purification units help maintain stringent purity standards. In environmental monitoring, membrane-based sensors and passive samplers provide rapid, sensitive detection of pollutants in water, air, and soil, while gas separation membranes enable compact solutions for air-quality management. Medical applications of membranes remain among the most impactful. Dialysis, membrane oxygenators, targeted drug-delivery systems, and bioartificial organs demonstrate

U. Balamurugan · S. Sugumar (✉)
Department of Biotechnology, College of Engineering and Technology, SRM Institute of Science and Technology, Kattankulathur, Tamil Nadu, India
e-mail: shobanas@srmist.edu.in

S. R. Natarajan
Department of Biotechnology, Bharathidasan University, Tiruchirapalli, Tamil Nadu, India

A. Mandpe et al. (eds.), *Membrane Filtration and Membrane Bioreactors for Wastewater Treatment*, Advances in Wastewater Research,
https://doi.org/10.1007/978-981-95-6706-5_7

how selective mass-transfer principles can directly support human health and therapeutic innovation. Overall, this chapter provides a concise synthesis of the wide-ranging applications of membrane filters, highlighting their technological significance and growing relevance in sustainable manufacturing, environmental stewardship, and advanced healthcare systems.

Keywords Bioprocessing · Environmental monitoring sensors · Membrane filtration · Medical applications · Membrane-based separation system · Next-generation membranes

Abbreviation

MF	Microfiltration
NF	Nanofiltration
HFM	Hollow-fiber membrane
ED	Electrodialysis
RO	Reverse osmosis
UF	Ultrafiltration
HRR	Higher release rates
TFF	Tangential flow filtration
NFF	Normal flow filtration
MCR	Medium-concentrated retentate
LPC	Liquid pre-cheese
BOD	Biochemical Oxygen Demand
WPC	Whey protein concentrate
WPI	Whey protein isolates
UHT	Ultra-High Temperature
HMW	High molecular weight
SS	Suspended solids
RO	Reverse osmosis
SPM	Semipermeable membrane
PV	Pervaporation
ED	Electrodialysis
MBR	Membrane bioreactors
MC	Membrane contactors
PPESK	Polyphthalazine ether sulfone ketone
O&G	Oil and Gas
AGMD	Air gap membrane distillation
MD	Membrane distillation
H2S	Hydrogen sulfide
GO	Graphene oxide
MCDI	Membrane capacitive deionization
OSN	Organic solvent nanofiltration

AOP	Advanced oxidation processes
CMP	Chemical mechanical polishing
Mxene	2D transition-metal carbides
IPA	Isopropyl alcohol
GPMCP	Gas-permeable membrane-based conductometric sensing device
OC	Open-chamber
PTFE	Polytetrafluoroethylene
DO	Dissolved oxygen
VOC	Volatile organic compounds
DAC	Direct air capture
ESRD	End-stage renal disease
kDa	Kilo Dalton
L	Liter
ECMO	Extracorporeal membrane oxygenation

7.1 Introduction

Membrane filters are synthetic filters composed of materials such as polyamide, polypropylene, and cellulose acetate that form a complex network of interconnected channels. They are employed in a variety of applications for filtration and exhibit a higher flow rate than other filters (George et al. 2017). The survival rate of tissues and layers of cells that are cultured is also guaranteed by the use of filter membranes as culture substrates. Porous filter membranes allow for the diffusion of culture medium and the transfer of cellular waste products from the cell mass to the membrane, in contrast to nonporous culture surfaces. Membrane filters are developed from a variety of fabricated raw materials like polyamide and polytetrafluoroethylene polypropylene. With microporous membranes, insulin, bovine serum albumin (BSA), and immunoglobulin G (IgG) can also be recovered (Mukherjee 2019). Our discussion in this chapter has encompassed the diverse applications of various membrane varieties and they are as follows:

7.2 Pharmaceutical and Biotech Industries

The waste generated during downstream processing in pharmaceutical manufacturing is significantly influenced by separation technologies and requires mitigation. To remediate effluent from pharmaceutical facilities, membrane filtration techniques such as nanofiltration and reverse osmosis are being used in conjunction with bioreactors and advanced oxidation processes. Nanofiltration with organic solvent-stable membranes can facilitate solvent exchange and catalyst recovery during the organic solvent-based drug synthesis of pharmaceutical compounds/intermediates.

Membrane technologies are frequently employed to facilitate separations, which are crucial at nearly every stage. The four essential steps that are used in microfiltration and are pertinent to this chapter are as follows:

7.2.1 Cell Repossession and Recovery

To achieve optimal cell number and efficiency, it is advisable to maintain the cells within the bioreactor or implement a recycling mechanism, regardless of whether the biotechnology process is conducted continuously or occasionally. Cell repossession and recycling can be achieved by size-based membrane filtration or sedimentation procedures. These techniques are necessary to maintain cell viability. However, centrifugation and depth filtration have emerged as the most often used methods for the final cell harvest. The separation of the membranes can be classified into two categories: external, which involves recycling the cells using a membrane section, and internal, which involves keeping the cells utilizing an immersed membrane (Carstensen et al. 2012).

Industrial-scale production commonly employs a sequence of culture flasks and bioreactors that progressively increase in size. To provide the final bioreactor with media containing a high concentration of cells, the second-to-last tissue culture fermentor is frequently operated in a partial perfusion mode. A hybrid perfusion approach, which involved a fed-batch operation after retaining cells using an HFM device, was demonstrated to increase productivity in a study. By simultaneously removing inhibiting substances, providing nutrients, and restoring optimal environmental conditions, perfusion culture guarantees uninterrupted cell growth and production.

7.2.1.1 Cell Harvesting

The cells must be harvested from the growth medium after the fermentation or cell culture has concluded. The cells frequently eradicate antibiotics, monoclonal antibodies, and small metabolites, while microbes generate recombinant proteins that are stored within the cells. The three primary methods employed to extract cells and refine culture broths are centrifugation, depth filtration, and microfiltration. MF has been employed as the initial stage in the recovery procedure for numerous biologics and monoclonal antibodies in the past. It is selective, and laboratory experiments can be conducted with minuscule quantities of it.

Conventional methods for clarifying mammalian cell culture harvests include centrifugation, tangential flow filtration (TFF), and normal flow filtration (NFF), either alone or in combination (Gupta et al. 2020). To explain things further, depth filtration is often used after centrifugation or microfiltration. This is because it would take a lot of expensive filter media to separate cells from high-density culture broth in the first place. Also, to get an internal product later on, you need a cell paste that has been compressed. This can be done by centrifuging or microfiltration (Tomic et al. 2015).

7.2.1.2 Product Fractionalization

For the first step, remove the product liquid from the cells once the drug of interest has been pushed out of the cells. This step is pretty much the same as the step where cells were collected and water was cleaned up in the last section. Keep in mind that the cell extract that comes after crossflow MF usually has a lot more liquid than 50% by volume. This is because the mix is very thick and needs to be able to move. To study an extracellular drug, which means it leaves cells, the first step is to get the resulting liquid out of the cells. Before this step, we talked about collecting cells and making the soup clear. This step is pretty much the same. The slurry is very thick and more than half of its volume is liquid, so it needs to be able to run. Also, it might be even more important to rinse or do diafiltration if the membrane doesn't let all of the soluble product through (i.e., the product transfer is less than 100%). The percolate stream, on the other hand, has less product because it has been cleaned or filtered more than once. This makes the next steps in the purification process harder (Aires-Barros and Azevedo 2017). Centrifugation provides the advantage of producing higher cell concentrations; however, microfiltration is straightforward to operate in a diafiltration mode for the rinsing and buffering phases (Lee et al. 2019). There are two categories of filtration procedures.

1. **Conventional filtration** is typically used for the following applications: sterile filtration (removing microorganisms), air, gases, medium, and buffer filtration; separation of cells/cell debris from extracellular products dissolved in the liquid (antibiotics, steroids); removal of cell debris from lysed cells (intracellular products). Particles with sizes ranging from 1000 to 0.1 mm are separated using the filtration method (Aires-Barros and Azevedo 2017) (Fig. 7.1).
2. **Tangential flow or Crossflow** with sizes ranging from 10 to 0.0001 mm, tangential filtration (membrane processes) is typically used for the concentration of cells and protein solutions, removal of viruses and extracellular products, and exchange or removal of salt or salts in a protein solution. To separate inclusion bodies from soluble proteins (host-cell protein, or HCP) and other compounds released during lysis, MF can also be utilized in conjunction with inclusion bodies. Once separated from the remaining cell detritus, the solubilized protein product in the inclusion bodies can be folded again and purified (Madsen et al. 2022).

7.2.1.3 Microbe Removal

The final application of MF membranes that is being investigated in this study is sterile filtering, which is used to eliminate microorganisms during the fabrication and recovery of biotechnology-based medications. The extraction and analysis of

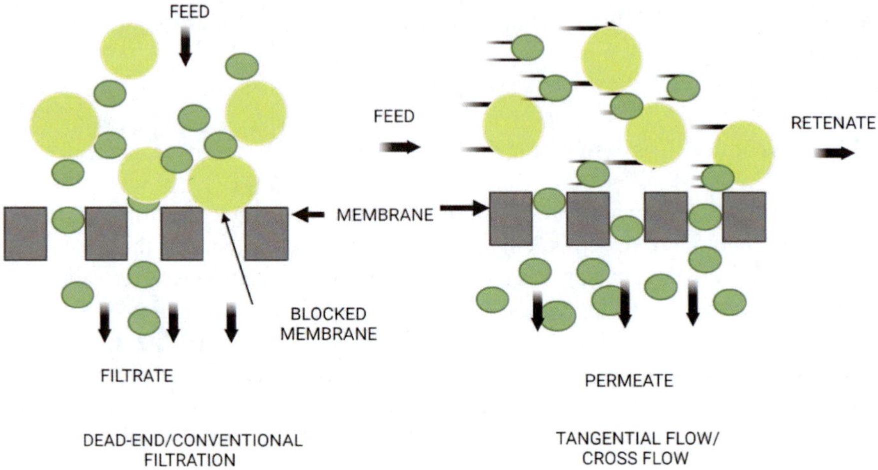

Fig. 7.1 Difference between the feed flow of dead-end/conventional versus tangential flow/crossflow

microorganisms from a variety of environments, such as solid surfaces, liquids, and biological samples. This phase is widely practiced in the pharmaceutical technological advancement industry, and it is frequently implemented on multiple occasions throughout the process (Norrrahim et al. 2021). To prevent the introduction of objectionable microorganisms into the fermentation media or cell culture, feed supplies and buffers are initially filtered to ensure sterility. Secondly, sterile filtering is occasionally conducted between intermediate phases. Sterile filtering is frequently employed after the restoration stages to eliminate any pathogens before the final formulation, thereby ensuring that the product is stable and free of contaminants.

7.2.2 Purification and Concentration of Proteins

A membrane is an interphase that functions as a barrier to the passage of molecular and ionic species in the liquids and/or vapors that come into contact with the two surfaces. Membranes have garnered the interest of chemists, chemical engineers, and biotechnical engineers over the past three decades as a result of their distinctive separation principle, which involves the selective transport and efficient separation of materials in comparison to other unit operations. In contrast to other thermal separation processes, membrane separations do not necessitate additives and can be conducted isothermally at low temperatures with reduced energy consumption. The up-scaling and downscaling of membrane separation processes, as well as their integration into other separation or reaction processes, are easily accomplished as a result of the distinctive properties of membranes (Saxena et al. 2009).

7.2.3 *Isolation and Purification of Nucleic Acids (DNA, RNA) Using Membrane Filters*

The objective is to ensure that molecular diagnostics are accessible to all by developing extraction methods that are based on paper or membranes and can be used in point-of-care (POC) devices. The capacity to multiplex, increased sensitivity, and the epidemiological monitoring of disease transmission and drift are among the numerous advantages of NA sequencing. A proof-of-concept system for DNA purification and concentration was developed in porous membrane substrates using the linear polysaccharide chitosan. It is widely acknowledged that polymers can be used to engineer or modulate surfaces by altering their properties, such as charge (Byrnes et al. 2015).

7.3 Food and Beverage

Membranes are widely utilized in the food and beverage sectors. The dairy industry, along with beverages such as wine, beer, and fruit juices, as well as sugar products, are the primary areas where these technologies are used. This chapter provides a discussion on various frequent uses (Table 7.1).

7.3.1 *Dairy Industries*

In dairy industries, different products are made from milk in different ways; Buttermilk, fresh cheese, Camembert, miniature sweet cheese, and other foods that include milk fat globule membrane (MFGM) fragments can be more concentrated by using an effective technique called microfiltration to extract the globules from whole milk (Manoni et al. 2020). A fat globule emulsion in an aqueous phase similar to milk.

Table 7.1 List of food products that utilize different types of membranes

Product	Membrane process	Process
Beer, wine, milk, fruit juices	MF	Dealcoholization, drying, thickening
Milk, cheese, whey	MF, NF, UF	Cold sterilization
Lactic acid, Citric acid	UF, ED	Product recovery
Aromas, flavors	PV, RO	Product improvement
Desalted cheese, soft water, potable water	RO, ED, NF	Desalination
Proteins (eggs, whey, blood),carbohydrates	UF	Fractionation

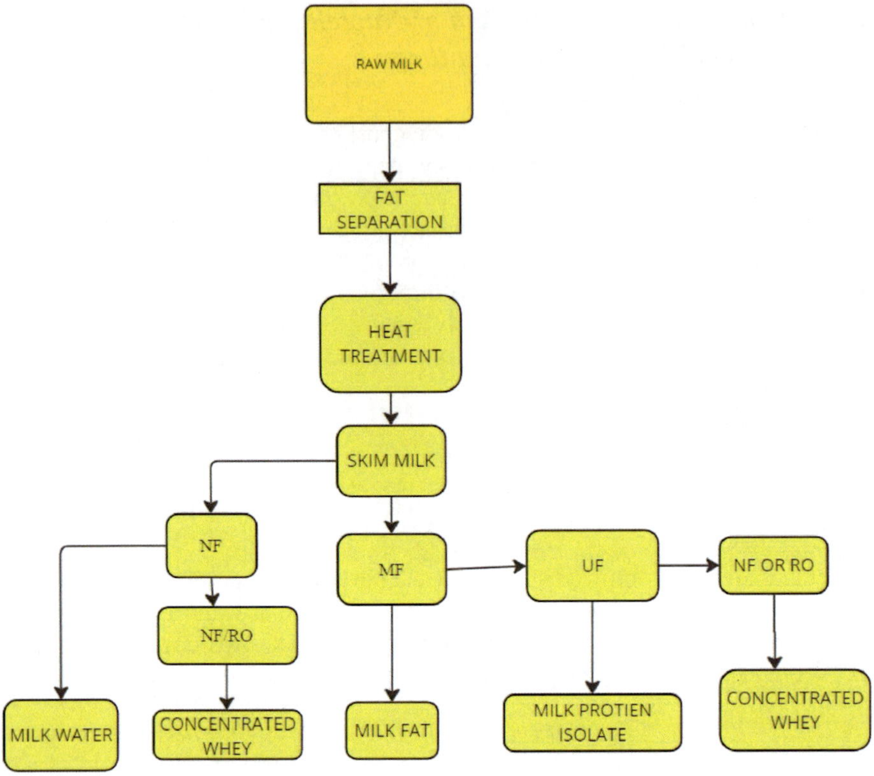

Fig. 7.2 Work flow of dairy products manufacturing from raw milk (Reig et al. 2021)

Aqueous phase constituents include lactose, casein micelles, serum proteins, and salts, among other suspended and dissolved substances. The primary component of milk fat globules with a diameter ranging from 0.1 to 15 μm. Small globules are typically less than 2 μm in diameter, while giant globules are defined as having a diameter greater than 2 μm. The fractionation of milk lipids according to pore size is facilitated by MF technology, which is capable of a multitude of applications (Postelmans et al. 2020) (Fig. 7.2).

7.3.1.1 Integrated Systems in Cheese-Making

One of the most consolidated membrane-based processes in the dairy industry is the UF of milk for cheese production. The primary benefits of milk concentration by UF are improved equipment operations, increased cheese yields, and the reduction of rennet consumption (Possas et al. 2021). Based on the specific cheese variants, milk ultrafiltered at LCR is produced. Despite the slight increase in protein content (4.5–5%) at these concentration factors, it is possible to completely standardize milk to determine the variations in milk composition. MCR can be produced from the preparation of feta cheese, gouda cheese, and Havarti cheese at concentration factors spanning from 2 to 6. The production of a liquid pre-cheese (LPC) at higher concentrations, which is equivalent to the final cheese's composition, prevents whey syneresis and reduces whey drainage (Faion et al. 2019). The UF aids in the permeation of lactose and soluble ions along with WPC in skimmed milk. Concentrated casein micelles and cream can be combined to produce cheese. NF is employed to concentrate it (Fadavi and Beglaryan 2015).

7.3.1.2 Whey Processing

Whey is a derivative of the cheese industry that is difficult and expensive to dispose of due to its high biological oxygen demand (BOD5 = 30–50 g/L for 1000 L of whey) and low solids content (up to 5–6%). Classified among the production of whey involves crucial applications in the dairy industry's membrane technology. A series of various membrane procedures can be employed to achieve whey concentration, fractionation, demineralization, and purification. Simultaneously, there is substantial interest in the creation of premium products derived from whey compounds. When moderate thermal treatment is applied to lipoproteins, calcium bridging results in the formation of aggregates. These agglomerated compounds can be removed using MF with a pore size of 0.14 μm (Baldasso et al. 2011).

7.3.1.3 Membrane Filtration Is an Alternative Efficient Way of Pasteurization

The skimmed milk needs to go through UHT and pasteurization procedures before homogenization. Although UHT is a more successful treatment than pasteurization, because it is performed at a higher temperature, it may also be more harmful to the characteristics of milk. As a result, MF is a better option than UHT for eliminating germs and spores from milk without changing the milk's chemical or organoleptic qualities. Membrane processes, in addition to classical processes, are employed in the dairy industry for a variety of applications, including the concentration of casein micelles, the separation and/or fractionation of fat globules from whole milk, the removal of bacteria and spores from skim milk (cold pasteurization), the separation of casein micelles from soluble proteins contained in skimmed milk, and the recovery of serum proteins from cheese whey (Song et al. 2018).

7.3.2 Fruit Juices

Fruit juice production is the most highly valued industry economically. The fruit juice industry's general production flow commences with the grinding or crushing of the fruits to an optimal and homogeneous particle size, followed by the extraction of the fruit mash. The fruit juice is concentrated to minimize the costs of transportation and storage after clarification/fining (Ahmad and Ahmed 2014). The consumption of fruit juices has significantly increased in recent years as a result of the perception that juices are a healthful, natural source of nutrients. The primary objective of MF in the fruit juice processing industry is to clarify the juice by removing suspended solids (SS), lipids, and high molecular weight (HMW) proteins. In addition, microfiltration can be employed to separate fruit juices into a fibrous concentrated purée and a clarified fraction that is free of spoilage microorganisms. The fruit juice processing industry extensively employs MF for beverage clarification (Bhattacharjee et al. 2017). Solutes including pigments and organic acids are present in fruit fluids with low molecular weight and High molecular weight solutes, including polysaccharides like hemicelluloses, lignin, protein, and starch (Echavarría et al. 2012).

7.3.2.1 Fruit Clarification

Key technologies in the food industry include pressure-driven membrane operations, including microfiltration (MF), ultrafiltration (UF), nanofiltration (NF), and reverse osmosis (RO). The basic properties of these processes render them suitable for the treatment of both food products and by-products. These processes offer a variety of advantages over conventional separation technologies, high selectivity, energy saving with environment-friendly and easy scale-up (Katibi et al. 2023). Other membrane processes, such as MC, PV, and MBRs, are employed in this food industry (Cassano 2016).

The desire for a clarification process depends on how the fruit juice will be used. For instance, the manufacturing of sweets by melting products, pastries made with natural essences, translucently made sauce from fruits, and clear beverages including soft drinks, aromatic waters, alcohol-based beverages, cold teas, etc., all require clarified fruit juices. As a part of conventional clarification methods sequential procedures such as diatomaceous earth clarification, and bentonite are used. In the 1970s as an alternative solution to conventional fruit clarification processes membrane filtration and ultrafiltration were first introduced (Charcosset 2021).

7.3.3 Wine Production

Grapes are fermented to produce wines, which are alcoholic beverages. Their makeup is dictated by the grape's makeup, which is influenced by genetic traits, vine growth circumstances, grape ripeness at harvest, and winemaking techniques. A significant element of winemaking, particularly in wine clarity, is played by suspended particles. Cell debris, colloidal aggregates, tartrate crystals, and microorganisms (such as yeast and bacterial chains) are all included in this group. Although the filters catch them, they could inadvertently end up in the finished wine (El Rayess et al. 2011).

In CFMF, a pressure drop allows the fluid to pass through the membrane while flowing parallel to its surface. To maintain a reasonably thin cake layer, the deposited particles may be swept toward the concentrate or retentate side by the shear that the feed solution's parallel flow to the membrane surface exerts. In the instance of CFMF, a layer of cake will progressively accumulate (Jordão et al. 2023). A quasi-steady state is eventually established, and a long-lasting quasi-steady filtrate flow is produced. The equilibrium between the movement of particles into the cake layer and the return movement of particles into the feed stream is responsible for this occurrence. The liquid amount that permeates the membrane is known as permeate, and the solvent and molecules that are retained make up the retentate, which gradually becomes concentrated throughout filtration cycles. In many food sectors today, including those that produce milk, cheese, wine, vinegar, beer, juice, and vegetable oils, this manner of operation is standard.

7.3.4 Sugar Production

When producing sugar, sugarcane and beets have been utilized as raw materials.

The process of producing sugar from beet sugar includes demineralization of beet juice, recycling of pulp, purification of raw juice, and pre-concentration of thin juice; however,

in the case of the purification of raw sugarcane juice, the concentration of clarified cane juice, the treatment of molasses, and the decolorization of remelted raw sugar in the sugarcane business are followed (Luo et al. 2016). Conventional purification treatment entails the addition of sulfur dioxide and lime, followed by the boiling of the treated juice. The resultant sludge is subsequently removed through vacuum filtration and sedimentation process. To resolve the aforementioned issues, alternative procedures were investigated, including electrodialysis, membrane filtration, activated carbon adsorption, and ion exchange. Membrane filtration, particularly ultrafiltration (UF), is regarded as the most promising method among these methods due to its ability to eliminate the use of chemicals, accomplish continuous and automated production, and produce juice of superior quality. The poly phthalazine ether sulfone ketone (PPESK) membrane is a thermostable material that is frequently employed in high-temperature applications. In general, the permeate permeability of the PPESK membrane increased as the temperature increased, while the retention began to decrease when the temperature exceeded 60 °C (Luo et al. 2016).

7.4 Oil and Gas Industries

Membranes have a diverse array of applications on a global scale. The processes and/or membrane material development described below are still in the early stages of development, and deployment will probably require an extended period.

7.4.1 Air Gap Membrane Filtrations

The oil and gas industry faces a significant challenge in obtaining copious amounts of pure water, particularly in arid and coastal regions. During the extraction of oil or gas, brine or saline water from subterranean formations is brought to the refinery. After treatment, it is used as an alternative and sustainable method in oil and gas.

AGMD is characterized by a variety of appealing attributes, including the ability to operate at lower temperatures than conventional processes; the primary solution, water, is not necessarily heated to the boiling point. Additionally, the HSP encountered in AGMD is less than that of pressure-driven membrane processes such as RO. AGMD is very appropriate for desalination and the removal of volatile compounds from aqueous solutions (Khayet and Cojocaru 2012).

7.4.2 Pressure-Retarded Osmosis with FO Membranes

Traditional phase inversion and thin-film composites (TFC) were employed to fabricate membranes through interfacial polymerization procedures. The reverse salt fluxes of FO membranes produced through the layer-by-layer method have been examined; however, they are generally very high. It is imperative to employ hydrophilic materials as substrates for FO membranes to improve water flux (Chung et al. 2015). During PRO operations, the majority of FO membranes may collapse or undergo significant deformation. Flow channels are preserved and mass transfer is enhanced by the use of spacers to support PRO flat-sheet membrane modules. In addition to causing hydraulic pressure losses along the flow channels, the feed separators inevitably deform PRO membranes during high-pressure operations (Adham et al. 2018). Therefore, this hypersalinity represents a high osmotic pressure. The fouling process in PRO membranes is more intricate than that in FO because the feed stream is exposed to porous substrates during PRO operations. Furthermore, the reverse salt flux has the potential to exacerbate fouling and confound fouling mechanisms. We are in urgent need of PRO membranes that possess superior antifouling properties, can withstand higher pressures, and have a higher power density (Adham et al. 2018).

7.4.3 H2S Removal with Hydrophobic Membrane Contactor

The removal of H2S from acidic water is one of the primary challenges in the O&G industry concerning water management. For a variety of reasons, such as environmental regulations, health and safety concerns, and corrosion, it is necessary to remove H2S, a toxic gas, from produced water before any treatment or reuse. To reduce the likelihood of reservoir souring, it may also be necessary to remove H2S before reinjection into disposal wells. Currently, the standard method for removing

H2S is the use of acidic water strippers. Membranes may be employed as an alternative technology for the remediation of acidic water. It is pH-dependent, as the speciation of H2S changes with pH. The removal rates were substantially higher at pH 4 or lower, as all sulfide species are present as H2S. The rate of H2S removal was greater than 98%, and no significant membrane contamination was observed. The concentrations of H2S in the feed and effluent were 98.4 mg/L and < 2 mg/L, respectively (Adham et al. 2018).

7.4.4 Next-Generation Membranes

7.4.4.1 UF Membranes

Graphene oxide (GO) nanosheet is one of the most highly regarded 2D constituents for the development of membranes with high permeability due to its extraordinary atomic thickness and superior amphiphilic properties. Ultrafiltration (UF) is a membrane filtration procedure that employs semipermeable membranes to separate particles according to their size. The separation through a semipermeable membrane is facilitated by pressure or concentration gradients in UF. The membrane functions as a barrier, permitting the passage of water like LMW solutes such as ions, while retaining suspended particulates and HMW solutes such as proteins in the retentate. UF is capable of removing macromolecules and particulates from fresh water, thereby generating potable water. It is employed in water treatment facilities or as a standalone system in remote regions. UF is utilized in the food and beverage industries, pharmaceutical industries, and dairy industries for recycling and value addition. UF is also employed in the procedure of blood dialysis (Yang et al. 2016).

7.4.4.2 Aquaporin Membranes

Aquaporin are proteins that serve as channels that enable the movement of water across the membranes. They are a vast family of proteins that facilitate the transportation of water across cell membranes. The thin-film coating that encases the aquaporin proteins ensures that their natural activity is preserved during water treatment. The co-occurrence of nano-oil droplets and small surfactant molecules continues to complicate the oil/water nano-emulsion separation process. The superior amphiphilic properties and exceptional atomic thickness of graphene oxide (GO) nanosheet render it one of the most exceptional 2D constituents for the development of membranes with high permeability. The development of novel graphene-based membranes was facilitated by the advancement in molecular simulation of graphene. The nano-pores that were examined exhibited salt rejection and water permeability that were 2–3 orders of magnitude greater than those of commercialized membranes, as determined by simulation (Hegab et al. 2024).

7.5 Chemical Industries

7.5.1 Desalination

In a variety of applications, including desalination, wastewater treatment, food processing, and fuel cells, ion-exchange materials, and membranes are now an established and integral element of separation processes. Brackish water has a salinity content of up to 10,000 ppm, while seawater typically has a salinity of 35,000–45,000 ppm in the form of total dissolved salts. Consequently, these waters are unsuitable for ingesting and the majority of domestic purposes (Ali et al. 2018). MCDI has endeavored to challenge benchmark methods, including reverse osmosis, which involves the application of a voltage between two electrodes that are covered with ion-exchange membranes to remove salt under ambient conditions. It has been demonstrated that the desalination performance is optimized in terms of energy efficiency and salt removal when ion-exchange materials are incorporated over electrodes in MCDI (McNair et al. 2021).

7.5.2 Organic Solvent Nanofiltration

Organic solvent nanofiltration is a membrane separation technology that works by applying pressure and is capable of separating small molecules from organic solvents. In ultrafiltration and reverse osmosis, OSN membranes are capable of identifying nanometer-sized molecules in organic solvents due to their pore size distributions of 0.5–2 nm. Three primary kinds of OSN membranes can be broadly classified:

1. Polymeric membranes exhibit exceptional selective permeability; however, they are frequently subject to expansion in organic solvents.
2. Inorganic membranes, which are predominantly ceramic membranes, are capable of withstanding high temperatures and organic solvent vents.
3. Nanocomposite membranes combine the benefits of both Inorganic and polymeric membranes; however, they may encounter difficulties associated with the compatibility of the filler agglomeration and polymer matrix (Gu et al. 2024)

In organic solvents, the nanofiltration is capable of differentiated molecules within a molecular weight range of approximately 100–1000 Da. Molecular separation in non-polar solvent systems, such as n-hexane, n-heptane, and toluene, has become of predominant significance due to their widespread utilization in the pharmaceutical, petroleum, and natural gas industries (Rayaroth et al. 2022).

7.5.3 *Catalytic Membrane-Based Oxidation-Filtration Systems*

The novel technology for the effective biological treatment of organic containments in water treatment is catalytic membranes, which can simultaneously achieve physical separation and chemical oxidation in a single integrated system. Not only does the catalytic membrane considerably improve the efficiency of pollutant removal when combined with advanced oxidation processes (AOPs), but it also prevents membrane fouling through self-cleaning. The mechanisms and efficacy of the various integrated catalytic membrane systems, which are combined with photocatalysis, Fenton oxidation, persulfate activations, ozonation, and electrocatalytic oxidation, are examined (Li et al. 2021).

7.5.4 *Cation Exchange Membranes*

The metal species are significantly used by the advancement of industrial society as a direct consequence of the rapid consumption of natural resources and the discharge of hazardous waste. Transition metallic ions are the primary source for metal species, which can be found in an extensive variety of sources, including waste acidic substances and electroplating procedures. The term "industrial metals" is frequently used to describe these ions, including Ag^{+}, Cu^{2+}, Ni^{2+}, and Fe^{3+}. They are identified by having a significant amount of metallic elements that are alkali e.g., k^{+}, Na^{+}, and alkaline e.g., Li^{+}., Ca^{2+}, and Mg^{2+}). CEMs have the potential to selectively separate a single metal ion from a variety of effluents in electrodialysis, regardless of whether they are of the same or distinct valences. The construction and operational parameters of the electrodialysis management, in addition to the inherent characteristics of membranes, all contribute to the specificity of metal cations. It is possible to manipulate the transport ratio of competing counter-ions by exploiting variations in the mass transport of ions at interfaces across the membrane (Tekinalp et al. 2023).

7.5.4.1 Polymer Blending in Membranes

Due to its straightforward preparation process, capacity to integrate advantageous characteristics into the membrane, and substantial capacity to concurrently alter the membrane's characteristics during the phase-change process, this blending approach has been extensively employed in the manufacturing of membranes made of polymers. Hydrophilic compounds, such as inorganic nanoparticles, hydrophilic polymers, and amphiphilic copolymers, were initially introduced to enhance the antifouling and flux performance (Nasrollahi et al. 2022).

7.6 Semiconductor

7.6.1 Mxene-Based Ceramic Nanofiltration

Semiconductor manufacturing is a technologically advanced industry that strictly enforces water quality standards and consumes a significant quantity of water. The semiconductor manufacturing process generates wastewater that contains toxic and hazardous compounds, necessitating appropriate treatment to prevent potential environmental pollution. In particular, the wastewater generated by the chemical mechanical polishing (CMP) process, which is employed for wafer polishing, contains elevated concentrations of dissolved and particulate silica, with a value of approximately 120 mg/L. The eliminating efficiency of organic pollutants in semiconductor wastewater was significantly improved by these MXene-coated ceramic NF membranes in comparison to the pristine ceramic membrane. It increased the removal efficiencies of dissolved silica by 4.2 times, dimethyl sulfoxide (DMSO) by 3.7 times, and isopropyl alcohol (IPA) by 2.5 times.

7.6.2 Performance of Integrated Membrane Filtration

Chemical Mechanical Polishing waste streams contain nano-scale particles, such as alumina (Al_3O_4) or silica (SiO_2), as well as compounds used to control surface etching and slurry dispersion. A modified ED system is employed to improve the adsorption efficacy for the adsorption of Cu-CMP heavy metals. This is because it can reduce the occurrence of issues such as blocked ion-exchange resin, biofilm formation through microbial activities, and membrane fouling. The feasibility of Cu removal and recovery from wastewater was demonstrated by the successful treatment of Cu-CMP wastewater by the combination of membrane systems, resulting in a level that readily satisfies the local discharge regulations (3 mg/l). The residual water quality of the UF/ED-treated Cu-CMP waste waters was found to be suitable for reuse as feed water for UPW production, thereby improving the water reuse efficiency of wafer fabrication processes (Su et al. 2014).

7.6.3 Integrated Membrane-Adsorption

The semiconductor industry necessitates the utilization of ultrapure water in the fabrication of silicon wafers to eliminate organic impurities, metals, and particles. Membrane filtration units are primarily employed for the treatment of wastewater and along with the recovery of silica, while GSCs are additionally employed for the polishing and treatment of semiconductor industry effluent. The rejection of various water quality parameters is used to assess the efficacy of the integrated membrane-adsorption system. GSC adsorption was conducted at three distinct bed heights 2 cm, 4 cm, and 6 cm, and ceramic ultrafiltration membranes with pore sizes of 0.1 μm and 0.02 μm were evaluated. The rejection rate of TSS is 99.25%, ammonium-nitrogen is 93%, and COD is 100% greater for a 0.02 μm membrane than for a 0.1 μm membrane.

7.7 Environmental Monitoring

The primary objective of the development of new BOD sensors with oxygen electrodes has been to enhance commercial viability and optimize sensor structures. This includes reducing operating and maintenance costs and enhancing the stability and correlation between biosensor data and standard BOD assessments. The cellular reagents can be injected and removed through a connecting conduit by a certain thickness of internal void between the polytetrafluoroethylene (PTFE) membrane of the DO probe and a dialysis membrane. The steady-state method and the dynamic transient method are the primary methods used to process oxygen electrode signals. The BOD is determined by the former, which compares the steady-state current values before and after the addition of the sample. The response time of these sensors is contingent upon the temperature, the number of microbes, and the thickness of the biofilm. Conversely, the sensor's functionality is contingent upon the concentration of DO (Zhou et al. 2024).

Membrane-based monitoring can be implemented through direct sampling where the synchronous analysis is gas chromatography, mass spectrometry, and spectroscopy for the detection of organic compounds. This is due to their capacity to separate and identify multiple substances from mixtures simultaneously and detection of contaminants in field conditions, or passive sampling and detection, which involves the steady-state collection of pollutants from the field environment and subsequent laboratory analysis. This section introduces these categories.

(A) **Direct sampling and detection**

Direct sampling and detection is a two-in-one monitoring method that integrates membrane technique with synchronous analysis technique to achieve accurate substance monitoring. This method involves the simultaneous sampling and detection of an analyte in a single monitoring device. To quantitatively and qualitatively identify and detect specific compounds in the environment, physical, chemical, and biological methods are frequently combined with a membrane technique.

(B) **Membrane-based electrochemical Sensors**

At present, these are the most frequently employed environmental monitoring devices. These devices primarily consist of membrane-based potentiometric sensing devices, amperometric sensing devices, and conductometric sensing devices. In this type of sensor, the shift of ion equilibrium across the membrane is represented by a specific membrane potential that can be quantitatively measured. This measurement allows for the calculation of the analyte concentration using the Nernst equation. In the past three decades, there has been a significant increase in the development and improvement of three primary categories of ion-exchange membrane sensors: crystal electrodes with solid membranes, ion-exchange membranes with liquid membranes, and carrier electrodes (e.g., enzyme electrodes). These sensors are capable of detecting a variety of environmental elements.

(C) **Passive sampling and detection**

The monitoring substances can be accumulated on-site using in situ passive sampling technology, without any impact on the surrounding environment. The concentration of the analyte that has been accumulated in the passive sampler can accurately represent the actual concentration or the time-weighted concentration (Gao et al. 2020).

(D) **Membrane-based amperometric detection**

It employs a conventional electroanalytical technique, as evidenced by voltammetric or potentiometric measurements, to precisely identify electroactive groups in a sample. It typically comprises an electrolyte, an auxiliary electrode, and a working electrode. This electrolyte facilitates a rapid, reversible, and redox reaction (Belluzo et al. 2008).

7.7.1 Air Quality Monitoring

Technologies based on membranes are used to monitor air pollutants such as gaseous species, particulate matter, and volatile organic compounds (VOCs). Real-time monitoring of air quality is made possible by membranes, which aid in the selective penetration of specific molecules. The threat or injury to the environment and humanity is mitigated by the prompt identification of pollution and its sources. Membranes are becoming a proficient technology for the rapid identification of polluting gases. It is possible to design them to selectively transmit the target analyte because their functionality and porosity can be customized. This provides a dual benefit: firstly, it decreases the volume of analyte that the sensors will detect; and secondly, it safeguards the sensors from deterioration when exposed to corrosive gases, thereby prolonging their service life. The diffusion of gases through membranes is the basic premise of sensing technology (i.e., membrane-based gas sensors) (Li et al. 2023) (Table 7.2).

7.7.1.1 Membrane-Based Gas Sensors

Conventional direct gas sensing methods have many with low selectivity or sensitivity. In such situations, membrane-based gas sensors are suitable substitutes. The

Table 7.2 Membrane-based technologies involved in air cleaning

Membrane technology	Description	
Hollow-fiber membranes	Used in air purification and also used in bioreactors for wastewater treatment	Ignatusha et al. (2024)
Nanofibrous membranes	Recent advancements indicate that the eradication of a variety of air pollutants can be achieved in an effective manner	Kang et al. (2024)
Gas separation membranes	Used in Direct air capture (DAC) technology, complementing with Sorbent-based approaches	Mata et al. (2022)

selective membrane enables gas molecules to enter the recognition unit matrix by passing through the membrane. The accuracy of the sensors may be compromised by membrane degradation, which is a potential disadvantage of these devices (Li et al. 2023).

Chemical and biosensors are two categories for membrane-based gas sensors, the recognition that can able to find which kind of reaction is occurring. The recognition unit interacts with the infiltrated gas molecules through appropriate chemical or biological interactions, leading to in a quantifiable signal that the transducers sense and send. Based on the type of transducer, we can divide the gas sensors into two categories: Biosensors and electrochemical.

7.7.1.1.1 Electrochemical-Based

The gaseous state molecules penetrate the membrane and reach the recognition unit in all electrochemical sensors. The matrix of this unit consists of a cathode, anode, and a reference solution (electrolytic solvent or gels). When the gas molecule enters the electrodes' presence, it reacts chemically with the solution. Despite their high sensitivity, optical sensors are not widely used because of their comparatively costly cost.

7.7.1.1.2 Biosensors-Based

Permeated gas molecules in the recognition unit of biosensors go through a biological reaction that produces an electrical/optical signal that is picked up by transducers. Due to the extremely selective biological interactions involved, this technique is relatively more sensitive. Their compactness, exceptional sensitivity, and selectivity are suitable for on-field environmental monitoring (Li et al. 2023).

7.7.2 Soil

To remove and examine pollutants from soil and groundwater samples, membranes are employed. They offer a quick and easy method for concentrating and separating analytes in preparation for further detection using analytical techniques.

7.7.2.1 Monitoring of Soil Ammonia

Soil ammonia emission flux ($JNH3$) detection is made possible by on-site soil ammonia monitoring device by the embedding of a gas-permeable membrane-based conductometric sensing device (OC-GPMCP) in an open-chamber deployment unit, eliminating the need for frequent calibration. The discharge of nitrogen fertilizer on aquatic systems can easily monitored. The advancement of GCMCP is accurate, rapid, and so sensitive providing the critical study of emission of ammonia into the soil (M. Zhou et al. 2021). The solution-diffusion model states that a single gas passes through a dense tubular membrane in several steps. Gas is first absorbed from space at the outer membrane. Adsorption of the gas molecule results in absorption or desorption, depending on surface energetics. Absorption—a dissolving process—is the rate-limiting stage compared to rapid adsorption. Gas molecules inside the membrane distribute along the radius according to the concentration gradient (Lazik et al. 2009) (Fig. 7.3).

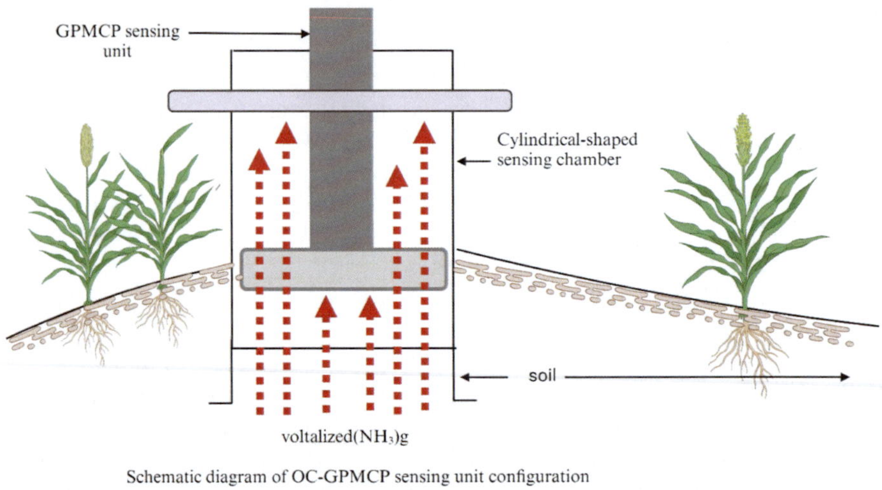

Fig. 7.3 Schematic diagram of OC-GPMCP

7.8 Membrane Applications in Medical Application

7.8.1 Dialysis

End-stage renal disease (ESRD) is characterized by the accumulation of metabolic waste products. The kidney eliminates metabolic-resulted toxins with molecular weights less than or equal to 58 kDa in a healthy human organism under normal conditions. The accumulation of fluids in the body leads to swelling of arms, legs, and even pulmonary edema, and in serious cases, organ failure is also possible. The objective of dialysis is to eliminate contaminants and excess water from the bodies of individuals with renal failure diseases (Lee and Son 2021). The renal replacement modality can be classified into three primary subsections based on the governing mechanism of solute removal from the main bloodstream: diffusion, convection, and hemodialysis. If diffusion is the controlling mechanism, the method is referred to as "hemodialysis." Renal dialysis involves two primary streams on opposite sides of the membrane: one that contains a greater quantity of targeted compounds and another that contains zero or lower concentrations (dialyzing fluid or dialysate). Same as dialysis, Hemofiltration involves a high ultrafiltration rate, equivalent to 20 L in a typical dialysis session (Mollahosseini et al. 2020).

7.8.2 Targeted Drug Delivery

It is also referred to as smart drug delivery, a technique that is designed to enhance the concentration of medication in specific regions of the body while simultaneously reducing its impact on healthy tissues. The osmotic-controlled drug delivery systems are used as the pulsed or delayed drug delivery of active constituents to the intestine based on the osmotic gradient, which is generated by the diffusion of water (Stamatialis et al. 2008).

The active agents are disseminated at the target site by being compressed out of the core through laser-drilled small holes due to increased hydrostatic pressure. The therapeutic benefits of OD delivery systems include the potential for HRR in comparison to conventional diffusion-controlled drug delivery systems, as well as zero-order delivery rates that are not affected by gastric pH or hydrodynamic conditions. This is followed by a continuous drug discharge that may persist for up to 24 h.

7.8.3 Membrane Blood Oxygenator

Blood oxygenators or artificial lungs are medical devices that perform the function of the lung by removing the carbon dioxide and oxygenating the blood. In open-heart surgery, they are frequently employed to provide temporal cardiopulmonary bypass. Additionally, they are occasionally employed for extracorporeal membrane oxygenation (ECMO) to provide short-term cardiopulmonary support for patients with respiratory failure, even though solid membrane-type oxygenators are the most prevalent in ECMO use (Wrisinger and Thompson 2022). The outer diameters of these hollow fibers range from 200 to 400 μm, and they are constructed with microporous walls that are 20–50 μm thick. For blood oxygenation, the membrane wall apertures remain gas-filled, allowing respiratory gases to diffuse easily across it, as hollow-fiber membranes are typically composed of hydrophobic polymers (such as polypropylene) (Zhang et al. 2007).

7.8.4 Bioartificial Pancreas

The pancreas' primary function is to regulate the balance of endocrine hormones like glucagon and insulin in the bloodstream, thereby ensuring glucose homeostasis. Diabetes is caused due to high blood sugar levels, as a result of pancreas failure, as the body is unable to regulate its insulin (Schweicher 2014). The primary cause of Type 1 diabetes, which predominantly affects children, is that the body's insulin secretion is insufficient to meet its needs. In a manner comparable to BAL, the bioartificial pancreas (BAP) technology integrates islet cell transplantation technology with membrane technology (Berney and Johnson 2010). A BAP is composed of islet cells that are enclosed in a semipermeable membrane capsule that functions as an immune system barrier. The SPM, which encases the islet cells, is capable of safeguarding the cells from immunoglobulin while simultaneously supplying them with the essential nutrients, oxygen, and glucose. The molecular weight (MW) of the smallest immunoglobulin, IgG, is 150 kDa. The optimal membrane MWCO (or pore size) to accomplish these objectives is approximately 50–150 kDa (it is important to note that solute MW alone should not be the sole determinant of selectivity) (Duy Nguyen et al. 2021).

7.8.5 Membranes for Artificial Liver

The membrane has the potential to be a critical component of artificial liver support systems. Hepatocytes are cultured in the majority of these systems, which are composed of flat sheet or porous fiber membranes that are made from a variety of polymers. To assist patients with liver failure, numerous artificial liver devices have been created. The objective of these devices is to temporarily supplant the detoxification functions of the liver. Maintenance of differentiated hepatocyte function for extended periods is the primary obstacle. The future challenge for membranologists will be the development of novel scaffolds and efficient bioreactors.

7.9 Conclusion and Future Gaps

Various industrial applications have increased as a result of the recent advancement in membrane technology, which has facilitated the efficient separation, purification, and filtration processes. The efficacy of membranes utilized in aqueous and organic fluxes could be enhanced by the integration of nanomaterials, including metal oxides, MOFs, (GO and CNT), COFs, zeolites, and nano-sized polymeric particles, with membrane fabrication methods, such as 3D printing. Furthermore, membrane technology provides a viable alternative to traditional separations and purifications in the following areas: energy production, environmental monitoring and quality control, food and beverage processing, fuel cells, and bioseparation systems. This is particularly true for chemicals and pharmaceutical processing (Fane et al. 2011).

References

Adham S, Hussain A, Minier-Matar J, Janson A, Sharma R (2018) Membrane applications and opportunities for water management in the oil & gas industry. Desalination 440:2–17. https://doi.org/10.1016/j.desal.2018.01.030

Ahmad S, Ahmed SkM (2014) Application of membrane technology in food processing. In Malik A, Erginkaya Z, Ahmad S, Erten H (eds) Food processing: strategies for quality assessment. Springer, New York, pp 379–394. https://doi.org/10.1007/978-1-4939-1378-7_15

Aires-Barros MR, Azevedo AM (2017) Fundamentals of biological separation processes. In: Current developments in biotechnology and bioengineering. Elsevier, pp 187–237. https://doi.org/10.1016/B978-0-444-63668-3.00007-X

Ali A, Tufa RA, Macedonio F, Curcio E, Drioli E (2018) Membrane technology in renewable-energy-driven desalination. Renew Sustain Energy Rev 81:1–21. https://doi.org/10.1016/j.rser.2017.07.047

Baldasso C, Barros TC, Tessaro IC (2011) Concentration and purification of whey proteins by ultrafiltration. Desalination 278(1–3):381–386. https://doi.org/10.1016/j.desal.2011.05.055

Belluzo MS, Ribone ME, Lagier CM (2008) Assembling amperometric biosensors for clinical diagnostics. Sensors 8(3):1366–1399. https://doi.org/10.3390/s8031366

Berney T, Johnson PRV (2010) Donor pancreata: evolving approaches to organ allocation for whole pancreas versus islet transplantation. Transplantation 90(3):238–243. https://doi.org/10.1097/TP.0b013e3181e25a40

Bhattacharjee C, Saxena VK, Dutta S (2017) Fruit juice processing using membrane technology: a review. Innov Food Sci Emerg Technol 43:136–153. https://doi.org/10.1016/j.ifset.2017.08.002

Byrnes SA, Bishop JD, Lafleur L, Buser JR, Lutz B, Yager P (2015) One-step purification and concentration of DNA in porous membranes for point-of-care applications. Lab Chip 15(12):2647–2659. https://doi.org/10.1039/C5LC00317B

Carstensen F, Apel A, Wessling M (2012) In situ product recovery: Submerged membranes vs. external loop membranes. J Membr Sci 394–395:1–36. https://doi.org/10.1016/j.memsci.2011.11.029

Cassano A (2016) Integrated membrane processes in the food industry. In: Basile A, Charcosset C (eds) Integrated membrane systems and processes, 1st edn. Wiley, pp 35–60. https://doi.org/10.1002/9781118739167.ch3

Charcosset C (2021) Classical and recent applications of membrane processes in the food industry. Food Eng Rev 13(2):322–343. https://doi.org/10.1007/s12393-020-09262-9

Chung T-S, Luo L, Wan CF, Cui Y, Amy G (2015) What is next for forward osmosis (FO) and pressure retarded osmosis (PRO). Sep Purif Technol 156:856–860. https://doi.org/10.1016/j.seppur.2015.10.063

Duy Nguyen BT, Nguyen Thi HY, Nguyen Thi BP, Kang D-K, Kim JF (2021) The roles of membrane technology in artificial organs: current challenges and perspectives. Membranes 11(4):239. https://doi.org/10.3390/membranes11040239

Echavarría AP, Falguera V, Torras C, Berdún C, Pagán J, Ibarz A (2012) Ultrafiltration and reverse osmosis for clarification and concentration of fruit juices at pilot plant scale. LWT Food Sci Technol 46(1):189–195. https://doi.org/10.1016/j.lwt.2011.10.008

El Rayess Y, Albasi C, Bacchin P, Taillandier P, Raynal J, Mietton-Peuchot M, Devatine A (2011) Cross-flow microfiltration applied to oenology: a review. J Membr Sci 382(1–2):1–19. https://doi.org/10.1016/j.memsci.2011.08.008

Faion AM, Becker J, Fernandes IA, Steffens J, Valduga E (2019) Sheep's milk concentration by ultrafiltration and cheese elaboration. J Food Process Eng 42(4):e13058. https://doi.org/10.1111/jfpe.13058

Fane AG, Wang R, Jia Y (2011) Membrane technology: past, present and future. In Wang LK, Chen JP, Hung Y-T, Shammas NK (eds) Membrane and desalination technologies. Humana Press, pp 1–45. https://doi.org/10.1007/978-1-59745-278-6_1

Gao L, Gao B, Xu D, Liu L (2020) DGT: A promising technology for in-situ measurement of metal speciation in the environment. Sci Total Environ 715:136810. https://doi.org/10.1016/j.scitotenv.2020.136810

George JH, Waller ST, Ye H (2017) 4.13 Use and manipulation of microporous membranes in mammalian cell cultures. In: Comprehensive membrane science and engineering. Elsevier, pp 272–292. https://doi.org/10.1016/B978-0-12-409547-2.12267-X

Gu S, Li S, Xu Z (2024) Organic solvent nanofiltration membranes for separation in non-polar solvent system. Green Energy Environ S2468025724000396. https://doi.org/10.1016/j.gee.2024.02.007

Gupta A, Amara JP, Gousseinov E, Cacace B (2020) Recent advances in harvest clarification for antibodies and related products. In: Approaches to the purification, analysis and characterization of antibody-based therapeutics. Elsevier, pp 117–136. https://doi.org/10.1016/B978-0-08-103019-6.00006-0

Hegab HM, Elmekawy A, Aubry C, Kallem P, Wadi VS, Banat F, Hasan SW (2024) 3D hierarchical aquaporin-like nanoporous graphene membrane with engineered tripartite nanochannels for efficient oil/water separation. NPJ Clean Water 7(1):10. https://doi.org/10.1038/s41545-024-00303-y

Ignatusha P, Lin H, Kapuscinsky N, Scoles L, Ma W, Patarachao B, Du N (2024) Membrane separation technology in direct air capture. Membranes 14(2):30. https://doi.org/10.3390/membranes14020030

Jordão A, Oliveira M, Correia A (2023) Crossflow and membrane plate filtration

Kang Y, Low Z-X, Zou D, Zhong Z, Xing W (2024) Multifunctional nanofibrous membranes for integrated air purification. Adv Fiber Mater. https://doi.org/10.1007/s42765-024-00427-3

Katibi KK, Mohd Nor MZ, Yunos KFMd, Jaafar J, Show PL (2023) Strategies to enhance the membrane-based processing performance for fruit juice production: a review. Membranes 13(7):679. https://doi.org/10.3390/membranes13070679

Khayet M, Cojocaru C (2012) Air gap membrane distillation: desalination, modeling and optimization. Desalination 287:138–145. https://doi.org/10.1016/j.desal.2011.09.017

Lazik D, Ebert S, Leuthold M, Hagenau J, Geistlinger H (2009) Membrane based measurement technology for in situ monitoring of gases in soil. Sensors 9(2):756–767. https://doi.org/10.3390/s90200756

Lee SY, Chew KW, Show PL (2019) Cell separation and disruption, product recovery, and purification. In Berenjian A (ed) Essentials in fermentation technology. Springer, pp 237–271. https://doi.org/10.1007/978-3-030-16230-6_8

Lee H-J, Son Y-J (2021) Prevalence and associated factors of frailty and mortality in patients with end-stage renal disease undergoing hemodialysis: a systematic review and meta-analysis. Int J Environ Res Public Health 18(7):3471. https://doi.org/10.3390/ijerph18073471

Li N, Lu X, He M, Duan X, Yan B, Chen G, Wang S (2021) Catalytic membrane-based oxidation-filtration systems for organic wastewater purification: a review. J Hazard Mater 414:125478. https://doi.org/10.1016/j.jhazmat.2021.125478

Li T, Zhou M, Wang Z, Xing C, Zhang S (2023) Environmental monitoring and membrane technologies: a possible marriage? In: Current trends and future developments on (bio-) membranes. Elsevier, pp 101–135. https://doi.org/10.1016/B978-0-12-824103-5.00009-7

Luo J, Hang X, Zhai W, Qi B, Song W, Chen X, Wan Y (2016) Refining sugarcane juice by an integrated membrane process: filtration behavior of polymeric membrane at high temperature. J Membr Sci 509:105–115. https://doi.org/10.1016/j.memsci.2016.02.053

Madsen E, Kaiser J, Krühne U, Pinelo M (2022) Single pass tangential flow filtration: critical operational variables, fouling, and main current applications. Sep Purif Technol 291:120949. https://doi.org/10.1016/j.seppur.2022.120949

Manoni M, Di Lorenzo C, Ottoboni M, Tretola M, Pinotti L (2020) Comparative proteomics of milk fat globule membrane (MFGM) proteome across species and lactation stages and the potentials of MFGM fractions in infant formula preparation. Foods 9(9):1251. https://doi.org/10.3390/foods9091251

Mata TM, Martins AA, Calheiros CSC, Villanueva F, Alonso-Cuevilla NP, Gabriel MF, Silva GV (2022) Indoor air quality: a review of cleaning technologies. Environments 9(9):118. https://doi.org/10.3390/environments9090118

McNair R, Szekely G, Dryfe RAW (2021) Ion-exchange materials for membrane capacitive deionization. ACS ES&T Water 1(2):217–239. https://doi.org/10.1021/acsestwater.0c00123

Mollahosseini A, Abdelrasoul A, Shoker A (2020) A critical review of recent advances in hemodialysis membranes hemocompatibility and guidelines for future development. Mater Chem Phys 248:122911. https://doi.org/10.1016/j.matchemphys.2020.122911

Mukherjee S (2019) Isolation and purification of industrial enzymes: advances in enzyme technology. In: Advances in enzyme technology. Elsevier, pp 41–70. https://doi.org/10.1016/B978-0-444-64114-4.00002-9

Nasrollahi N, Ghalamchi L, Vatanpour V, Khataee A, Yousefpoor M (2022) Novel polymeric additives in the preparation and modification of polymeric membranes: a comprehensive review. J Ind Eng Chem 109:100–124. https://doi.org/10.1016/j.jiec.2022.02.036

Possas A, Bonilla-Luque OM, Valero A (2021) From cheese-making to consumption: exploring the microbial safety of cheeses through predictive microbiology models. Foods 10(2):355. https://doi.org/10.3390/foods10020355

Postelmans A, Aernouts B, Jordens J, Van Gerven T, Saeys W (2020) Milk homogenization monitoring: fat globule size estimation from scattering spectra of milk. Innov Food Sci Emerg Technol 60:102311. https://doi.org/10.1016/j.ifset.2020.102311

Rayaroth MP, Aravindakumar CT, Shah NS, Boczkaj G (2022) Advanced oxidation processes (AOPs) based wastewater treatment—unexpected nitration side reactions—a serious environmental issue: a review. Chem Eng J 430:133002. https://doi.org/10.1016/j.cej.2021.133002

Reig M, Vecino X, Cortina JL (2021) Use of membrane technologies in dairy industry: an overview. Foods 10(11):2768. https://doi.org/10.3390/foods10112768

Saxena A, Tripathi BP, Kumar M, Shahi VK (2009) Membrane-based techniques for the separation and purification of proteins: an overview. Adv Coll Interface Sci 145(1–2):1–22. https://doi.org/10.1016/j.cis.2008.07.004

Schweicher J (2014) Membranes to achieve immunoprotection of transplanted islets. Front Biosci 19(1):49. https://doi.org/10.2741/4195

Song H, Xie F, Chen W, Liu J (2018) FO/MD hybrid system for real dairy wastewater recycling. Environ Technol 39(18):2411–2421. https://doi.org/10.1080/09593330.2017.1377771

Stamatialis DF, Papenburg BJ, Gironés M, Saiful S, Bettahalli SNM, Schmitmeier S, Wessling M (2008) Medical applications of membranes: Drug delivery, artificial organs and tissue engineering. J Membr Sci 308(1–2):1–34. https://doi.org/10.1016/j.memsci.2007.09.059

Su Y-N, Lin W-S, Hou C-H, Den W (2014) Performance of integrated membrane filtration and electrodialysis processes for copper recovery from wafer polishing wastewater. J Water Process Eng 4:149–158. https://doi.org/10.1016/j.jwpe.2014.09.012

Tekinalp Ö, Zimmermann P, Holdcroft S, Burheim OS, Deng L (2023) Cation exchange membranes and process optimizations in electrodialysis for selective metal separation: a review. Membranes 13(6):566. https://doi.org/10.3390/membranes13060566

Tomic S, Besnard L, Fürst B, Reithmeier R, Wichmann R, Schelling P, Hakemeyer C (2015) Complete clarification solution for processing high density cell culture harvests. Sep Purif Technol 141:269–275. https://doi.org/10.1016/j.seppur.2014.12.002

Wrisinger WC, Thompson SL (2022) Basics of extracorporeal membrane oxygenation. Surg Clin North Am 102(1):23–35. https://doi.org/10.1016/j.suc.2021.09.001

Yang X, He Y, Zeng G, Zhan Y, Pan Y, Shi H, Chen Q (2016) Novel hydrophilic PVDF ultrafiltration membranes based on a ZrO_2–multiwalled carbon nanotube hybrid for oil/water separation. J Mater Sci 51(19):8965–8976. https://doi.org/10.1007/s10853-016-0147-6

Zhang J, Nolan TDC, Zhang T, Griffith BP, Wu ZJ (2007) Characterization of membrane blood oxygenation devices using computational fluid dynamics. J Membr Sci 288(1–2):268–279. https://doi.org/10.1016/j.memsci.2006.11.041

Zhou M, Li T, Liu P, Zhang S, Liu Y, An T, Zhao H (2021) Real-time on-site monitoring of soil ammonia emissions using membrane permeation-based sensing probe. Environ Pollut 289:117850. https://doi.org/10.1016/j.envpol.2021.117850

Zhou Y, Zheng S, Qin W (2024) Electrochemical biochemical oxygen demand biosensors and their applications in aquatic environmental monitoring. Sens Bio-Sens Res 44:100642. https://doi.org/10.1016/j.sbsr.2024.100642

Chapter 8
Circular Economy Driving Sustainable Effluent Solutions

Van Toi Pham, Dinh Pham Van, and Tuan-Dung Hoang

Abstract This chapter looks at membrane technology's role in waste management as we move toward a circular economy. The circular economy seeks to: (1) extend the life cycle of various materials and waste products; and (2) make reuse easier by approaching them as manageable resources. The manufacturing of pharmaceuticals, the food sector, wastewater treatment, wastewater product recovery, etc. are the principal applications for membrane technologies. These days, membrane filtration technologies—such as ultrafiltration membrane (UF), microfiltration membrane (MF), reverse osmosis membrane (RO), and nanofiltration membrane (NF)—are some of the most accessible and advantageous sophisticated technologies for society. The food sector has made extensive use of membrane technology (MB) in its manufacturing processes. This technique combines the benefits of high efficiency and energy conservation with the ability to separate, concentrate, and purify food. Among the principal uses are fruit and vegetable juices, dairy products, egg white concentration, food additive separation and concentration, and so forth. The pharmaceutical sector is using membrane separation technology more and more frequently as a result of the technology's quick development. Numerous uses in pharmaceutical operations and wastewater treatment have been discovered by researchers. Reasonable and scientifically based membrane separation technology applications in the pharmaceutical field can ensure the progressiveness of pharmaceutical processes, boost production efficiency, reduce pollution, and promote the pharmaceutical industry's growth in an eco-friendly, efficient, and green direction. Furthermore, membranes are extensively employed in biotechnology, mostly for purifying bioproducts. Membranes are also utilized in drug delivery, biological applications like blood oxygenation and dialysis, and laboratory filtration and analysis as parts of membrane bioreactors. MB is one of the wastewater treatment innovations that has advanced significantly

V. T. Pham · D. P. Van
Faculty of Environmental Engineering, Hanoi University of Civil Engineering, Hanoi, Vietnam
e-mail: toipv@huce.edu.vn

T.-D. Hoang (✉)
VNU School of Interdisciplinary Studies and Arts, Vietnam National University, 144 Xuan Thuy, Cau Giay, Hanoi, Vietnam
e-mail: tuandung@vnu.edu.vn

A. Mandpe et al. (eds.), *Membrane Filtration and Membrane Bioreactors for Wastewater Treatment*, Advances in Wastewater Research,
https://doi.org/10.1007/978-981-95-6706-5_8

during this time. The advantages membrane technology provides for treating water and wastewater have led to its substantial growth in the last few decades. Membrane technology offers numerous prospects in wastewater treatment due to its low initial cost, energy requirement, and significant equipment size reduction. The membrane technology has the potential to close the gap between affordability and sustainability because of its universal accessibility, low to no chemical consumption, and environmental friendliness. It has been proven successful to treat textile wastewater using membrane technologies. UF, NF, MF, and RO textile wastewater have often been treated using membranes alone, but a sequence of filtrations has also been used. Membrane bioreactors (MBRs) combine biological processes (typically activated sludge) with membrane technologies to produce extremely high-quality effluents when needed, such as for water reuse. Some of the special benefits of membrane separation technology include its high efficiency, broad applicability, and low environmental effect. As a result, this technology will be extensively utilized in the future in the circular economy for the processing of food, desalination, wastewater treatment, and other industries.

Keywords Membrane technology · Wastewater treatment · Environmental protection · Waste management · Membrane bioreactor

8.1 Membrane Filtration in Food and Beverage Industry

Over the past few decades, membrane processes have grown in importance in the food business. Next to the dairy industry, the common fields of use for this technology are drinks such as fruit juices, sugar products, wine, etc. MF, UF, NF, and RO are examples of common MB processes.

8.1.1 Dairy Industry

Milk includes water, salts, fat globules, serum proteins, lactose, casein micelles, and other suspended and dissolved substances. Season, climate, and nutrition affect the content of milk, which varies depending on the breed, age, and lactation stage of each cow (Brans et al. 2004). In the dairy industry, MB is used in addition to traditional methods for a number of tasks, such as fat globule fractionation and separation, bacterial and spore removal, casein micelle concentration, and casein micelle separation (Brans et al. 2004).

8.1.1.1 Elimination of Spores and Bacteria

Human harmful microorganisms including Listeria, Brucella, Mycobacterium, or Salmonella may be present in milk collected by the dairy facility (Charcosset 2021). Consequently, it is imperative to achieve a reduction in bacteria and spores without compromising the composition of milk proteins (especially for cheese production). Unwanted bacterial and fungal growth can cause a late ripening blowout in cheese. This is especially true for hard and semi-hard cheese types, where factors like high moisture content and ripening temperatures, low salt content, etc. create an environment that encourages the formation of the spores.

One potential technique for lowering the quantity of bacteria and spores is milk centrifugation, also known as bactofugation (Brans et al. 2004; Charcosset 2021). However, this method uses a lot of energy, and bactofugation alone removes only 90–95% of the spores. Higher temperatures can alter the milk's flavor and extend its shelf life compared to pasteurization. These days, MF is used to get rid of germs and spores in the dairy sector. This method's benefit is that it retains the milk's flavor while reducing the amount of microbes and spores. Moreover, MF can be used to pretreat skim milk in order to produce milk cheeses and decrease spores in acid cheese milk, in addition to producing milk with a prolonged shelf life (Brans et al. 2004). The method is widely recognized. There are, as far as we know, only a few recent articles that address highly specific parts of this subject. Griep et al. (2018) showed that spore size and surface characteristics could influence spore retention. Besides, they found that a MB with a hole size of 1.2 μm might be superior to the conventional 1.4 μm hole size selected. Nevertheless, a drop in the concentration of milk protein and an increase in membrane fouling can result from a larger retention of casein micelles due to the reduced pore size. These days, a Sterilox® membrane (hole size of 1.4 μm) is used to circulate heated skim milk at a speed of 7.2 m/s (Luciana and Jean-Louis 2000). Membrane fouling happens during skimming milk by MF, much like in all membrane processes. Therefore, backflushing procedures or high cross-flow speed (4–8 m/s) are required to reduce the creation of the fouling layer.

8.1.1.2 Separation of Casein Micelles

The aqueous phase of skimmed milk contains lactose, minerals, and soluble proteins suspended in a complex dispersion of casein micelles (Jiao et al. 2004). Negatively charged colloids in equilibrium with the aqueous phase are casein micelles. Therefore, the changes in ionic strength modify the composition of skim milk by modifying the properties of casein micelles, the protein and mineral (Jiao et al. 2004).

Cross-flow MF, which has a pore size of 0.1 μm, is used to separate soluble proteins from casein micelles. Subsequently, those casein micelles are employed to produce a lactose, mineral, and soluble protein-rich permeate and a retentate abundant in casein micelles (Jiao et al. 2004). Concentrated casein micelles can be reconstituted by cream in order to make cheese. Furthermore, skimmed milk MF permeate is a great starting point for the purification of certain proteins (Brans et al. 2004). However, casein micelles could adhere to the membrane's surface, generating a fouling layer and lowering permeability and selectivity in the process (Charcosset 2021). The process's overall performance may be significantly impacted by the physicochemical circumstances. For instance, casein micelles may form a tight, difficult-to-remove gel at high ionic strength (Jiao et al. 2004).

8.1.1.3 Separation and Fractionation of Fat Globules

Distributed fat globules, which average 3.4 μm in diameter and range in size from 0.1 to 15 μm, are the main constituent of milk fat (Henri et al. 2000). Given that MF is known to not harm the membranes of fat globules, fractionation by MF may be a better option than other methods. Using a 2-μm MF ceramic membrane, Henri et al. (2000) reported the fractionation of milk fat globules. Goods manufactured with huge fat globules collected in the retentates tend to be more firm, with the exception of butter, than goods made with milks containing little fat globules discovered in the permeate.

Usually, centrifugation is utilized to separate fat. Milk fat globules typically separate at a temperature of about 50 °C to prevent clumping. About 40% of the cream that is left over after centrifugation is fat. When relatively modest cross-flow velocities are employed, membrane separation has the advantage of requiring less energy and causing less damage to shear-sensitive components, such as the membranes of milk fat globules (Brans et al. 2004). Better cream stability and increased consumer product sensory qualities could come from this.

8.1.2 Wine

Among the most widely consumed alcoholic beverages worldwide is wine, which is produced in significant quantities in France, Italy, Spain, etc. Wine is made by fermenting grapes, and the final product's composition is influenced by the grape's makeup, maturity at harvest, growing conditions, and winemaking methods. Additionally, the metabolism of yeast and bacterial strains affects their makeup.

Polyphenols, polysaccharides, and proteins are components of wine colloids (Rayess et al. 2011). The two most common phenolic compounds found in red wine are tannins and anthocyanins; in white wine, one of the main polysaccharide macromolecule groups is hydroxycinnamic acids. Depending on where they come from, polysaccharides can be categorized into three groups: fungal polysaccharides like beta-glucan, yeast polysaccharides, and grape polysaccharides (pectin, pectin compounds). Grape proteins and, to a lesser extent, proteins from autolyzed yeast make up wine proteins. Because tannins precipitate proteins, red wines have relatively little of them. Furthermore, the protein content of white and rosé wines, which mostly come from grapes, might vary by a few hundred (10–500 mg/L).

Wine filtering is utilized in the entire production process to enhance microbial stability and reduce turbidity (Rayess et al. 2011). The presence of haze caused by suspended particles and macromolecules is the cause of turbidity. Turbidity is expressed in nephelometric turbidity units (NTU) following measurement with a turbidimeter. The initial NTU may approach 2000 NTU prior to therapy (Rayess et al. 2011). Wines must have turbidity levels of less than 2 NTU following flotation. Conventional filtration methods involve utilizing diatomaceous earth and filtration on sheets. One potential substitute to improve the stability and limpidity of wine is cross-flow MF. Because of its benefits, which include removing the need for diatomaceous earth and the hazards to human health and the environment that go along with it, as well as its ability to combine sterile filtering, stabilization, and clarifying into one process, it is being used more and more in the wine industry.

When wine is fouled during cross-flow mechanical fining (MF), the permeation flux decreases, which affects the process's viability economically, and the component retention increases, which may have an effect on the organoleptic qualities of the wine (Rayess et al. 2011). The complexity of the fouling image is influenced by the intricate composition of wine, the membrane's effect on wine colloids and particles, and the operating conditions. Wine polysaccharides have been shown in numerous studies to have a negative effect on the permeation flow. Membrane fouling is more closely associated with the structure, composition, and balance of the various polysaccharide groups than with the overall amount of polysaccharides present (Charcosset 2021). Moreover, proteins may cause extremely significant and nearly permanent fouling. Interactions between proteins and the membrane material through hydrophobic interactions are the main processes (Ulbricht et al. 2009). Additionally, yeasts and other particles may potentially have a major impact based on the size of the particles. Furthermore, cake formation may even offer a membrane's partial defense against internal pore obstruction (Ulbricht et al. 2009). For instance, Boissier et al. (Boissier et al. 2008) examined the effects of wine particles on cross-flow MF performances. Reversible deposits were only generated by yeast, according to their hydrophilic nature and uncharged characteristics. On the other hand, over a specific TMP, fines created a cohesive and adherent cake.

The type of membrane used in wine cross-flow MF greatly affects fouling. In actuality, interactions between wine components and the membrane surface account for a sizable amount of membrane fouling. For instance, Ulbricht et al. (2009) utilized MF to clarify a white wine and found that although both membranes had the same pore size (0.2 μm), polypropylene membranes produced significantly more fluxes than polyarylsulfone membranes. They demonstrated that polyphenols and polysaccharides were widely absorbed by polyethersulfone membranes but only very little by polypropylene membranes using this membrane for wine. Furthermore, it was suggested that the main process involved the interaction of polyphenols with the membrane surface, and that the aggregation of red wine's polyphenols and polysaccharides significantly affected adsorptive fouling.

8.1.3 Beer

Brewing is one of the main players in the food sector. Following milk, tea, and coffee in popularity worldwide, beer comes in at number five. One of the most crucial processes in the brewing sector is beer clarification (Sman et al. 2012). To protect the beer's biological stability, it involves harsh beer filtering to get rid of the yeast and colloidal particles that cause haze. Maltose, which is produced from starch, is fermented by yeast cells to produce alcohol and carbon dioxide. Consequently, fermented beer is an intricate blend of macromolecules, cells, aggregates, and fluids. Haze particles, which are aggregates of proteins and polyphenols, and yeast cells are the sources of turbidity. The industry standard for clarifying beer is dead-end filtration, which uses filter aids like diatomaceous earth. However, cross-flow MF has been created and employed in certain industrial facilities because of limitations in technology, funding, and environment.

Severe membrane fouling during MF beer clarifying results in low permeate flows (often less than 100 L/h m^2), the retention of protein and aroma components, and labor-intensive cleaning processes (Czekaj et al. 2000). Membrane fouling may result from the buildup of particles, like macromolecules or macromolecular aggregates, on the upper surface of the membrane, pore size constriction, or the deposition of cells and/or cell detritus. Furthermore, macromolecules included in beer, such as proteins, polyphenols, polysaccharides, etc., can nevertheless considerably contribute to fouling even though their pore sizes are far smaller than those of MF membranes.

The size of the membrane pores is an important factor in controlling flux and filtrate quality throughout the beer-clarifying process. Instead of 0.2 and 1.3 μm, the optimal pore diameter is 0.45 μm to achieve a balance between flux and selectivity (Gan et al. 2001). Because ceramic membranes can tolerate strong cleaning conditions, they are chosen over polymeric membranes. The obtained fluxes are typically significantly smaller, though. Since surface fouling and the transfer of macromolecular solutes are primarily regulated by in-pore fouling, improved surface hydrodynamics have little influence on boosting flow and retention. On the other hand, strong flux enhancement was demonstrated to result from inverted membrane architecture and high-frequency backflushing (Gan et al. 2001). By altering the configuration, feeding the fluid to the substructure side, and removing the permeate from the typical retentate side, the pore morphology of the membrane can be reversed. Next, the TMP is applied in the reverse manner.

Beer clarification is done in a number of commercial MF plants (Sman et al. 2012). To create cool haze, the raw beer is first chilled to 0 °C. Additionally, the membrane holds onto this haze, keeping the beer from becoming turbid when it is bottled. Modules for hollow fiber or flat sheets are often configured using hydrophilic polyethersulfone membranes in the 0.45–0.65 μm pore diameter range in commercial systems. Before MF, some commercial systems remove the yeast cells and big aggregates using a centrifuge separation step. However, most industrial operations have not included a centrifugation phase since the expenses of this stage are larger than those of the MF stage.

In the brewing business, tank bottom recovery is a significant application of membrane technologies. Many industrial facilities use polysulfone (0.6 μm) or ceramic (0.4–0.8 μm) MF membranes (Fillaudeau et al. 2006). Beer can be extracted from tank bottoms using a filter press, centrifugation, or natural sedimentation. Alternatively, MF can provide a superior beer with less color and bitterness loss that still has flavor and haze. The creation of beers with low or no alcohol content is one of the more recent trends. Beers classified as non-alcoholic typically have an ethanol content of less than 0.5% vol. Many methods, including thermal ethanol removal, membrane processes, using specific yeast strains that either absorb alcohol or do not create it, or biological processes that limit fermentation to limit the quantity of ethanol produced, can be used to de-alcoholize beers. The primary benefit of membrane techniques over conventional thermal technologies for dealcoholization of beers is their ability to function at moderate temperatures. These membrane methods include pervaporation, RO, and dialysis. Low flux for fouling, dialysis, and high energy use for RO, and volatile chemical loss due to pervaporation are some of the drawbacks of these techniques. Other methods, such as forward osmosis distillation and non-porous membrane distillation, are suggested (Charcosset 2021). In forward osmosis, as opposed to RO, the difference in osmotic pressure between a semi-permeable membrane's two sides drives the solution with less fouling. Ethanol and water are extracted concurrently, and the alcohol concentration is decreased by adding water through diafiltration.

8.1.4 Fruit Juices

One of the most significant areas of the fruit juice industry is fruit juice production. Consumption of fruit juice has skyrocketed in recent years because of the perception that juices constitute a natural, high-nutrient food source (Charcosset 2021). Fruit juices include both large and low-molecular-weight solutes, and microorganisms. The fruit juice business has been using membrane techniques more and more in recent decades. Their low operating temperatures and excellent selectivity allow these methods to create high-quality fruit juices. They are mostly used for concentration and clarity, as well as the separation and purification of certain components (like scent and bioactive compounds).

8.1.4.1 Fruit Juice Clarification

Certain fruit juices naturally exhibit turbidity because they contain insoluble components such as pectin, starch, and juice cells (Charcosset 2021). The need for a clarification process depends on how the fruit juice will be used. For example, fruit liquids that have been cleared are used in the production of candies, pastries, and transparent drinks. Sequential batch procedures like enzymatic pretreatment, pasteurization, or diatomaceous earth clarity are typically used in conventional clarification. The conventional clarifying techniques that were first developed in the late 1970s were replaced by MF and UF. Higher juice yields, one-step operation, shorter working and enzyme utilization times, no need for filter aids, elimination of pasteurization needs, and continuous process production are some advantages of MF and UF (Rouquié et al. 2019).

The effectiveness of utilizing UF or MF for fruit juice clarification is significantly limited by concentration polarization and fouling processes. Membrane characteristics, juice properties, and operating factors like as temperature, TMP, and recirculating flow rate are crucial in averting membrane fouling, which therefore impacts process efficiency. Many studies have looked into how operating circumstances affect membrane fouling in cross-flow MF and UF of various fruit juices (Laorko et al. 2010; Cassano et al. 2007). The permeate flux, similar to other MF and UF apps, reaches a limiting value at increased pressure and exhibits a linear connection with pressure below a TMP threshold. Increased flow rates cause the concentration polarization layer to shrink, which is linked to an increase in permeation flux, by raising the membrane surface's wall shear stress. Higher juice temperatures also result in greater permeability because they decrease juice viscosity and increase macromolecule diffusion coefficients (Cassano et al. 2007).

When using MF or UF to clarify fruit juice, pretreatment is crucial. Juice filtering can yield low permeate filtrates because of its comparatively high pectin, lignin, and hemicellulose content. Two popular preparation techniques are juice centrifugation and enzymatic treatment. Prior to passion fruit juice MF, Domingues et al. (2014) examined the effects of centrifugation, chitosan coagulation, and enzymatic liquefaction as pretreatments. Centrifugation reduced color and turbidity, whereas the enzymatic preparation reduced juice viscosity. The most promising pretreatment was chitosan coagulation, which produced the most permeate flux as well as the greatest reduction in color and turbidity. Additionally, it was found that the pretreatment influences the main membrane fouling mechanism. Fruit juice that has been centrifuged or treated with enzymes will mostly foul due to cake formation; fruit juice that has been treated with chitosan will foul primarily due to internal pore clogging.

8.1.4.2 Fruit Juice Concentration

Fruit juices are typically concentrated to improve preservation, streamline processes, and lower expenses associated with distribution, storage, and transportation. The traditional techniques for concentrating juice, such as thermal evaporation, often use high temperatures to evaporate water. Heat, however, can result in unfavorable modifications to the sensory and nutritional qualities of the product, including color and flavor alterations as well as a decrease in nutritional content. Processes for membrane concentration are substitutes for thermal evaporation. Juices can be concentrated at ambient temperature with little to no harm to the final product (Jiao et al. 2004). In this field, membrane technology advancements and innovations have been examined by Jiao et al. (2004).

The fruit processing sector has been using RO-based fruit juice concentration for approximately 40 years (Bhattacharjee et al. 2017). Thick membranes that function at a TMP higher than the osmotic pressure of the solution are the foundation of reverse osmosis, which permits water to pass through from the side with a high solute concentration to the one with a low one. The attainment of superior fruit juices typically leads to a notable retention of advantageous, fragrant, and taste-enhancing compounds.

Because RO operates at moderate temperatures, it has several advantages over thermal evaporation, including less harm to the fruit juice, less energy use, and reduced capital costs. Additionally, phase change is not used in the process of removing water. Numerous fruit juices, such as those from apples, citrus fruits, grapefruits, kiwis, pineapples, watermelon, passion fruit, and tomatoes, have had their RO concentrations studied (Bhattacharjee et al. 2017). Numerous studies have shown how the membrane and operating conditions affect the penetration and retention of juice components, particularly sugars and fragrance chemicals. Elevated pressures may lead to decreased processing time, increased flux, and increased aroma compound rejections. Both penetrating fluxes and volatiles' rates of permeation increase with temperature (Butterfield et al. 2001). The retention of beneficial substances in various membrane topologies, like the plate-and-frame structure and the spiral-wound system, may also have an impact on the concentration of fruit juice.

In fruit juices, some macromolecules, particularly polysaccharides, have a tendency to create a gel layer close to the membrane surface. Enzymatic hydrolysis has been suggested as a solution to mitigate this issue by breaking down the methoxylated pectin. As a result, mass transport across the membrane is encouraged and juice viscosity is decreased. Despite their excellent capacity for solute retention and selectivity, RO membranes have a significant downside. Fruit juice's high osmotic pressure restricts its degree of concentration to the necessary threshold. According to several studies, fruit juices have a final concentration of less than 25–30°Brix, which is the point at which flux and solute recovery are most effective (Bhattacharjee et al. 2017). Consequently, in order to save energy and boost production capacity, RO can be employed as a preconcentration step before other procedures like freeze concentration or evaporation. A further practical industrial solution is reverse osmosis with many stages. A study on industrial full-scale plant-based spiral wound membrane modules for apple juice concentration was carried out by Al-Obaidi et al. (Al-Obaidi et al. 2017). They determined the ideal conditions for making an apple juice product with a high concentration.

8.1.5 Sugar

In the sugar industry, membrane processes have been suggested as a replacement for or addition to conventional main operations (Lipnizki 2013). These applications pertain to the sugar industries that produce cane and beets. Beet sugar is used in the demineralization of beet juice, pulp recycling, raw juice purification, and others. These include the concentration of clarified cane juice, the purification of raw sugar cane juice, the decolorization of remelted raw sugar, and the molasses treatment in the sugarcane business. Additionally, sugar vinasse therapy has drawn increased attention lately (Charcosset 2021).

Several downstream unit processes are involved in the conversion of sugarcane juice into sucrose (Abels et al. 2013). The classic purification method known as double-sulfurization involves adding sulfur dioxide and calcium hydroxide (lime) to purify the unrefined juice in a sugar cane mill. Consequently, calcium sulfate and the raw juice's suspended colloids coagulate and precipitate (Lipnizki 2013; Abels et al. 2013). Following a second sulfurization the juice that has been clarified is concentrated by evaporation and bleached. Ultimately, white sugar crystals are obtained by crystallizing the product stream. Nevertheless, the ineffective extraction of materials from the raw juice, such as gum, ashes, silica, etc., as well as the ineffective extraction of sulfur from the product stream, are problems with the double-sulfurization method. In order to purify sugarcane juice, MF and UF have been studied as a potential replacement for the second sulfurization phase. Ceramic tube filtration systems or spiral wound or flat-sheet polymeric membranes can be used for the UF of sugarcane juice (Jegatheesan et al. 2009). Compared to the rise of 0.5–1 unit obtained by the liming–sulfurization procedure, the fructate obtained had an increase of 1.5–3 units of juice purity. Furthermore, juice clarity improved with membrane clarification, resulting in a roughly 60% color reduction and a decrease in the juice's inorganic content (Fig. 8.1).

In both lab and industrial trials, the clearing of sugarcane juice by UF and MF has been thoroughly investigated. For instance, in a white sugar plant in northern India, a pilot study on the ultrafiltration membrane of sugarcane juice was conducted by Ghosh and Balakrishnan (2003). Using polymeric spiral-wound membrane modules,

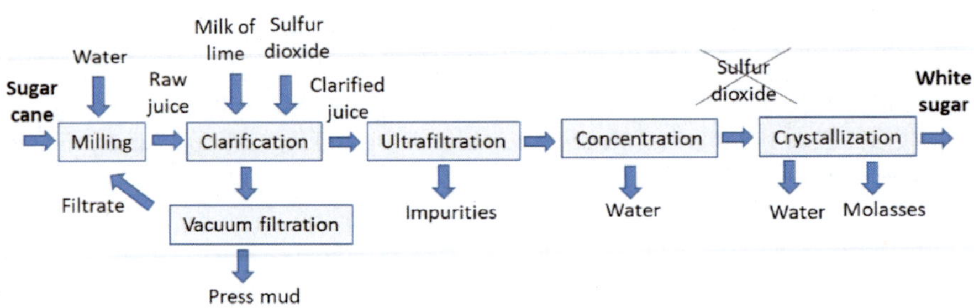

Fig. 8.1 Application of UF in cane sugar refining (Abels et al. 2013)

sugarcane juice was cleared at 91–97 °C in a trial facility with a 10 m^3/h output capacity. The permeate had an average purity increase of 0.9 units, 47% less color, and 31% less turbidity. The membrane modules were found to be temperature-resistant and to demonstrate good separation. The average flux was only 7 L/m^2 h, though. Membrane fouling is a significant restriction in the UF of cleared sugarcane juice, as it is in other food membrane processes (Saha et al. 2009). Sugarcane juice has between 0.85 and 1.45% soluble solids in organic, non-sugar forms such proteins, polysaccharides, and waxes, and between 3 and 5% soluble solids in the form of colorants, color precursors, etc. Sugarcane juice's polysaccharide content is known to function as a membrane foulant. Furthermore, the creation of gels by polysaccharides is enhanced by the presence of divalent cations, such as Ca^{2+}. Recently, Shi et al. (2019) emphasized how crucial it is to pretreat factory sugarcane juice prior to UF in order to improve juice quality and permeate flux while lowering membrane fouling. Flocculation lowers contaminant loading and membrane fouling before sugarcane juice is clarified. However, research into the flocculants being assessed in the sugar industry is still ongoing (Shi et al. 2019). Moreover, intriguing options for sugarcane juice purification are integrated processes. Integrated membrane approaches have been demonstrated by Luo, Hang (Luo et al. 2016) to be a viable substitute for attaining superior permeate flux and high color removal in the refining of sugarcane juice. The three components of the integrated membrane process were spiral-wound NF, spiral-wound tight UF, and tubular loose UF.

8.2 Membrane Applications in Pharmaceutical and Biotechnology Processes

8.2.1 Membrane Ultrafiltration and Microfiltration

The membrane-separation techniques of ultrafiltration and microfiltration are widely recognized (Charcosset 2006). Tubular, hollow fiber, spiral wound, flat plate, and rotating devices are examples of module configurations. Deadend and cross-flow designs are two typical modes of operation. The filtered fluid is forced through the membrane by a pressure differential operating parallel to its surface in the cross-flow mode. To keep the level low, the cross-flow lessens the filter cake's ability to form.

Product concentration—the elimination of solvent or buffer—is a popular use for ultrafiltration in downstream processing. Diafiltration can be used for buffer exchange and desalting, which involves washing out the original buffer and replacing it with a fresh buffer. Microfiltration is most frequently used for sterile filtration, or the elimination of bacteria, which comes before many goods are finalized. It's also used for fermentation broths' first clarifying, to get rid of suspended cell mass and other particle debris. It has been demonstrated that sterile filtering in a dead-end arrangement with 0.2 μm pore-size membranes may totally eliminate Brevundimonas diminuta (Charcosset 2006). Under some process conditions, these sterilizing filters can allow the passage of extremely microscopic microbes. Thus, to improve sterility assurance in pharmaceutical operations, some users use membranes with hole sizes of 0.1 μm (Sundaram et al. 1999).

Microfiltration is a useful technique for removing viruses from cell cultures. It can be used to separate viruses, which have a size range of around 12–300 nm, from proteins, which have a size range of 4–12 nm (Reis and Zydney 2001). Recombinant proteins with medicinal, preventive, or diagnostic uses are commonly produced using mammalian cell lines. On the other hand, viruses or particles like viruses could contaminate cell lines. Viruses can also enter a fermentation process through the addition of supplements and other ingredients, handling, and other processing-related procedures, in addition to endogenous contamination. In order to separate viruses from proteins, membrane producers are creating membrane filters with higher resolution (Reis and Zydney 2001). Because there is a chance that cell cultures will become contaminated with parvovirus, this is crucial for the biotechnology sector. The small size of parvoviruses (approximately 20 nm in diameter) and their strong resistance to different forms of inactivation, including chemical and thermal ones, make them exceptionally challenging to eradicate. Membrane integrity is checked both before and after use to make sure that viral eradication is consistent with validation investigations.

Membrane filtration is also very effective in the manufacture of antibiotics. As building blocks for semisynthetic antibiotics like erythromycin, medmycin, and benzylpenicillin (penicillin G), the majority of antibiotics are commonly utilized. Following fermentation, the traditional methods of filtration (biomass removal), solvent extraction (isolation and purification), and crystallization (polish) are used to remove them from their broths. These fermentation broths are first clarified by membrane filtration (Charcosset 2006). Combining filtration and diafiltration processes to get a very high yield is one of the key benefits of employing membrane systems for the initial recovery of antibiotics from a fermentation broth. Antibiotics can almost completely pass through membranes with hole widths as small as 0.2–0.45 μm, and as long as the membrane's nominal molecular cut-off is more than 20,000 g/mol, all of the cells and particulate debris will be preserved (Charcosset 2006).

The development of novel membrane modules with enhanced mass-transfer properties for ultrafiltration and microfiltration processes continues to receive a lot of attention. This comprises conically shaped rotors, cylindrical Taylor vortex devices, spinning disk filters (Jaffrin et al. 2004), and helical coiled Dean vortex systems (Charcosset 2006). Since centrifugal flow instabilities result in very high mass-transfer rates, Dean vortex devices have these characteristics. Protein transmission and capacity are significantly increased by these devices; however, fouling is still an issue in many applications. Levesley and Hoare (Levesley and Hoare 1999) provide an alternate method of continuously cleaning the membrane surface: high-frequency back-pulsing. It has been demonstrated that high-frequency back-pulsing enhances flow, lessens fouling, and promotes protein transmission during the conjugated vaccine product purification process. In addition, pulsative flow, electric fields, combination electric/ultrasonic fields, etc. are useful methods for increasing the effectiveness of ultrafiltration and microfiltration. Jaffrin et al. (2004) looked at how various hydrodynamic parameters affected the permeate flux produced by two separate dynamic filtering systems using the same fluids and membrane materials. This work can be cited in this subject. The devices that were tested included a pilot for the VSEP using a circular vibrating membrane and two homemade rotating disk modules. Larger vanes on the disk, faster rotation, or higher vibration amplitude can all be used to boost the flux to extremely high levels. It was demonstrated that the flux is primarily determined by the maximum shear rate rather than the specifics of the internal flow. Industrial systems may be scaled up using these findings. The work of Schutyser et al. (2002) provides another illustration in this area. They improved cross-flow microfiltration and diafiltration for the concentration and purification of rBDNF inclusion bodies from E. coli cell suspensions by using Dean vortex microfiltration with controlled centrifugal instabilities (Dean vortices produced in helical flow). The helical module outperformed the linear module in the microfiltration tests employing feeds such as cell and homogenate suspensions, improving flow by roughly 50 and 70%, respectively. When the helical module, rather than the linear module, was employed for diafiltration with homogenate suspensions as feed, solute transfer increased from 100 to 40% after 40 and 100 min, respectively.

Further advancements in our comprehension of the effects of the solution environment on molecules, particle retention, and fouling have also led to improved performance in microfiltration and ultrafiltration. For instance, several recent studies have shown that adjusting the ionic strength or pH of the solution can regulate the pace at which proteins are transported through membranes (Charcosset 2006). The procedure must be performed at the pI of the protein that is transmitted and close to the pI of the protein that is retained. When the ionic strength is kept low to facilitate separation, the diffuse double layer of the charged solute is thicker and has higher retention than that of the uncharged solute, which enters the membrane more quickly. Protein separation can also be achieved by utilizing these electrical interactions.

8.2.2 *Membrane Bioreactors*

It has been previously reviewed that membrane bioreactors can be found in Giorno and Drioli (2000) and Giorno et al. (2003). These approaches serve as alternatives to the traditional ones for immobilizing biocatalysts, which involve encasing them in a reaction vessel separated by a membrane or immobilizing them inside the membrane matrix itself. These biocatalysts include things like microorganisms, enzymes, and antibodies. A membrane-separation unit might be integrated into a conventional stirred tank reactor in the first method; in the second, the membrane serves as both a catalyst support and a separation unit. The biocatalyst can be immobilized on the membrane by being trapped, gelified, physically adsorbing, ionic, covalent, or cross-linked; alternatively, it can be flushed along a membrane module, segregated inside one, or both. Immobilizing enzymes is said to have several benefits, such as better product quality and purity, decreased waste, and enhanced productivity and stability of the reactor (Giorno and Drioli 2000). Numerous research endeavors focus on examining the optimal working settings and enhancing the diverse attributes of these membrane bioreactors. The effectiveness of the system as a whole is influenced by the geometric (morphology, pore size distribution, etc.), hydrodynamic (flow velocity, transmembrane pressure, etc.), and biochemical (reaction kinetics, concentration, etc.) characteristics (Giorno et al. 2003).

Successful tests have also been conducted with immobilized whole cell membrane bioreactors (Belfort and Heath 1994). Increased cell densities are possible in membrane bioreactors for immobilized entire cells, which results in higher product titre. The cells are continuously supplied with medium through a membrane, which allows them to receive nutrients and oxygen while eliminating wastes and desirable products. Typically, a membrane barrier keeps the cells inside the bioreactor. Numerous membrane topologies, including spinning bioreactor setups and flat sheet designs, have been investigated; of these, the hollow fiber configuration is the most intriguing. Cell development can be accomplished in one of two ways: either outside the fibers with the media developing inside the cells, or inside the fibers with the medium passing through themIt was demonstrated that there was an amass transfer constraint for both glucose and oxygen, and that one way to lessen this phenomena was to have consistent, well-defined spacing between the fibers. To develop the cells in the annulus between the two threads, a number of arrangements were suggested, including hollow fibers placed into another (Charcosset 2006). The distance required for the diffusion of medium components, such as oxygen, glutamine, and other nutrients, can be reduced to about 50 μm for the distant cells by carefully selecting the diameters of the inner and outer fibers. When producing antibodies from hybridoma cells, conventional and concentric reactors were compared. It was found that the concentric reactor can support a much greater cell concentration in addition to having superior specific productivity and cell durability.

The chemistry of enzyme immobilization is crucial to membrane bioreactors that employ immobilized enzymes. It is frequently carried out via spacer arms or directly on the membrane, frequently via the protein's lysine residues' ε-amino activity (Butterfield et al. 2001). The immobilized enzyme's activity is frequently greatly reduced as a result of substrate accessibility being blocked to the active site, multiple point-binding occurring, or the enzyme being denatured. Randomly immobilizing enzymes onto polymeric microfiltration-type membranes generally results in increased stability. Numerous techniques have been devised to enable the immobilization of enzymes with various structural features at specified sites. These include site-directed mutagenesis to introduce distinct cysteines to enzymes, post-translational modification to add a single biotin moiety, and gene fusion to add a peptidic affinity tag at the N- or C-terminus of the enzyme (Butterfield et al. 2001).

Membrane bioreactors have been around for more than thirty years, and their primary application has been in the industrial treatment of wastewater, which includes wastewater from residences, companies, and local governments. Membrane bioreactors have had some success in the field of biotechnology. The primary technological challenges associated with membrane bioreactors in the industrial setting are connected to rate-limiting factors and scale-up issues, as well as the enzyme's lifespan, the cost-effectiveness of pure enzyme availability, etc.

8.2.3 Membrane Chromatography

Research on adsorptive MB has been conducted for about twenty years as a substitute for conventional resin-based chromatography columns (Ghosh 2002). Because the interactions between molecules and active sites on the membrane take place in convective through-pores rather than in stagnant fluid inside the pores of an adsorbent particle, adsorptive membranes exhibit shorter diffusion times than those achieved in resin-based chromatography. As a result, large macromolecules with low diffusivities can be effectively employed with adsorptive membranes to sustain high flow rates. Reduced denaturation and protein breakdown are also associated with their use.

Chromatographic membranes have been employed in many different ways, including as spiral wound membranes, hollow fibers, anchored membranes, and adsorptive processes like hydrophobic, reversed-phase, ion-exchange, etc.. Numerous ligands, including immunoaffinity ligands, Protein A and G, low-molecular-mass ligands, etc. have been evaluated with affinity membranes (Ghosh 2002). Membrane materials investigated for chromatographic applications include hydrazide, cellulose, polyamide, polysulfone, and composite membranes. Commercial membranes include flat sheet systems, hollow-fiber modules, radial flow cartridges, and membrane stacks (Sartorius, Göttigen, Germany). Membranes come in two varieties: affinity (protein A and Cu^{2+}) and ion-exchange (cation and anion) (Charcosset 2006).

Chromatographic membrane purification techniques have been documented for a variety of substances, including DNA, viruses, and proteins (enzymes, serum antibody, monoclonal antibody, etc.). Thiophilic membranes are utilized in various applications such as the extraction of monoclonal antibodies from medium used in cell culture, the isolation of antibacterial peptides from lactoferrin using ion-exchange membranes, and so on (Charcosset 2006). Particularly intriguing is the early research on membrane chromatography conducted by Brandt et al. (1988). It was proposed to employ a hollow-fiber device to separate fibronectin from blood plasma and to support protein A with a hollow-fiber membrane in order to separate IgG. This device's quick throughput rate and effective ligand utilization allowed for quick bind-elute cycle times. Even though a typical agarose affinity system had volumes that were 100–1000 times larger, the affinity-membrane device only needed 0.1% more ligand to achieve the same mass-transfer efficiency and throughput. An illustration of a successful use in the pharmaceutical production industry was given by Belanich et al. (1996). It has been shown that scalable strong anion-exchange membrane adsorbers can remove endotoxins from bacterial extracts without compromising the enzyme activity of the protein combination.

Due to the massive diameter-to-length ratio of the modules, one of the main drawbacks of membrane chromatography is the membrane's uneven dispersion of flow. Process mathematical modeling brought this constraint even more into focus by highlighting the need to consider the system dispersion curve in order to provide a reliable comparison between computed and experimental concentrations. This can frequently be a severe problem, lowering membrane efficiency to that of closely spaced beds. However, research has shown that a suitable flow distributor design mitigates this problem (Ghosh 2002).

It is undeniable that the method of membrane chromatography has not achieved the anticipated level of success (Millipore Inc. no longer supplies membrane discs for chromatography). Most likely, this is because prospective users of this new technology are hesitant. Furthermore, preparative chromatography—which was first developed by Sepracor Inc. (Brandt et al. 1988)—makes membranes for chromatography particularly appealing for the purpose of purifying vast quantities of molecules. More so than flat-sheet membrane modules, hollow fibers (Brandt et al. 1988) are especially well adapted in this aspect. Lastly, compared to standard chromatographic supports, analytical chromatography membranes offer less advantages than preparative chromatography membranes.

8.2.4 Membrane Contactors

Membrane contactors have garnered increased attention in recent times. For instance, they have been documented in the process of creating emulsions (also known as "membrane emulsification") and precipitates (Charcosset 2006). The joining of phases A and B via membrane pores is referred to as a "membrane contactor." In a membrane contactor, a scattered phase A is pushed by pressure to penetrate a membrane and change into a continuous phase B that approaches the membrane surface tangentially. As a result, membrane pores function as parallel capillaries when phase A and phase B are introduced. Membrane emulsification is the process by which a reaction (precipitation preparation) or none at all may occur between the two stages.

Over the last fifteen years, membrane emulsification has changed (Joscelyne and Trägårdh 2000). The unique characteristic is that the membrane selection, not the incidence of turbulent droplet break-up, determines the final droplet size. Compared to conventional emulsification systems, the apparent shear stress is smaller because the dispersed phase breaks into small droplets through micropores more quickly than it breaks into huge droplets in high-energy density zones. It is possible to create emulsions with narrow droplet size distributions in addition to using shear-sensitive components. Additionally, compared to traditional mechanical methods (106–108 J/m^3), membrane emulsification technologies enable the creation of emulsions with reduced energy input (104–106 J/m^3).

It has been possible to manufacture both single and multiple emulsions with droplet sizes ranging from 0.8 to over 100 μm and a typical coefficient of variation of 10–15%. A range of microspheres, ranging in diameter from 2 to 100 μm, have been created through the use of the droplet swelling approach, subsequent solvent evaporation, or a combination of membrane emulsification and suspension polymerization (Ma et al. 2003). Metal solder particles, integrated TiO_2 microcapsules, bichromal particles, solid microcarriers, and solid lipid nanoparticles—which are obtained by chilling the lipid emulsion produced by membrane emulsification to room temperature—are among the other particles that are primarily produced by membrane emulsification (Charcosset 2006). Various researchers have utilized microporous membranes for membrane emulsification, such as polypropylene, polyamide, anodic porous alumina, polytetrafluoroethylene (PTFE) membranes, and Shirasu Porous Glass (SPG) membranes (Joscelyne and Trägårdh 2000).

Other uses cited are the production of nanosized $BaSO_4$ and $CaCO_3$ particles, as well as polymeric nanoparticles through interfacial polymerization or nanoprecipitation processes between an organic and an aqueous phase (Charcosset 2006). Phase A of these processes passes via the membrane pores, while phase B flows tangentially to the membrane surface. For instance, a membrane contactor was recently described (Charcosset and Fessi 2005) for the synthesis of nanoparticles. Using the filtrate side, the organic phase was forced through the membrane pores. The nanoparticles that had gathered at the pore outputs were washed away by the aqueous phase as it traveled throughout the membrane module. It was shown that nanoparticles as small as 260 nm may be produced with a 1000-Da nanofiltration membrane, 3 bar of transmembrane pressure, and 1.7 m s^{-1} of cross-flow velocity. In just 4 min, 1.8×10^{-3} m^3 of nanoparticles with an average diameter of 360 nm could be created thanks to the ultra-high fluxes (1.6 $m^3/h\ m^2$) produced by the microfiltration membrane with pore size of 0.1 μm. It was shown that the capacity to scale up this membrane contactor in comparison to alternative methods for nanoparticle synthesis and the ability to regulate nanoparticle size through suitable membrane selection are its main advantages.

The membrane's maximum disperse phase flux, which is relatively modest, ranges from 0.01 to 0.1 $m^3/(m^2\ h)$. This phase flux is a possible downside of direct membrane emulsification, but it is necessary to prevent steric obstruction among droplets that may develop simultaneously at the neighboring pores and to prevent the transition from a "size-stable" to a "continuous outflow" zone. Various operating methods were suggested to circumvent this potential constraint, including recurrent membrane extrusion of coarsely pre-emulsified feeds (Vladisavljević et al. 2004) and revolving membranes. For example, Vladisavljević used multi-stage (repeated) premix emulsification to create different w/o/w emulsions using SPG membranes with a mean pore size of 10.7 μm (Vladisavljević et al. 2004). To achieve additional droplet homogenization and size reduction, a coarse emulsion containing droplets with a mean particle size of around 100 μm was homogenized five to six times using the same membrane at a constant pressure differential of 20–300 kPa. To become a standard in industrial practice, membrane contactors will need to compete with alternative emulsification and particle preparation techniques.

8.3 Textile Industry Wastewater Treatment with Membrane Systems

The textile industry is one of the largest water-consuming industries in the world, and its wastewater contains complex pollutants such as dyes, degradable organics, detergents, stabilizing agents, de-sizers, inorganic salts, and heavy metals. In the context of environmental protection and circular economy, water-recovery treatment in the textile industry is an urgent requirement for the world to develop sustainably. Membrane systems have emerged as an efficient alternative for treating textile industry wastewater due to their ability to selectively remove a wide range of contaminants while offering advantages such as high efficiency, compact design, scalability, and environmental impact. Most of the review research done so far has been conducted in the mention of single membrane without showing its existence in a system. Thus, this review provides a comprehensive view of the treatment of textile wastewater by membrane systems. For this purpose, single membrane systems and combined membrane systems are presented separately. In fact, a single membrane system does not provide the desired water quality enabling reuse or require high energy consumption; thus, combined membrane systems can be used to increase the removal efficiency, reduce energy costs, and be eco-friendly.

8.3.1 Introduction

Today, the textile industry is one of the world's greatest economic sectors in terms of output, employment, and GDP (Siliņa et al. 2024). The global textile sector is estimated to be worth US$1 trillion, accounting for 7% of total global exports and employing around 35 million people (Global Market Report on Sustainable Textile 2010). At the same time, it is thought to consume more water than any other industry, resulting in a great environmental impact on people and the earth (Christian et al. 2023). According to the World Fund for Nature, growing 1 kg of cotton (equivalent to making one pair of blue jeans) requires approximately 8500 L of water (Desore and Narula 2018). An average-sized textile industry can consume 200 L of water for every kg of fabric processed within the industry. Therefore, the textile industry consumes approximately three trillion gallons of fresh water worldwide to produce 60 billion kilograms of fabric (Christian et al. 2023). Textile industries consume 80% of the total production of 130,000 tons of dyes which have a yearly growth rate of 8.13% between 2016 and 2023 (Samsami et al. 2020). Moreover, according to an estimation of industries in the world, more than 35% of chemicals released into the environment are a result of various textile treatment and dyeing processes (Desore and Narula 2018). With emerging challenges such as climate change (Bandh, Suhaib A., et al., 2025) resource scarcity (Pham Van Toi et al., 2025), rigorous regulatory regimes, environmental security (Nguyen, To The, et al. 2025), psychological health (Ta Qui Thanh, et al., 2025) and demand for sustainable textiles, as well as increasing textile consumption, the issue of environmental damage caused by textile production must be addressed promptly.

Many different methods have been researched and applied to deal with textile wastewater including physical, chemical, and biological methods (Christian et al. 2023; Kallawar and Bhanvase 2024; Siddique et al. 2017; Halepoto et al. 2022). Traditional biological treatment methods are often ineffective due to the complex characteristics of textile industry wastewater (TIWW) containing many toxic substances as well as difficult-to-biodegrade substances. While physical and chemical processes require a high cost of chemical consumption and a negative impact on the environment (Keskin et al. 2021; Adane et al. 2021). Thus, membrane systems have emerged as an efficient alternative for treating TIWW due to their capacity to manage a wide range of contaminants while offering advantages such as high efficiency, compact design, scalability, low operational cost, and environmental impact (Keskin et al. 2021; Adane et al. 2021).

This paper aims to review the application of membrane systems to deal with TIWW. The structure of this paper is organized into four parts: Sect. 8.1 is an introduction; Sect. 8.2 describes characteristics of textile wastewater; Sect. 8.3 shows the application of single-membrane; Sect. 8.4 presents integrated membrane systems; Sect. 8.5 introduces operational cost; and Sect. 8.6 presents conclusions and recommendations.

8.3.1.1 Characteristics of TIWW

Wastewater from the textile industry contains metals, dyes, and other contaminants. The features of textile effluent vary based on the type of textile, production process, fabric generated, chemicals used, and so on (Adane et al. 2021; Yaseen and Scholz 2019; Othman 2021). Textile industries are classified based on the fabrics they generate, including cellulosic materials derived from plants, animal protein fibers, and synthetic fabrics manufactured artificially (Othman 2021). To form products, these raw materials must go through a series of stages such as sizing, de-sizing, sourcing, bleaching, mercerizing, dyeing, printing, and finishing techniques (Kallawar and Bhanvase 2024; Adane et al. 2021). Each of these stages creates wastewater sources with different characteristics as shown in Fig. 8.2.

Overall, textile wastewater is a complicated mixture. Recent statistics on the characteristics of TIWW are shown in Table 8.1 with notable features:

- Textile processing relies heavily on organic ingredients such as chemicals, dyes, and sizing agents. Textile wastewater frequently has a chemical oxygen demand (COD) that is significantly greater than the biological oxygen demand (BOD) (see Table 8.1). This imbalance indicates that the wastewater includes complex organic compounds that are difficult for bacteria to break down;
- Textile operations commonly use sodium-based salts, leading to high TDS levels in wastewater. This salinity can kill freshwater creatures and render water unsuitable for irrigation;

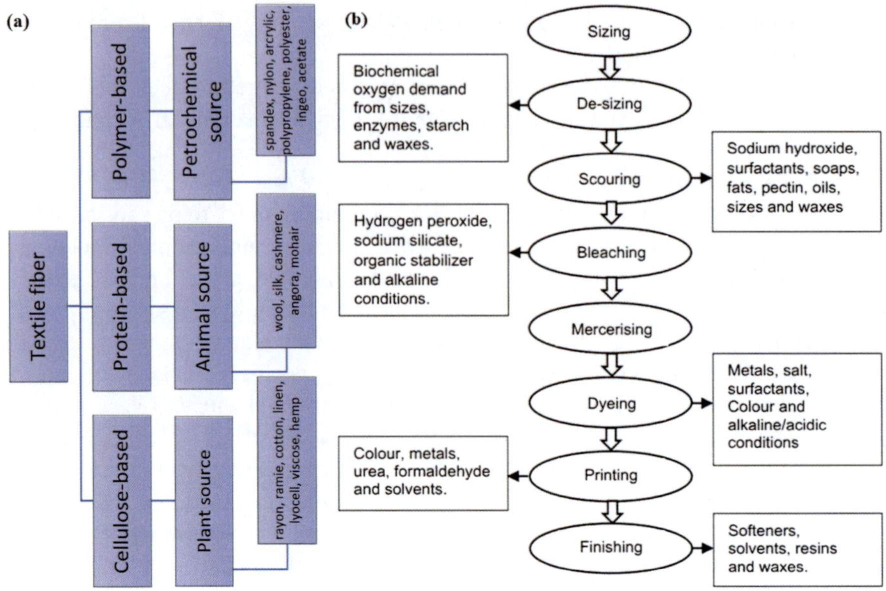

Fig. 8.2 **a** Classification of textile fiber; **b** textile processing, adapted from Yaseen and Scholz (2019); Othman 2021)

- Textile effluent may contain heavy metals like chromium, copper, and arsenic, which can build in the food chain, as well as acids and alkalis used in the dyeing and printing stages, which can harm aquatic life and modify the pH of the wastewater.

The textile industry not only generates wastewater with complex properties but also generates a huge amount of wastewater, accounting for 20% of global industrial wastewater (Keskin et al. 2021). Therefore, biological, chemical, or physicochemical methods either cannot treat thoroughly or have very high costs (Adane et al. 2021; Othman 2021). In that context, membrane technology is considered potential because of: (i) the ability to selectively remove substances from wastewater; (ii) its low cost; and (iii) its small system.

8.3.1.2 Membrane Technology for Treatment of TIWW

Membrane technology has emerged as an effective solution for treating TIWW due to its ability to handle a wide range of contaminants. Classification of membrane technology can be separated into two groups as shown in Fig. 8.3. In which, physical technologies are sub-classified as microfiltration (MF), ultrafiltration (UF), nanofiltration (NF), and reverse osmosis (RO) (Keskin et al. 2021). Moreover, a more complex technique that combines a biological treatment process with a membrane

Table 8.1 Characteristics of TIWW

References	Ghaly et al. (2014)	Kehinde and Aziz (2014)	Hussein (2013)	Upadhye and Joshi (2012)	Alcaina-Miranda et al. (2009)	Eswaramoorthi et al. (2008)
Temp. (°C)	35–45	21–62	33–45	35–45		35–45
pH (−)	6–10	6.95–11.8	5.5–10.5	6–10	7–9	6–10
Color (Pt–Co)	50–2500	50–2500		50–2500	50–2500	
COD (mg/l)	150–12,000	150–30,000	150-10,000	150-10,000	150–12,000	1000–1500
BOD (mg/l)	80–6000	80–6000	100–4000	100–4000	80–6000	300–500
TSS (mg/l)	15–8000	15–8000	100–5000	100–5000	15–8000	200–400
TDS (mg/l)	2900–3100	2900–3100	1500–6000	1800–6000	2900–3100	8000–12,000
Chlorine (mg/l)	1000–6000				1000–1600	
Chlorides (mg/l)			200–6000	1000–6000		3000–6000
Free chlorine (mg/l)	< 10					< 10
TA (mg/l) as CaCo3		17–22	500–800	500–800		
TH (mg/l) as $CaCo_3$						
TKN (mg/l)	70–80	70–80	70–80	70–80	70–80	
TNK(mg/l)	10–30					10–30
NO_3-N (mg/l)	< 5					< 5
Free ammonia (mg/l)	< 10					< 10
Na_2CO3 (mg/l)						
NaOH (mg/l)						
NaCl (mg/l)						
Sulphates (mg/l)	600–1000		500–700			600–1000
Sulphides (mg/l)			5–20			

(continued)

Table 8.1 (continued)

References	Ghaly et al. (2014)	Kehinde and Aziz (2014)	Hussein (2013)	Upadhye and Joshi (2012)	Alcaina-Miranda et al. (2009)	Eswaramoorthi et al. (2008)
Sulphur trioxide (mg/l)						
Oil and grease (mg/l)	10–30	5–5.5	10–50			10–30
Dye (mg/l)						
Zink (mg/l)	< 10		3–6			< 10
Nickel (mg/l)	< 10					< 10
Manganese (mg/l)	< 10					< 10
Iron (mg/l)	< 10					< 10
Copper (mg/l)	< 10		2–6			< 10
Boron (mg/l)	< 10					< 10
Arsenic (mg/l)	< 10					< 10
Silica (mg/l)	< 15					< 15
Mercury (mg/l)	< 10					< 10
Fluorine (mg/l)	< 10					< 10
Chromium (mg/l)			2–5			
Potassium (mg/l)			30–50			
Sodium (mg/l)	7000		400–2175	610–2175		7000

Notes *COD*—chemical oxygen demand, *BOD*—5-day biochemical oxygen demand, *TOC*—total organic carbon, *EC*—electrical conductivity, *TS*—total solids, *TSS*—total suspended solids, *TDS*—total dissolved solids, *TVS*—total volatile solids, *TA*—total alkalinity, *TH*—total hardness, *TKN*—total kjeldahl nitrogen, *TN*—total nitrogen, NO_3-*N*—nitrate-nitrogen, Na_2CO_3—sodium carbonate, *NaOH*—sodium hydroxide, and *NaCl*—sodium chloride

filtration process, called a bioreactor (MBR), is also playing an increasingly important role in textile wastewater treatment.

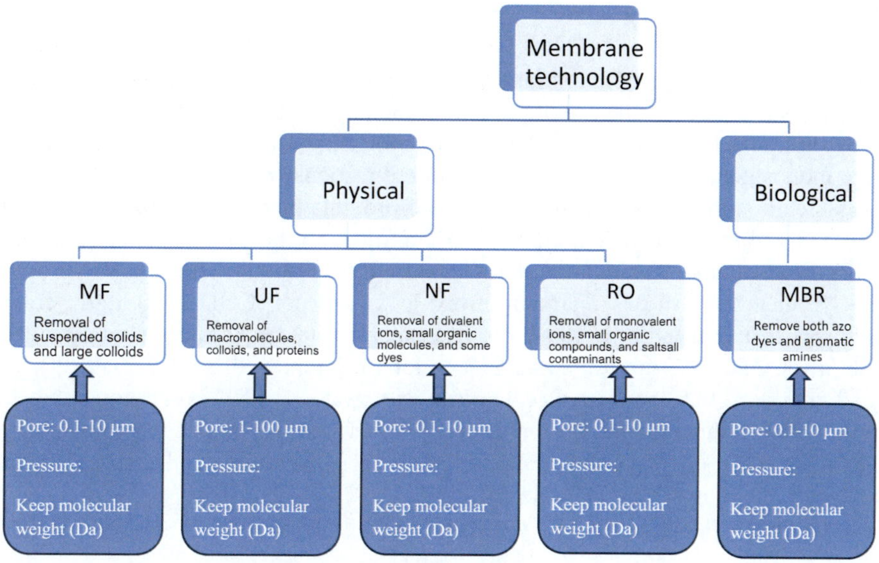

Fig. 8.3 Classification of membrane technologies and their basic applications in treatment of TIWW (Samsami et al. 2020; Periyasamy 2018)

8.3.1.2.1 Microfiltration

Microfiltration is considered a cost-effective membrane operation capable of handling ample liquid volume at low membrane pressure (Kallawar and Bhanvase 2024). With pore size within 0.1–10 µm, MF can remove suspended solids and large colloids in the water stream but cannot remove dissolved substances; therefore, it is suitable to apply for treatment of TIWW at the stages producing highly suspended solid materials such as fiber manufacturing, spinning/weaving, mercerization, dyeing, and printing, or it could play the role of pretreatment in a complete treatment system (Keskin et al. 2021**)**. Reports on employing only MF membranes for textile wastewater treatment are limited since MF membranes cannot remove dye molecules via size exclusion due to dye molecules being smaller than the average MF membrane pore size (Othman 2021). To improve the effectiveness of MF, ceramic MF has been recently developed. Saini, Bulasara (Saini et al. 2019) designed a new ceramic MFM based on kaolin with a porosity of 40.2%, pore size of 0.27 µm, and permeability of 1376 L/m^2.h.bar. The treatment effectively removed 25% COD, 21% turbidity, and 26% color as well as 100% TSS from textile effluents. Using natural magnesite, Manni et al. (2020) created a ceramic membrane with excellent thermal stability. A membrane with 48.15% porosity, 1.14 µm pore size, and 922 L/m^2 h bar water permeability was used to remediate effluent from the textile sector. It was claimed that 69% COD and 99% turbidity were successfully removed upon use.

8.3.1.2.2 Nanofiltration

Nanofiltration is a type of membrane separation technique that separates the matter from the wastewater by the use of pressure (much higher than in the MF technique). Due to pole size in the range of 1–10 nm, low-molecular-weight organic compounds, large monovalent and divalent ions, hydrolyzed responsive dyes, and dyeing auxiliaries are all stored in NF membranes. Therefore, this kind of membrane has been used widely for the management of color control from the TIWW (Siddique et al. 2017). For example, Tavangar et al. (2019) tested the effectiveness of a polypropylene NF membrane in the treatment of textile wastewater (COD 2690 mg/L, color 2647 pt–Co, and turbidity 1500 NTU). The experimental result in this study showed that 74% COD, 99% turbidity, and 87% color were removed. Aouni et al. (2012), Bes-Piá et al. (2009) also used real textile wastewater for NF experiments, but at a much lower contaminant concentration (COD 708 mg/L). The test obtained very high removal levels of contaminations (94–99% COD removal). Details of these experiments are presented in Table 8.2.

8.3.1.2.3 Ultrafiltration

Ultrafiltration can be used to remove particles larger than 10 nm and utilized as a secondary or pretreatment step in textile wastewater processing. Ultrafiltration membranes have larger pores than nanofiltration membranes. The use of UF membranes in textile wastewater treatment has been well-known in the literature. For instance, Barredo-Damas et al. (2012) performed ultrafiltration of wastewater from yarn production having characteristics of COD 1240–1740 mg/L, conductivity 5950–8720 μS/cm. Upon the separation process, the removal efficiency was obtained by a membrane with 85% COD, 88.5% color, and 12.5% conductivity. Aouni et al. (2012) examined a performance comparison between UF and NF membranes in textile wastewater treatment with its characteristics: COD of 708 mg/L, the conductivity of 3840 μS/cm. The treatment by UF was 75% COD and 30% conductivity removal. While higher efficiency was achieved in NF with 90% COD and 80% conductivity removal. These results are due to UF membranes having larger pores than NF membranes. This leads to higher flux and improved mobility of the multivalent salt ions, making it more appropriate for the separation of dye and salt in textile wastewater (Christian et al. 2023). More similar results are presented in Table 8.2.

8.3.1.2.4 Reverse Osmosis

Reverse osmosis is a widely used technology for treating textile wastewater. RO works on the principle of applying pressure to overcome osmotic pressure, forcing water molecules through a semi-permeable membrane while rejecting most dissolved salts, organics, bacteria, and particulates. Therefore, the RO membrane can decolorize and remove chemical auxiliaries from dyehouse wastewater all in one step

Table 8.2 Illustrative applications of the physical membrane in the treatment TIWW

No	Type of production/ wastewater source	Characteristic	Membrane characteristics	Filtration capacity	Treatment performance	References
1	Dyes and chemical substances production Finishing units	pH 5.5–12 Cond. 1850–6400 μS/cm Turbidity 51–176.5 NTU COD 1530–2750 mg/L	**NF1:** Spiral wound, NF270, Dow Chemical, Polyamide **NF2:** Spiral wound, Duraslick NF, GE Osmonics, Polyamide	**NF1:** 104 L/ m2·h at 10 bar **NF2:** 88 L/ m2·h at 10 bar	**NF1:** 91% COD, 94% Color, 69% Conductivity, 96% Turbidity Removal **NF2:** 98% COD, 98% Color, 50% Conductivity, 98% Turbidity Removal	Alcaina-Miranda et al. (2009)
2	Textile and nontextile dye production Dyeing, printing and rinsing units	pH 7.11 and Color 0.12 Pt Co Conductivity 3840 μS/ cm Turbidity 4.02 NTU COD 708 mg/L	**UF:** Flat sheet, 1 kDa, Synder Filtration, Polyethersulfone, 1–100 nm **NF:** Flat sheet, NF270, Dow/ Filmtec (USA), Polyamide, 0.84 nm	**UF:** 120 L/ m2·h at bar **NF:** 145 L/ m2·h at 15 bar	**UF:** 75% COD, 90% Color, 30% Conductivity Removal **NF:** 90% COD, 90% Color, 80% Conductivity Removal	Aouni et al. (2012)
3	Yarn production Dyeing, printing and finishing units	pH 8–12 Cond. 5950–8720 μS/cm Turbidity 38–65 NTU COD 1240–1740 mg/L	**UF:** Multichannel tubular shape of ceramic membrane, Tami Industries, Inside-C´ eRAM™	**UF:** 120 L/ m2·h bar at 2 bar	**UF:** 85% COD, 88.5% Color, 12.5% Conductivity, 96% Turbidity, 91% SS Removal	Barredo-Damas et al. (2012)
4	**Yarn and fiber production** **Dyeing, printing, washing and rinsing units**	**pH 7.2–11.4** **Color 0.01–1.76 Pt Co** **Conductivity 300–6110 pS/cm COD 34–1111 mg/L**	**NF:** NF270, Dow/Filmtec (USA), Polyamide, **RO:** XLE, Dow/Filmtec (USA), Polyamide	**NF:** 45 L/ m2·h at 6 bar **RO:** 26 L/ m2·h at 20 bar	**NF:** 96% COD, 99% Color, 71% Conductivity Removal **RO:** 95% COD, 99% Color, 96.4% Conductivity Removal	Kurt et al. (2012)

(continued)

Table 8.2 (continued)

No	Type of production/ wastewater source	Characteristic	Membrane characteristics	Filtration capacity	Treatment performance	References
5	Cotton and polyester fabrics production Dyeing and finishing units ok	pH 7.03 Conductivity 14.95 mS/cm Turbidity 1500 NTU COD 2690 mg/L	**NF:** Flat sheet, support layer polypropylene, NADIR NP010	**NF:** 50 10L/m bar2·h at	**NF:** 74% COD, 99% Turbidity, 87% Color Removal	Tavangar et al. (2019)
6	Textile dye production Dyeing, bleaching, printing and finishing units	pH 7.76 Conductivity 976 µS/cm Turbidity 148 NTU COD 2830 mg/L	Hydrophobic and Hydrophilic-Coated Polytetrafluoroethylene (PTFE)	**MD:** 30 L/ m2·h	**MD hydrophobic:** 98.2% COD, 98.9% Conductivity, 98.9% Salt Removal **MD hydrophilic:** 93.4% COD, 99.5% Conductivity, 99.5% Salt Removal	Villalobos García et al. (2018)
7	Textile dye production Dyeing and printing units	pH 8.04 Conductivity 8620 µS/cm Turbidity 0.948 NTU COD 329.4 mg/L	**NF:** NF270 Flat sheet, polyamide, 2 nm	**NF:** 85 L/ m2·h at 11 bar	**NF:** 93% Color, 65% Conductivity, 60% SS, 91% Turbidity Removal	Fersi and Dhahbi (2008)

(Siddique et al. 2017). Apply RO membrane and NF membrane in the treatment of wastewater from dyeing and printing units, Li et al. (Li et al. 2020) observed the same ability of color removal in both membranes. However, the conductivity removal capacity of the RO membrane (96.4%) was significantly higher than that of the NF membrane (71%). Sahinkaya et al. (2019) obtained the results of 93% color removal and 98% conductivity removal when treatment wastewater (color of 1000 pt–Co and conductivity of 5.5 mS/cm) from dyeing units. This result is often difficult to observe with other types of membranes. However, the water recovery rate in this study was only 70%. Low water recovery rate and energy-consuming operation (high pressure) are the two main disadvantages that lead to consideration when using RO membranes.

8.3.2 Membrane Bioreactor

The membrane bioreactor (MBR) differs from typical biological treatment processes in that it separates treated wastewater from activated sludge. The MBR process has the following advantages: low maintenance, less sludge production, a small footprint, constancy in final treated water quality regardless of sludge conditions in the bioreactor, and high organic and nutrient removal (Al-Asheh et al. 2021). Furthermore, one of the primary advantages of MBR treatment over traditional dye removal procedures is its high efficacy in keeping microorganisms/biocatalysts (Samsami et al. 2020). This process is divided into different stages: (i) the wastewater enters the reactor and comes into contact with the biomass, (ii) the mixture of feed wastewater and biomass is fed to the membrane unit, and the retentate is recycled after the permeate is discharged into the separation unit (Siddique et al. 2017). MBR systems use membranes with pore sizes ranging from 10^{-4} to 10^{-3} nm, which can be referred to as MF or UF. MBRs are frequently classed according to their aeration strategies/steps (Samsami et al. 2020).

Aerobic MBR (AeMBR) which requires aeration to supply oxygen, is widely used for textile wastewater treatment, achieving COD removal efficiencies > 90%. For example, 94.8% COD removal and 72.9% dye removal were reported (Yun et al. 2006). Hoinkis et al. (2012) tested COD concentration in the range of 600–1250 mg/L and ascertained the efficiency of an MBR system. The removal efficiency for COD, ammonium, and color was > 90%, > 90%, and in the range of 60–75%, respectively. (Samsami et al. 2020). However, AeMBR systems are generally recommended to operate effectively for wastewater streams with COD concentrations < 1000 mg/L (Al-Asheh et al. 2021).

In contrast to AeMBR, anaerobic MBR (AnMBR) uses anaerobic microorganisms that do not require oxygen to decompose organic matter. So, compared with AeMBR, AnMBR has the advantage of lower operating costs, less sludge generation, and even energy generation. And also AnMBR is recommended for use with the waste stream having COD > 4000 mg/L when it's not suitable to apply for AeMBR (Al-Asheh et al. 2021). Yurtsever et al. (2015) evaluated the use of MBR (both anaerobic and aerobic) to treat synthetic textile wastewater. The aerobic MBR's dye removal effectiveness was only 30–50%, whereas the anaerobic MBR removed 100% of the color. However, this study also shows that AeMBR treated organic matter more thoroughly than AnMBR as known in the literature. In addition, AnMBR is less effective in removing NH4 + (Al-Asheh et al. 2021). Therefore, many recent studies suggest that AnMBR and AeMBR should be used in the same system to meet discharge quality (Al-Asheh et al. 2021).

The performance of MBR systems depends not only on factors such as wastewater characteristics, membrane type, and membrane configuration but also on conditions related to microbial growth such as mixed liquor suspended solids (MLSS) concentration, food-to-mass ratio (F/M) ratio, organic loading rate (OLR) and hydraulic retention time (HRT) (Keskin et al. 2021; Al-Asheh et al. 2021). Some illustrations of MBR applications for textile wastewater treatment are presented in Table 8.3.

Table 8.3 Illustrative applications of the MBR for the treatment of TIWW

No	Type of wastewater	Characteristics	Type of membrane	Operation	Removal	References
1	Clothing production industry Dyeing units	pH 9.4 Color 153 mgPt/L Conductivity 1340 μS/cm Turbidity 20.8 NTU COD 840 mg/L	**MBR:** Norit X-flow Airlift **RO:** Spiral Wound	**MBR:** 20 L/m2·h **RO:** 5 L/m2·h at 12.6 bar	**MBR:** 75% COD, 28.6% Color, 94% Turbidity, 28.6% Conductivity Removal **RO:** 87% COD, 95% Color, 97% Turbidity, 93% Conductivity	Jager et al. (2014)
2	Cotton fabric industry Dyeing, bleaching, printing and finishing units	pH 10.05 Color 141.5 mg Pt/L Conductivity 2993 μS/cm Turbidity 67.41 NTU COD 815 mg/L	**MBR:** Submerged flat sheet UF, Polyvinylidene fluoride, Sinap, SINAP-25, < 100 nm **NF:** Spiral wound, NF90, Dow, 1 nm,	**MBR:** 60 L/m2·min	**MBR:** 75.4% COD, 99% Turbidity, 59.2% Color **NF:** 97.3% COD, 99.6% Turbidity, 96.5% Color	Li et al. (2020)
3	Woolen fabric production Spinning, dyeing and finishing units	pH 7.5 Color 435 Pt Co Conductivity 3463 μS/cm Turbidity 148 NTU COD 918 mg/L	modules (GEMINI-RF-500 × 4	Capacity 6 m^3/d; Flux 10 L/m^2-h Pressure 1.5–2.0 bar	11.4% conductivity, 70% COD, 62% color, and 99% turbidity	Cinperi et al. (2019)

8.3.3 Assessments

Figure 8.4 shows a general performance treatment of textile industrial wastewater using different membranes based on the data in Table 8.1 and the statistical data in the report of Christian et al. (2023). Accordingly, with the characteristics of high substrate concentration as shown in Table 8.1, MF or UF membranes cannot process TIWW to achieve quality; they can only play a pretreatment role. Both NF and RO membranes have great performance, but if used directly, they will encounter fouling problems. Therefore, they can act as main processing along with pretreatment or as post-treatment. It can be seen that the COD and color removal efficiency between NF and RO are equivalent. So, under normal conditions, it is recommended to use NF instead of RO because of lower energy costs and higher pure water ratio. In cases where it is necessary to remove additional inorganic ions or metals, NF cannot be performed; then RO must be used.

8.4 Integrated Membrane Systems

In some cases, because a single membrane process does not provide the desired water quality for reuse, a hybrid/combined membrane system can be used to increase the removal efficiency of organics, dyes, and pollutants. colors and ions. Therefore, many recent studies have paid more attention to how to combine different types of membranes to limit their disadvantages and combine their advantages to form

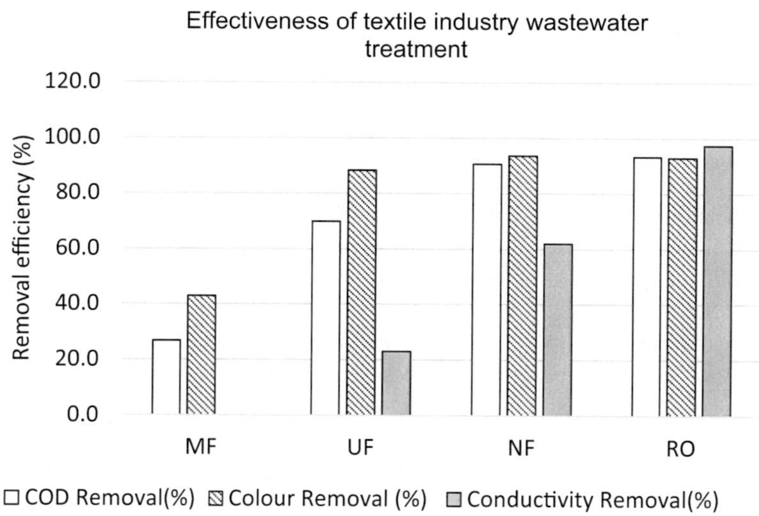

Fig. 8.4 Comparing performance between membrane types in textile industrial wastewater treatment

a system that can treat complex wastewater like TIWW. For the same purpose, membrane technology can also be combined with other types of technology in the same system.

8.4.1 Combined Membrane-Membrane Systems

The TIWW treatment process can be divided into sequential stages including suspended matter removal, dissolved organic removal, and ions/salt removal. Accordingly, the functional allocation of each type of membrane into a treatment system is clearer and shown in Fig. 8.1.

Tahri et al. (2012) performed the treatment of dyeing wastewater from the dyeing unit of the cotton fabric process by the MF + NF system. Color, turbidity, SS, conductivity, and COD removal efficiency were obtained at 97.2%, 99.5%, 100%, 42.4%, and 75.5%, respectively. According to the findings of this investigation, the MF + NF system was found to be effective in treating textile effluent and reusing it as process water. More complicated, Sahinkaya et al. (2019) set up a configuration of UF + RO system to deal with wastewater from dyeing units of cotton and polyester fabrics production due to the low quality of wastewater in the current treatment system. In this system, the UF membrane plays the role of removing particulates and reducing the load injected. The result obtained by the system was to remove 98% conductivity, 93% color, and 76% COD. The flux of RO membrane was kept at 19 $L/m^2{\cdot}h$ at 8 bar.

Cinperi et al. (2019) tested the treatment of TIWW in woolen fabric production textile mill industry by different processes. Only the MBR process was employed, the treatment system removed 11.4% conductivity, 70% COD, 62% color, and 99% turbidity. So the MBR method alone has been shown to not provide sufficient quality of permease as required. Because of that, BF was installed after MBR. The treatment system could achieve removal of 35% conductivity, 97% COD, 80% color, and 98% turbidity. In the next experiment, the previous NF membrane was replaced by an RO membrane. At that time, treatment effect removal was of 97% conductivity, 99% COD, 85% color, and 99% turbidity removal. Following the results, the NF or RO process should be used after the MBR process for a complete treatment process. Using another configuration of UF + MBR, De Jager et al. (2014) had a removal efficiency of 75% COD, 94% turbidity, 28.6% color, and 28.6% conductivity. Although the effluent from the current system met discharge regulations, RO membrane was added after MBR to further remove residual color and conductivity. At that time, removal efficiencies of COD, turbidity, color, and conductivity were 87%, 97%, 95%, and 93%, respectively (Table 8.4).

Table 8.4 Illustrations of the combined membrane-membrane systems

No	Type of wastewater	Characteristics	Combination	Removal	References
1	Clothing production industry Dyeing units	pH 9.4 Color 153 mgPt/L Conductivity 1340 μS/cm Turbidity 20.8 NTU COD 840 mg/L	UF + MBR	75% COD, 94% turbidity, 28.6% color and 28.6% conductivity	Jager et al. (2014)
			UF + MBR + RO MBR: Norit X-flow Airlift RO: Spiral Wound	87% COD, 97% turbidity, 95% color and 93% conductivity,	
2	Woolen fabric production Spinning, dyeing and finishing units	pH 7.5 Color 435 Pt Co Conductivity 3463 μS/cm Turbidity 148 NTU COD 918 mg/L	MBR + NF: 10 L/m2·h at 5–7 bar	97% COD, 80% Color, 98% Turbidity and 35% Conductivity	Cinperi et al. (2019)
			MBR + RO: 13 L/m2·h at 8–12 bar	99% COD, 85% Color, 99% Turbidity and 97% Conductivity	
3	Cotton fabric industry Dyeing unit	pH 9.23–10.65 TDS 9.54–11.42 mg/l COD 2160–2450 mg/l	MF: 163.5 L/m2·h at 2 bar NF: 45 L/m2·h at 6 ba	**73–85% % COD** 100% color	Lafi et al. (2018)
4	Cotton and polyester fabrics production Dyeing units	pH 9 Color 1000 Pt.Co Conductivity 5.5 mS/cm COD 1000 mg/L	UF: 50 LMH RO:19 L/m2·h at 8 bar	76% COD, 93% Color, 98% Conductivity	Sahinkaya et al. (2019)

8.4.2 Combine Membrane and Other Processes

The simplest combination is to use another simple physical filtration process as a pretreatment before using the membrane. For instance, Marcucci et al. (2002) compared sand filtration (SF) and MF as a pretreatment followed by NF process in wastewater treatment from dyeing departments. In this study, there was no significant difference in removal of SS (100%) and turbidity (78%) between SF and MF. Efficiency in pretreatment was to completely remove COD, conductivity, and color by NF. Marcucci et al. (2003) employed sand filtration (SF) before using MF/UF to deal with textile industry wastewater. The result achieved nearly 99% TSS removal and 80% turbidity removal. However, the configuration of SF + UF was significantly better than SF + MF in removing COD and color from wastewater. Both cases had to use post-treatment by RO and NF to completely remove color.

More complicated, Hassanzadeh et al. (2017) applied coagulation process as for pretreatment purpose prior to NF process treating woolen textile wastewater. The feed stream characteristics were COD (1560–3420 mg/L), BOD5 (558–1450 mg/L), turbidity (41–65 NTU). The study found that the optimal pretreatment conditions were 600 mg/L FeSO4 at pH 8. At that time, the removal effectiveness of turbidity and COD could reach 98%. In fact, coagulation/flocculation has been a widespread and important physicochemical treatment used in textile wastewater treatment. Advantage has been discovered to be a cost-effective, easy-to-use, and energy-saving treatment option. However, the main drawbacks of the coagulation/flocculation method include sludge generation and ineffective color removal of some soluble dyes (Othman 2021). In the textile industry, the majority of colors used are negatively charged. As a result, cationic polymer coagulants are frequently used for effective dye removal and degradation. The common coagulants are poly-aluminum chloride, ferric chloride, and magnesium chloride (Othman 2021). The dosage is determined by the type of pollutants, the pH of the water, the turbidity, and the temperature (Kallawar and Bhanvase 2024).

Tavangar et al. (2019) found that electrocoagulation (EC) or membrane filtration alone were insufficient to achieve the requisite treatment performance, therefore, NF was applied after the electrocoagulation procedure. The EC process eliminated 99% turbidity, 64% COD, and 94% color, whereas the NF method achieved removal efficiencies of 99%, 74%, and 87%. And the result from EC + NF system was 99.6% turbidity, 82.4% COD, 89% color removal. Another study, Lafi et al. (2018) combined UF and electrodialysis processes for the treatment of cotton fabric primary treated textile wastewater. UF process was used to remove macromolecules, colloids, and color; electrodialysis was used to separate the ionic species. Two types of operation mode with only UF process and UF + ED combination were investigated in this study. With UF + ED system, conductivity, turbidity, and COD removal efficiencies were found as 97%, 100%, and 77%, respectively.

Some recent studies have combined membrane filtration with chemical processes, typically advanced oxidation processes (AOP). For example, Sisira and Sheena (2020) applied MF followed by the heterogeneous fenton to treat textile wastewater (COD 1210 mg/L, turbidity 60 NTU) which could not be treated to meet the requirements. The heterogeneous fenton plays the role of oxidation of residual chemicals. The removal of COD and color could reach 80.4% and 88.6%, respectively. Lebron et al. (2021) compared two systems including MF-NF-AOP and MF-AOP-NF for treatment of wastewater collected after the rinsing step of the dyeing process. In both systems, the MF successfully separated 99.1% of the indigo blue, which can be reused. The MF also removed 65.4% COD and 99.8% color. The performance of NF-AOP step in the first system was 69.8% COD removal but did not meet the reuse criteria. While AOP-NF step in the second system could remove 78.5% COD, resulting in the treated water meeting recycling requirements.

Besides the combined use of membrane technology with physical, chemical, physicochemical, and electrochemical processes, membrane applications can be employed as a follow-up to a biological treatment procedure in case of over-pollutant load or insufficient treatment. Bes-Piá et al. (2009) used an activated sludge technique, followed by NF membrane to treat textile wastewater. Because the activated sludge process did not produce the needed treated wastewater quality for reuse in the textile mill, an NF unit was added to the system as a post-treatment step (Bes-Piá et al. 2009). In another study, the biological activated sludge process was employed as a pretreatment method, followed by membrane filtration processes (UF + RO) to treat wastewater from dyeing and finishing units in a textile industry. In the study, a hybrid system of UF and RO was used to reuse treated textile wastewater in the textile plant. The hybrid process achieved 99% conductivity, 76% COD, and 95% color removal efficiency (Ciardelli et al. 2001) (Table 8.5).

8.5 Economic Perspective

From the data collected in this study, it is feasible to apply membrane technology to treat and recover textile industrial wastewater according to a technical perspective. Even superior to other technologies. That is why research on membrane application for wastewater treatment has been increasing in recent years (Nthunya et al. 2024). Recently, Adane et al. (2021) conducted a comparative assessment of the costs of textile wastewater treatment using different types of technology. They claimed that a combination of MF + NF membrane filtration with the AOP process resulting an operating cost of 0.42–0.74 USD/m^3 which are much lower than physical, electrochemical, and even chemical processes. Some estimates of the operating costs of membrane systems in TIWW treatment are shown in Fig. 8.5. Ciardelli et al. (2001) investigated the water recovery from textile wastewater using a combination of SF/MF and NF. The author showed that to completely remove COD, the treatment cost was about 1 USD/m^3. Thus, by using it in conjunction with the AOP process, membrane filtration system costs can be reduced. Li et al. (2020) conducted an MBR-NF system having a capacity of 2.4 m^3/d for the treatment process of the textile mill. By applying this technology to a capacity scale of 3000 m^3/day, the estimated cost was 0.9 USD/m^3. More complicated, Cinperi et al. (2019) applied an MBR combined with an RO system to recover water from textile wastewater. By this way, turbidity and color were reduced by 97% and 74%, respectively. The analysis was conducted for a full-scale application showing an operation cost of 0.78–1.0 USD/m^3.

Although these studies were not estimated under the same conditions of wastewater characteristics, treatment capacity or treatment efficiency, they still partly reflect actual operating costs. Similar evaluation studies need to be performed to obtain sufficient data to draw firm conclusions.

Table 8.5 Illustrations of the membrane system combined with other processes

No	Type of wastewater	Characteristics	Combination	Removal	References
1	From the dyeing of cotton fabric process	pH 8.6 Conductivity 7850 mS/Cm TDS 2980 mg/L COD 220 mg/L	UF: 250 L/h.m^2.bar; 6 CFV m/s; 2.05 bar ED: 64 cm^2; d 0.5 mm; 15 V	94.2% TDS 97.1% conductivity77% COD	Lafi et al. 2018)
2	Woolen textile wastewater	pH 5.3–9.2 Turbidity 41–65 NTU COD 1560–3420 mg/L	Coagulation—flocculation: 600 mg/L FeSO4 at pH 8 and NF: spiral wound polyamide nano filter (TFC)	98% turbidity 98% COD	Hassanzadeh et al. 2017)
3	From dyeing and finishing of cotton and polyester fabrics	pH 7.03 Conductivity 14.95 mS/cm Color 2100 Pt–Co Turbidity 1500 NTU COD 2690 mg/L	**EC: 35cm x 18cm x 7cm; area 282 cm^2;** **NF:** 50 10L/m bar2·h at	EC: 99% turbidity, 64% COD, and 94% color NF: 99% turbidity, 74% COD and 87% color EC + NF: 99.6% turbidity; 82.4% COD; 89% color	Tavangar et al. 2019)
4	Cotton fabric industry Dyeing unit	pH 8.6 Conductivity 7850 μS/cm Turbidity 7.8 NTU COD 220 mg/L	**ED:** 10 L/h, 1.98A, 40 min **UF:** 150 L/ m2·h at 5 bar	77% COD, 100% Turbidity, 97% Conductivity	Lafi et al. 2018)
5	Natural and synthetic fiber production Dyeing and finishing units		Activated sludge + **UF**: 25 L/m2·h at 4 bar **RO**: 9.26 L/m2h at 8 bar	99% conductivity, 76% COD, and 95% color	Ciardelli et al. 2001)

(continued)

Table 8.5 (continued)

No	Type of wastewater	Characteristics	Combination	Removal	References
6	Textile and nontextile dye production Dyeing, bleaching, printing and finishing units	pH 7.11 Color 0.12 Pt Co Conductivity 3840 µS/cm Turbidity 4.02 NTU COD 708 mg/L	Activated sludge + **NF1:** 28 L/m2·h **NF2:** 32 L/m2·h **NF3:** 27 L/m2·h	**NF1:** 98.8% COD, 69.2% Conductivity, 65% Salt **NF2:** 98.4% COD, 49.2% Conductivity, 45% Salt Removal **NF3:** 94% COD, 11.5% Conductivity, 20% Salt Removal	Bes-Piá et al. 2009)
7	Cotton production Dyeing units	pH 10.9 COD 1210 mg/L Turbidity 60 NTU	MF + heterogeneous Fenton	80.4% COD 88.6% color	Sisira and Sheena 2020)
8	The rinsing stage of the dyeing process	pH 7.7 Color 7129 COD 1771 mg/L	MF-NF-AOP	69.8% COD > 98.5% color (NF-AOP)	Lebron et al. 2021)
			MF-AOP-NF	78.5% COD > 92.3% color (AOP-NF)	

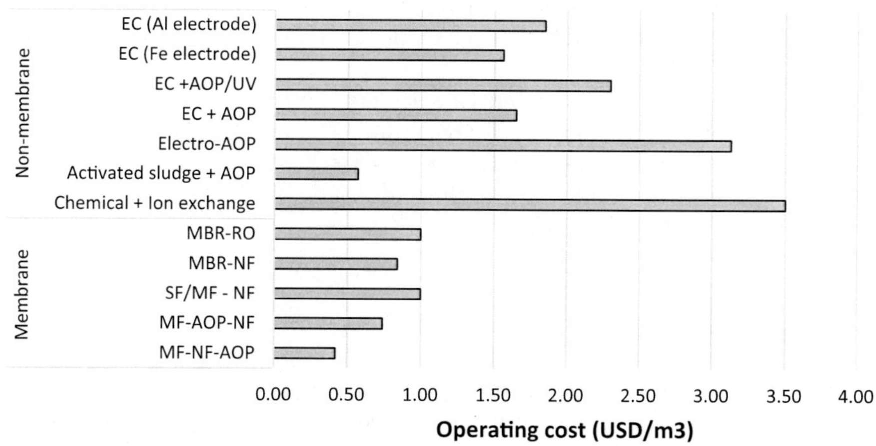

Fig. 8.5 Operating cost of the different systems for treatment of TIWW. Data compiled from Adane et al. (2021); Li et al. 2020; Cinperi et al. 2019; Ciardelli et al. 2001)

8.6 Conclusions and Recommendations

Membrane systems have high selectivity for pollutants, so they can be applied to treat heavy metal(oid)s (Tran et al. 2023), nanoparticles and recover substances in the textile industry. They also have a variety of configurations to respond to the complex nature of textile wastewater easily;

In textile wastewater treatment, membrane technology offers potentially lower operating costs than other technologies. This is the core condition for its large-scale application. Integrating different membrane technologies in the same system allows for thorough treatment of textile wastewater easily. But it can lead to high investment and operating costs. One notable potential solution is the use of membrane systems combined with AOP or coagulation.

Challenges

Costs: Initial investment costs for membrane filtration systems can be significant, especially for large-scale industrial applications. Operating costs including energy consumption and membrane replacement contribute to the overall expenses.

Membrane Fouling: Fouling happens when particles, microorganisms, or organic matter accumulate on the membrane surface, reducing efficiency and require frequent clean-up or membrane replacement activities; this is a major challenge in membrane filtration processes.

Energy Consumption: Energy-intensive processes such as membrane fouling control, backwashing, and membrane cleaning contribute to energy consumption. Minimizing energy usage while maintaining process efficiency is a significant challenge.

Regulatory Compliance: Meeting regulatory standards for treated water quality and discharge limits poses a challenge, particularly in industries with stringent environmental regulations.

Membrane Material and Durability: Selecting suitable membrane materials, such as membranes from bio-based sources (Sezali et al. 2024, Khedulkar, Akhil Pradiprao, et al. 2025) carbon-based materials (Hoang et al. 2023), with high durability, chemical resistance, and fouling resistance is essential for long-term performance and cost-effectiveness.

Future Prospects

Technological Advances: Ongoing research and development efforts are focused on enhancing membrane materials, module design, and system integration to improve performance, reduce fouling, and lower operational costs.

Integration with other advanced treatment processes: Integration of membrane filtration with advanced treatment processes, e.g. ozonation, UV disinfection, and advanced oxidation can enhance the removal of contaminants and improve overall process efficiency.

Smart Monitoring and Control Systems: Implementation of smart monitoring and control systems utilizing IoT, AI (Hoang 2022; Fernando, et al. 2022), blockchain technology (Thuy, Trang Le, et al., 2025) and machine learning (Hoang, Quy Nhan, et al., 2026) can enable real-time monitoring of process parameters, predictive maintenance, and optimization of system performance. Resource Recovery: Membrane technologies offer opportunities for resource recovery, including the extraction of valuable compounds from wastewater streams and the production of high-quality water for reuse, thereby contributing to sustainability and circular economy goals (Hoang et al. 2023; Fernando et al. 2022; Malla et al. 2023; Bandh et al. 2024).

Decentralized Treatment Systems: Membrane filtration systems, particularly MBRs, are increasingly being adopted in decentralized wastewater-treatment applications, offering flexibility, scalability, and reliability for various industrial and municipal settings.

Market Expansion: Growing awareness of the environmental and economic benefits of membrane filtration technologies is driving market expansion across diverse industrial sectors, including food and beverage, pharmaceuticals, chemicals, and electronics manufacturing.

Policy Support: Supportive policies and incentives promoting the adoption of membrane filtration technologies, such as subsidies for wastewater reuse and stricter discharge regulations, are expected to drive market growth and innovation. Good governance of water from the sources (Hoang et al. 2024; Anh and Hoang 2024) should be promoted to prevent the source water from being polluted and treated.

In summary, while membrane filtration and MBRs face challenges namely membrane fouling, high costs, and energy consumption, ongoing technological advancements, integration with advanced treatment processes, and supportive policies are expected to drive their continued adoption and growth in various industrial applications.

References

Abels C, Carstensen F, Wessling M (2013) Membrane processes in biorefinery applications. J Membr Sci 444:285–317

Adane T, Adugna AT, Alemayehu E (2021) Textile industry effluent treatment techniques. J Chem 2021(1):5314404

Al-Asheh S, Bagheri M, Aidan A (2021) Membrane bioreactor for wastewater treatment: a review. Case Stud Chem Environ Eng 4:100109

Alcaina-Miranda M et al (2009) Nanofiltration as a final step towards textile wastewater reclamation. Desalination 240(1–3):290–297

Al-Obaidi MA, Kara-Zaïtri C, Mujtaba IM (2017) Optimum design of a multi-stage reverse osmosis process for the production of highly concentrated apple juice. J Food Eng 214:47–59

Aouni A et al (2012) Reactive dyes rejection and textile effluent treatment study using ultrafiltration and nanofiltration processes. Desalination 297:87–96

Bandh SA et al (2025) Track to reach net-zero: progress and pitfalls. Energy Environ 36(5):2345–2370

Bandh SA et al (2024) Track to reach net-zero: progress and pitfalls, p 0958305X241260793

Barredo-Damas S et al (2012) Application of tubular ceramic ultrafiltration membranes for the treatment of integrated textile wastewaters. Chem Eng J 192:211–218

Belanich M et al (1996) Reduction of endotoxin in a protein mixture using strong anion exchange membrane adsorption. Pharm Technol 20(3):142–150

Belfort G, Heath CA (1994) New developments in membrane bioreactors. In: Crespo JG, Böddeker KW (eds) Membrane processes in separation and purification. Springer, Dordrecht, pp 127–148

Bes-Piá A et al (2009) Comparison of three NF membranes for the reuse of secondary textile effluents. Desalination 241(1–3):1–7

Bhattacharjee C, Saxena VK, Dutta S (2017) Fruit juice processing using membrane technology: a review. Innov Food Sci Emerg Technol 43:136–153

Boissier B et al (2008) Particles deposition during the cross-flow microfiltration of red wines—incidence of the hydrodynamic conditions and of the yeast to fines ratio. Chem Eng Process 47(3):276–286

Brandt S et al (1988) Membrane-based affinity technology for commercial scale purifications. Bio/technology 6(7):779–782

Brans G et al (2004) Membrane fractionation of milk: state of the art and challenges. J Membr Sci 243(1):263–272

Butterfield DA et al (2001) Catalytic biofunctional membranes containing site-specifically immobilized enzyme arrays: a review. J Membr Sci 181(1):29–37

Cassano A, Marchio M, Drioli E (2007) Clarification of blood orange juice by ultrafiltration: analyses of operating parameters, membrane fouling and juice quality. Desalination 212(1):15–27

Charcosset C (2006) Membrane processes in biotechnology: an overview. Biotechnol Adv 24(5):482–492

Charcosset C (2021) Classical and recent applications of membrane processes in the food industry. Food Eng Rev 13(2):322–343

Charcosset C, Fessi H (2005) Preparation of nanoparticles with a membrane contactor. J Membr Sci 266(1):115–120

Christian D et al (2023) Recent techniques of textile industrial wastewater treatment: a review. Mater Today Proc 77:277–285

Ciardelli G, Corsi L, Marcucci M (2001) Membrane separation for wastewater reuse in the textile industry. Resour Conserv Recycl 31(2):189–197

Cinperi NC et al (2019) Treatment of woolen textile wastewater using membrane bioreactor, nanofiltration and reverse osmosis for reuse in production processes. J Clean Prod 223:837–848

Czekaj P, López F, Güell C (2000) Membrane fouling during microfiltration of fermented beverages. J Membr Sci 166(2):199–212

De Jager D, Sheldon MS, Edwards W (2014) Colour removal from textile wastewater using a pilot-scale dual-stage MBR and subsequent RO system. Sep Purif Technol 135:135–144

Desore A, Narula SA (2018) An overview on corporate response towards sustainability issues in textile industry. Environ Dev Sustain 20:1439–1459

Domingues RCC et al (2014) Microfiltration of passion fruit juice using hollow fibre membranes and evaluation of fouling mechanisms. J Food Eng 121:73–79

El Rayess Y et al (2011) Cross-flow microfiltration applied to oenology: a review. J Membr Sci 382(1):1–19

Eswaramoorthi S, Dhanapal K, Chauhan D (2008) Advanced in textile waste water treatment: the case for UV-ozonation and membrane bioreactor for common effluent treatment plants in Tirupur, Tamil Nadu, India. Environment with People's Involvement & Co-ordination in India. Coimbatore, India

Fernando WAM et al (2022) Artificial intelligence in wastewater treatment systems in the era of industry 4.0: a holistic review, pp 45–85

Fersi C, Dhahbi M (2008) Treatment of textile plant effluent by ultrafiltration and/or nanofiltration for water reuse. Desalination 222(1–3):263–271

Fillaudeau L, Blanpain-Avet P, Daufin G (2006) Water, wastewater and waste management in brewing industries. J Clean Prod 14(5):463–471

Gan Q et al (2001) Beer clarification by microfiltration—product quality control and fractionation of particles and macromolecules. J Membr Sci 194(2):185–196
Ghaly A et al (2014) Production, characterization and treatment of textile effluents: a critical review. J Chem Eng Process Technol 5(1):1–19
Ghosh R (2002) Protein separation using membrane chromatography: opportunities and challenges. J Chromatogr A 952(1):13–27
Ghosh AM, Balakrishnan M (2003) Pilot demonstration of sugarcane juice ultrafiltration in an Indian sugar factory. J Food Eng 58(2):143–150
Giorno L, Drioli E (2000) Biocatalytic membrane reactors: applications and perspectives. Trends Biotechnol 18(8):339–349
Giorno L, De Bartolo L, Drioli E (2003) Membrane bioreactors for biotechnology and medical applications. In: Bhattacharyya D, Butterfield DA (eds) Membrane science and technology. Elsevier, pp 187–217
Griep ER, Cheng Y, Moraru CI (2018) Efficient removal of spores from skim milk using cold microfiltration: Spore size and surface property considerations. J Dairy Sci 101(11):9703–9713
Halepoto H, Gong T, Memon H (2022) Current status and research trends of textile wastewater treatments—a bibliometric-based study. Front Environ Sci 10:1042256
Hassanzadeh E et al (2017) An efficient wastewater treatment approach for a real woolen textile industry using a chemical assisted NF membrane process. Environ Nanotechnol Monit Manag 8:92–96
Henri G, Jacques F, Jean-Louis M (2000) Fractionation of globular milk fat by membrane microfiltration. Lait 80(1):93–98
Hoang T-D et al (2022) Artificial intelligence in pollution control and management: status and future prospects, pp 23–43
Hoang T-D et al (2024) Regional collaboration for water development projects for strengthening the governance of transboundary rivers in the Mekong countries. The mekong river basin. Elsevier, pp 211–235
Hoang T-D et al (2023) Carbon-based synthesized materials for CO_2 adsorption and conversion: its potential for carbon recycling. Recycling 8(4):53
Hoinkis J et al (2012) Membrane bioreactor (MBR) technology–a promising approach for industrial water reuse. Procedia Eng 33:234–241
Hussein FH (2013) Chemical properties of treated textile dyeing wastewater. Asian J Chem 25(16)
Hoang QN et al (2026) Artificial Intelligence (AI) in agriculture water management. In: Applications of AI in water resources management and sustainability. Springer Nature Switzerland, Cham, pp 83–97
Jaffrin MY et al (2004) A hydrodynamic comparison between rotating disk and vibratory dynamic filtration systems. J Membr Sci 242(1):155–167
Jegatheesan V et al (2009) Performance of ceramic micro- and ultrafiltration membranes treating limed and partially clarified sugar cane juice. J Membr Sci 327(1):69–77
Jiao B, Cassano A, Drioli E (2004) Rcccnt advances on membrane processes for the concentration of fruit juices: a review. J Food Eng 63(3):303–324
Joscelyne SM, Trägårdh G (2000) Membrane emulsification—a literature review. J Membr Sci 169(1):107–117
Kallawar GA, Bhanvase BA (2024) A review on existing and emerging approaches for textile wastewater treatments: challenges and future perspectives. Environ Sci Pollut Res 31(2):1748–1789
Kehinde F, Aziz HA (2014) Textile waste water and the advanced oxidative treatment process, an overview. Int J Innov Res Sci Eng Technol 3(8):15310–15317
Keskin B et al (2021) Pilot and full-scale applications of membrane processes for textile wastewater treatment: a critical review. J Water Process Eng 42:102172
Khedulkar, AP et al (2025) Bio-based nanomaterials as effective, friendly solutions and their applications for protecting water, soil, and air. Mater Today Chem 46:102688

Kurt E et al (2012) Pilot-scale evaluation of nanofiltration and reverse osmosis for process reuse of segregated textile dyewash wastewater. Desalination 302:24–32

Lafi R et al (2018) Treatment of textile wastewater by a hybrid ultrafiltration/electrodialysis process. Chem Eng Process-Process Intensif 132:105–113

Laorko A et al (2010) Effect of membrane property and operating conditions on phytochemical properties and permeate flux during clarification of pineapple juice. J Food Eng 100(3):514–521

Le Anh T, Hoang T-D (2024) Trans-boundary water governance in the Mekong countries. The Mekong river basin. Elsevier, pp 237–272

Lebron YAR et al (2021) Integrated photo-Fenton and membrane-based techniques for textile effluent reclamation. Sep Purif Technol 272:118932

Levesley JA, Hoare M (1999) The effect of high frequency backflushing on the microfiltration of yeast homogenate suspensions for the recovery of soluble proteins. J Membr Sci 158(1):29–39

Li K et al (2020) Influence of nanofiltration concentrate recirculation on performance and economic feasibility of a pilot-scale membrane bioreactor-nanofiltration hybrid process for textile wastewater treatment with high water recovery. J Clean Prod 261:121067

Lipnizki F, Ruby-Figueroa R (2013) Membrane operations in the brewing and sugar production, pp 163–195

Luciana VS, Jean-Louis M (2000) Current developments of microfiltration technology in the dairy industry. Lait 80(6):541–553

Luo J et al (2016) Refining sugarcane juice by an integrated membrane process: filtration behavior of polymeric membrane at high temperature. J Membr Sci 509:105–115

Ma G-H et al (2003) Synthesis and characterization of polyurethaneurea–vinyl polymer (PUU–VP) uniform hybrid microspheres by SPG emulsification technique and subsequent suspension polymerization. J Appl Polym Sci 89(1):163–178

Malla FA et al (2023) Circular economy to decarbonize electricity. Renewable energy in circular economy. Springer, pp 71–87

Manni A et al (2020) New low-cost ceramic microfiltration membrane made from natural magnesite for industrial wastewater treatment. J Environ Chem Eng 8(4):103906

Marcucci M et al (2002) Experimental campaigns on textile wastewater for reuse by means of different membrane processes. Desalination 149(1–3):137–143

Marcucci M et al (2003) Membrane technologies applied to textile wastewater treatment. Ann N Y Acad Sci 984(1):53–64

Nguyen TT et al (2025) The environment security with nontraditional security management approaches: case study of Hanoi. Soils Sustain Agric: Interplay Soil, Plant, Water Environ Syst Sustain Agric 945–964

Nthunya LN et al (2024) Progress in membrane distillation processes for dye wastewater treatment: a review. Chemosphere 142347

Othman MHD et al (2021) Advanced membrane technology for textile wastewater treatment. In: Membrane technology enhancement for environmental protection and sustainable industrial growth, pp 91–108

Periyasamy AP et al (2018) Sustainable wastewater treatment methods for textile industry. In: Sustainable innovations in apparel production, pp 21–87

Pham VT et al (2025) Water resources and bioenergy. In: Soils and sustainable agriculture: interplay of soil, plant, water and environmental systems for sustainable agriculture. Springer Nature Switzerland, Cham, pp 1041–1068

Rouquié C et al (2019) New prospects for immersed hollow-fiber membranes in fruit juices microfiltration: case of grapefruit juice. J Food Eng 246:75–85

Saha NK, Balakrishnan M, Ulbricht M (2009) Fouling control in sugarcane juice ultrafiltration with surface modified polysulfone and polyethersulfone membranes. Desalination 249(3):1124–1131

Sahinkaya E et al (2019) Performance of a pilot-scale reverse osmosis process for water recovery from biologically-treated textile wastewater. J Environ Manage 249:109382

Saini P, Bulasara VK, Reddy AS (2019) Performance of a new ceramic microfiltration membrane based on kaolin in textile industry wastewater treatment. Chem Eng Commun 206(2):227–236

Samsami S et al (2020) Recent advances in the treatment of dye-containing wastewater from textile industries: overview and perspectives. Process Saf Environ Prot 143:138–163
Schutyser M et al (2002) Dean vortex membrane microfiltration and diafiltration of rBDNF E.coliInclusion bodies. Biotechnol Progress 18(2):322–329
Sezali NAA et al (2024) Bio-based nanomaterials for energy application: a review
Shi C et al (2019) Ceramic membrane filtration of factory sugarcane juice: effect of pretreatment on permeate flux, juice quality and fouling. J Food Eng 243:101–113
Siddique K et al (2017) Textile wastewater treatment options: a critical review. In: Enhancing cleanup of environmental pollutants, vol 2: non-biological approaches, pp 183–207
Siliņa L, Dāboliņa I, Lapkovska E (2024) Sustainable textile industry—wishful thinking or the new norm: a review. J Eng Fibers Fabr 19:15589250231220360
Sisira S, Sheena K (2020) Heterogeneous fenton and microfiltration for the treatment of textile waste water. Int Res J Eng Technol (IRJET) 7(7)
Sundaram S et al (1999) Application of membrane filtration for removal of diminutive bioburden organisms in pharmaceutical products and processes. PDA J Pharm Sci Technol 53(4):186
Ta QTH et al (2025) The impact of waste on psychological health in urban environments. Soils Sustain Agric Interplay Soil, Plant, Water Environ Syst Sustain Agric 1011–1039
Tahri N et al (2012) Coupling microfiltration and nanofiltration processes for the treatment at source of dyeing-containing effluent. J Clean Prod 33:226–235
Tavangar T et al (2019) Toward real textile wastewater treatment: Membrane fouling control and effective fractionation of dyes/inorganic salts using a hybrid electrocoagulation–nanofiltration process. Sep Purif Technol 216:115–125
Thuy TL et al (2025) Blockchain-enabled water quality monitoring: a comprehensive review of digital innovations and challenges. Water 17(17):2522
Tran T-K et al (2023) Review on fate, transport, toxicity and health risk of nanoparticles in natural ecosystems: emerging challenges in the modern age and solutions toward a sustainable environment, p 169331
Tran T-K et al (2024) Applications of engineered biochar in remediation of heavy metal (loid) s pollution from wastewater: current perspectives toward sustainable development goals, p 171859
Ulbricht M et al (2009) Fouling in microfiltration of wine: The influence of the membrane polymer on adsorption of polyphenols and polysaccharides. Sep Purif Technol 68(3):335–342
Upadhye V, Joshi SS (2012) Advances in wastewater treatment. a review. Int J Chem Sci Appl 3(2):264–268
van der Sman RGM et al (2012) Review of hypotheses for fouling during beer clarification using membranes. J Membr Sci 396:22–31
van Reis R, Zydney A (2001) Membrane separations in biotechnology. Curr Opin Biotechnol 12(2):208–211
Villalobos García J et al (2018) Membrane distillation trial on textile wastewater containing surfactants using hydrophobic and hydrophilic-coated polytetrafluoroethylene (PTFE) membranes. Membranes 8(2):31
Vladisavljević GT, Shimizu M, Nakashima T (2004) Preparation of monodisperse multiple emulsions at high production rates by multi-stage premix membrane emulsification. J Membr Sci 244(1):97–106
Yaseen D, Scholz M (2019) Textile dye wastewater characteristics and constituents of synthetic effluents: a critical review. Int J Environ Sci Technol 16:1193–1226
Yun M-A et al (2006) Characterization of biofilm structure and its effect on membrane permeability in MBR for dye wastewater treatment. Water Res 40(1):45–52
Yurtsever A et al (2015) Performances of anaerobic and aerobic membrane bioreactors for the treatment of synthetic textile wastewater. Biores Technol 192:564–573

Chapter 9
Comprehending Membrane Fouling: Causes, Effects, and Mitigation Strategies

Saimah Khan, Zeba Ali Mumtaj, Abdul Rahman Khan, Amadur Rahman Khan, and Roohul Abad Khan

Abstract Membrane technology has emerged as a critical component in water treatment, wastewater reclamation, and food processing due to its efficiency and cost-effectiveness. Membrane fouling is one of the ongoing issues preventing its widespread use, though. Membrane fouling is the term used to describe the buildup of undesirable materials on the membrane's surface or inside its pores, which results in lower performance and higher operating expenses. The objective of this study is to present a thorough analysis of membrane fouling, emphasizing its causes, effects, and preventative measures. First, it discusses the various mechanisms that contribute to membrane fouling, such as physical deposition, adsorption, and biological growth. Comprehending these mechanisms is essential to formulating efficacious fouling control tactics. Second, the effects of membrane fouling on system efficiency and viability from an economic standpoint are explained. Fouling can have a number of negative effects, including decreased flux rates, higher energy consumption, and frequent membrane replacement or cleaning cycles. These effects highlight the necessity of taking preventative action to minimize fouling. Finally, a variety of mitigation techniques to stop membrane fouling are examined. These tactics cover surface modifications, chemical cleaning, backwashing, and sophisticated monitoring and control methods, among other preventive and corrective measures. Furthermore, new developments are examined, including the application of nanomaterials and bio-inspired coatings for fouling mitigation.

Keywords Membrane foulants · Mitigation strategies · Gas bubbling · Backwashing

S. Khan (✉) · Z. A. Mumtaj · A. R. Khan
Department of Chemistry, Integral University, Lucknow, India
e-mail: saimah@iul.ac.in

A. R. Khan
Department of Architecture, Aligarh Muslim University, Aligarh, India

R. A. Khan
Department of Civil Engineering, King Khalid University, Abha, Saudi Arabia

A. Mandpe et al. (eds.), *Membrane Filtration and Membrane Bioreactors for Wastewater Treatment*, Advances in Wastewater Research,
https://doi.org/10.1007/978-981-95-6706-5_9

9.1 Introduction

The main step regarding the desalination of water consists of the elimination of salts as well as impurities which are contained in brackish or salt water and result in the water being suitable for domestic or industrial use. Membrane technologies including reverse osmosis, microfiltration, ultrafiltration amongst others represent a group of highly accepted water treatment methodology as they occupy little space, are highly energy efficient, can be automated and their design is compact. For example, RO, MF, NF, UF, and MD are some of the most famous filtration methods based on the membrane (Prihasto et al. 2009; Henthorne and Boysen 2015). The limiting and troublesome thing of the accumulation and storage of the materials which are not needed and cause unpleasantness is called fouling. It is mostly happens in the external or inner part of a host, for instance, boilers, heat exchangers, and membranes. There inclusions could thus be, dissolved particles, partially soluble organic and/or inorganic macromolecules and/or living biological micro-microorganisms. Hence, among the membrane-based water treatment processes, the membrane fouling is an unavoidable phenomenon that can greatly undermine the operation of the processes and also pose challenges on sustainability and economics. Fouling phenomena are the products of the complex interactions that occur between a feedstock's chemical and physical features and the membrane's surfaces (Al-Juboori and Yusaf 2012; Ladewig and Al-Shaeli 2017). The major contributing factors to membrane fouling include (Guo et al. 2012): • Solids and liquids ratio, examples are: pH, ionic strength and fouling intensity. • Concentration polarization (CP): In a narrow sense, the CP can be defined as the precipitation of disliked solutes on the surface of the membrane, and the area close to the membrane where the solute concentration shows spatial variations is known as the polarized bed. This added resistance leads to a reduction of the osmotic pressure difference across the membrane and thus the driving force for the process trans membrane pressure (TMP), permeate flux rate, observed solute rejection, and increases the likelihood of membrane fouling (Ladewig and Al-Shaeli 2017; Field 2010) • Membrane characteristics include the type of material, porosity, hydrophobicity, charge, • Several process parameters like temperature, pressure, aeration, permeate flux and several other hydrodynamic conditions of the operating process.

This chapter aims at discussing types of fouling that occur in RO system, factors that affect fouling of RO membranes and some measures adopted to minimize this problem.

9.2 Membrane Foulants

In membrane fouling, the classification is usually done to the nature of foulants or determines the area for fouling. Foulants, which have both particulate and non-particulate components, such as biofoulants, organics, and inorganics are the substantial collectants in water flow systems and surfaces across built and natural environments. Relating to placement, this problem takes place inside the system or on the membrane's outside wall. In proceedings, initial clogging of the membrane pores is more likely to happen when the low-pressure membrane-based separation processes, like UF and MF that suffer from internal fouling by adsorption are involved. Furthermore, membranes with high density and compaction, even though usually semipermeable, such as those used in NF and RO eventually experience surface fouling. This part of the discussion will be further expanded in the next few paragraphs where the types of membrane fouling are categorized.

9.2.1 Particulate Fouling or Colloidal Dispersions

Are solutions of the two substances, commonly ranging in size from a few nanometers to micrometers, are evenly mixed in a mixture (AlSawaftah et al. 2021). Product colloids demo can be given in the description of size. It can be settled colloids where particles size is greater than 100 μm, supra-colloids that consists of particles having the size from 1 to one hundred micrometers and colloidal solids that are formed by particles in such small sizes from. They are as small as 0.001 microns and all the dissolved solids less than 0.0001 microns. The diameters of PM 2.5 and PM 10 particulates are respectively 0.01 and 0.001 mm (Yiantsios et al. 2005). Adding to this, aquatic colloids can also be classified as organic and inorganic based on their content material. The organic colloids referred to as proteins, fats, oils, greases, carbohydrates, surfactants, and bio-colloids are the first ones; the second ones are the inorganic colloids represented by mud slimes, clays, silt (the remnants of the microorganisms), crystals, silica sediments, as well as aluminum and iron precip Particulate fouling is a complicated plan, the start of which is the pore-blocking phase where an obstruction occurs near the pore opening and the aperture begins to narrow. Over time, the particles being deposited compel the membrane's obstruction and absorption increase as deeper and deeper layers get deposited. The third stage is the addition the deposition causing thick "cake layer" inhibiting the membrane's function and preventing contaminants removal. The cake layer resort that develops professional to higher hydraulic resistance (termed cake resistance) also invite complex CP reactions which as it is known for may involve decrease in permeate flux and also some unanticipated side effects of the foulants (Ladewig and Al-Shaeli 2017). The foulant particle occurrence on the membrane surface is determined by particular surface properties of the membrane itself, undefined from a physical perspective (e.g., morphology and topography), the fouling agents present in the feed water along with the ions and charges of the feed water constituents are involved. Furthermore, when it comes to the chemistry of the feed, such as pH, ionic strength and charge interactions, as well as the operating conditions, all of these factors should also be considered undefined, the temperature, pressure, flux and cross-flow velocity (CFV)) are the dominant factors for the growth of nuisance algae. Smoothness and multiple hydrophilic characteristics of surfaces with low chargedness are the factors considered to determine if the surface will be damaged by particles, and in very unsteady flow conditions such as high flux membranes or low CFV, the membrane may get damaged directly (Tang et al. 2011).

9.2.2 *Organic Fouling*

In the context of organic matter, it pertains to the settlement and buildup on the inner lining of the pipes of the denser organic materials. The can be described as polysaccharides, proteins, humid substances, nucleic acids, lipids, and amino acids. The dissolved organic matter dominates both in surface water and in the wastewater and generally falls into three categories: The main category of compounds that have been identified to cause fouling of the RO membranes is the natural organic matter (NOM) the synthanthropic compounds and the soluble microbial products (SMPs). NOM is not a single substance but a complex, heterogeneous mixture of organic compounds formed during the decay of plant and animal matter. DOMs of syndetic nature are the group of molecules which are synthesized in clean water systems and have been added or formed during disinfection process. Finally, Bio solids are less treated biological residues that are susceptible to biological decompositions of organic compounds. In line with the previous research studies that involve RO membrane fouling, the nature of foulants dominating the feed water seems to determine how the presence of organic matter affects RO membrane fouling, through the variety of the foulant-surface, and foulant–foulant interactions (Ladewig and Al-Shaeli 2017; Guo et al. 2012).

9.2.3 *Inorganic Fouling*

Inorganic fouling better known as scaling happens when inorganic compounds form a layer on the surface of the membrane or within the pores of the membrane. They could be inorganic elements that are sparingly soluble in water or more solutes which exist in the relatively larger volumes of water. They are able to make solid solutions with a concentration above the solubility and then crystallize (separate from the solution) on the surface of the membrane. This process may form scale on the surface of the membrane in two fashions where one of them is the crystal form and the other one is a particulate foul. That means the various types of filtration methods obey the first principle of ionic precipitation and deposition on the surface of the membrane. On the other hand, the convective transport of particles from the mass of the solution to the membrane surface is used in the second principle. As for the types of water including rivers and groundwater, seawater, and municipal wastewater, the main inherent got accumulated are hydroxides, sulfates, carbonates, calcium, magnesium, iron, ortho-phosphates, silicic acid and silica (Ladewig and Al-Shaeli 2017; Guo et al. 2012).

9.2.4 *Biofouling*

Here, biofouling means microorganisms take shelter on the solid surface and therefore leads to the generation of biofilm, which can reach up to 40 percent fouling (Al-Juboori and Yusaf 2012; Kalafatakis et al. 2020). In Table 9.1, microorganisms such as pathogens, fungi, and some bacteria present the attack of membranes are listed (Maddah and Chogle 2017). Slimy biofilm is the agglomeration of population of microorganisms that finds its place on the surface and is enveloped with the extracellular polymeric substance (EPS) matrix. It is composed of the chemical compound called EPS which is used to guard the microbial endospores against biocides as well as toxins. The composition of the matrix will be based on the environment in which biofilm forms, that is commonly translucent in nature and smart materials have also been used to augment the functionality, such as proteins, glycoproteins, lipoproteins, nucleic acids, lipids, and polysaccharides. Mineral crystals, clay, silt, or other particles such as tarnish and corrosion may also be part of the matrix (Al-Juboori and Yusaf 2012; Ladewig and Al-Shaeli 2017; Donlan 2002). The cell membrane surface roughness and hydrophobicity among the significant criteria of biofilm formation are the most prominent ones. However, biofouling is more common among hydrophobic membranes as a result of the close microbial interactions, which allow bio-contamination to occur. Whereas the rough surface membranes possess a larger surface area compared to the smooth surface ones, more active sites are present for microbial growth and adherence; hence, more biofouling conditions might be presented (Maddah and Chogle 2017) (Fig. 9.1).

The community of biofouling undergoes several effects such as reducing the performance of membrane systems and hampering its components. In detail, these impacts are illustrated (Maddah and Chogle 2017).

• Flux decline: Just like the fact that the biofouling, which is the deposition of organic material, increases the resistance and lowers the pure of the water, the biofilm formation also results in similar outcomes of higher resistance and lower permeates flux. • Membrane biodeterioration: destruction of the boundary barrier by the acid being created as a by-product of the biological activity of the bacteria. • Deteriorated salt retention: the presence of AP imposed on the cell membrane and thus the transport mechanisms that rely merely on diffusion is associated with the accumulation of dissolved salts and ions on the cell surface leading to the increase in AP. • Increased differential pressure: it is because of the synthesis of extra layer of material onto the microbes making them resistant to certain drugs that I consists

Table 9.1 Indicates types of membrane fouling and their related characteristics

Type of fouling	Characteristics	References
Particulate fouling	• Caused by colloidal dispersions ranging from nanometers to micrometers • Organic and inorganic colloids • Initial pore-blocking, followed by cake layer formation • Influenced by membrane properties, feed water chemistry, and operating conditions	AlSawaftah et al. (2021), Yiantsios et al. (2005), Ladewig and Al-Shaeli (2017), Tang et al. (2011)
Organic fouling	• Accumulation of dense organic materials likes polysaccharides, proteins, nucleic acids, etc • Categories: natural organic matter (NOM), synthanthropic compounds, soluble microbial products (SMPs) • Affects reverse osmosis (RO) membrane fouling	Ladewig and Al-Shaeli (2017), Guo et al. (2012)
Inorganic fouling	• Precipitation of inorganic compounds on or within the membrane • Can form through crystal formation or particulate fouling • Common foulants include hydroxides, sulfates, carbonates, etc	Ladewig and Al-Shaeli (2017), Guo et al. (2012)
Biofouling	• Microorganisms form biofilms on the membrane surface • Biofilms composed of EPS matrix with various organic and inorganic materials • More common on hydrophobic and rough surface membranes • Impacts: flux decline, membrane bio deterioration, deteriorated salt retention, increased differential pressure, higher energy requirements, frequent cleaning	Al-Juboori and Yusaf (2012), Kalafatakis et al. (2020), Maddah and Chogle (2017), Donlan (2002)

of bacteria, this is known as biofilm formation. • Higher energy requirements: the thick membrane-approaching section and permeate flux drop are the reasons why the energy consumption is so high. • Frequent chemical cleaning: water acting as cleaning agent increases amount and frequency of membrane damage that eventually shortens membrane lifespan. The Table 9.1 shows the fouling types and their characteristics.

9.3 Characterization Methods

Until now, several of characterization methods have been performed or applied to investigate the character of fouling layers and the mechanisms of fouling, and validate or verify the things to help breaking down fouling. For example, such techniques as

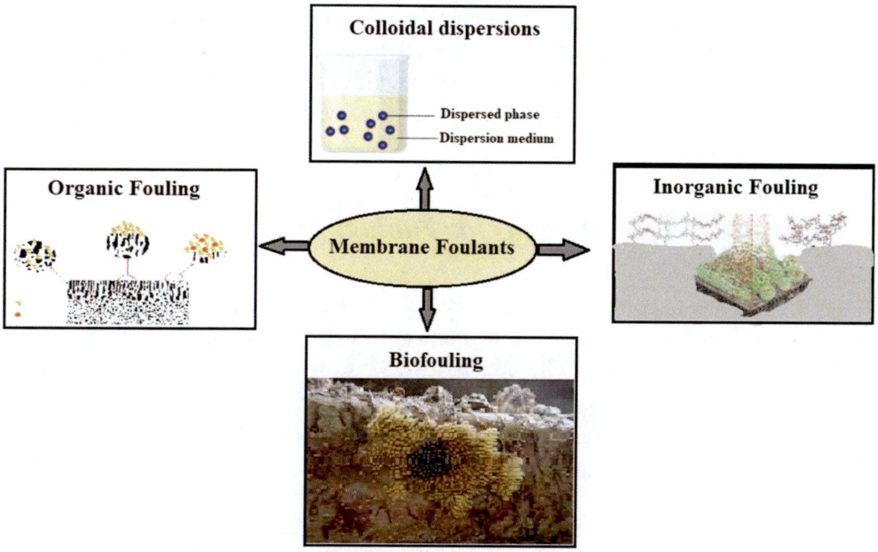

Fig. 9.1 Shows the different pretreatment methods for various types of fouling

scanning electron microscopy with energy dispersive X-ray spectroscopy23, atomic force microscopy24 provide important data about the fouling film buildup, surface morphology, pore size, etc. As an example, optical coherence tomography provides vivid, non-destructive, and in real-time details about the development of biofilms living on membranes 20 but the lack of particularly specific data about these contaminants is still preserved. Obviously, the methods stipulated in this case do not recognize the chemical composition of diverse foulants, which is the most vital element for chemically defining them and identifying the role they play in membrane-fouling process. Because of the nature of molecular entities, e.g., the organic matter from natural sources (NOM) and the extracellular polymeric substances (EPSs), having different chemical structures, absorption, or emission characteristics, spectroscopic methods using UV–vis, fluorescence, IR and Raman are good tools for the determination of chemical structure of foulants. Certain of them, for example, microscopic spectroscopic methods, can yellow them, and indicate the presence, location (in space), and movement (through time) of specific foulants (Zator et al. 2007; Virtanen et al. 2017). There is, however, some shortcoming predominantly emanating from some techniques such as labeling, sensitivity, and time-intensive nature, which are some of the challenges being grappled with. This notwithstanding, spectroscopic methods and the ultrafast development of spectroscopic instrumentation make the use of chemometric approaches for spectral data treatment from being the most efficient platform tools for characterizing membrane fouling. In this section, we would like to spotlight the newest techniques of molecular spectroscopy of membrane fouling, which would concentrate on the present-day spectroscopic imaging methods. This includes discussion of the main drawbacks and restrictions of the current methods of fouling feature recognition, the consequences of such restrictions, and the need for approaches which overcome these. We then envisage new methods that would apply the current approaches in fouling feature recognition; thus, significantly advancing membrane-fouling evolution characterization (Chen et al. 2018).

9.3.1 Spectroscopic Methods for Foulant Identification

Amongst the most popular techniques for characterizing membrane foulants (membrane-modified substances) and microbial organisms (MIC) in water, UV–vis spectroscopy is included (Chen et al. 2018). This is an accurate technique by measuring the amount (absorbance) of light at various wavelengths; it can give a detailed report about the structural changes of the foulants. These are for example, UV254 and SUVA254 parameters that indicate aromatic contents in high-molecular-weight substances (HSs) (Wei et al. 2009). SUVA280, SUVA365, and SUVA436, on the other hand, are associated with double bonds, molecular size, and chromophoric groups in dissolved organic carbon (DOC, respectively) (Uyguner and Bekbolet 2005).

Besides, the E2/E3 ratios and E4/E6 reflect humification degree and the structures of NOM through providing information about its aromaticity and molecular weight distribution (Yan et al. 2013). Wavelength specific spectra, which represent the exact wavelength ranges, provide more information about NOM and EPS changes and thus elucidate the fouling mechanisms. Indeed, a quadratic regression of differential absorbance spectroscopy (DAS) data with Gaussian modeling can reveal even the slightest changes triggered by changes in physico-chemical parameters (Wang et al. 2016).

On the other side, UV–Vis spectroscopy is limited by a lot of factors including—overlapping spectra—no identification of fluorescent foulers which are polysaccharides. A chemometric analysis or other correlated techniques are hence obligatory in order to authenticate and present complex analysis (Chen et al. 2018).

Fluorescence spectroscopy, as a means of the study of the fouling of membranes is empowered by its rapidity and great sensitivity (Chen et al. 2018). The hyperspectral fluorescence diagnostic techniques, for instance, excitation and emission spectra, synchronous fluorescence, and fluorescence excitation-emission matrix (EEM) can determine the sources and ingredients that lead to biofouling and humic fouling as well as screen the biofouling from the humic fouling (Henderson et al. 2009).

EEM-PARAFAC provides a method for obtaining qualitative data, which shed light on the dynamics of NOM and EPS mix by decomposing EEM into independent fluorescence components (Chen et al. 2018). The fluorescence techniques have successfully been used in different membrane process applications like in MBR and UF to determine the degree of fouling by the industrial and natural contaminants and how they affect the overall performance (He and Hur 2015).

Therefore, native full constructs might not be frequent fluorescent components with NOM and EPSs because only these small amounts of NOM and EPSs are fluorescent. Along with fluorescent ones, non-fluorescent ones also remain unidentified considering their invisibility that could be considered as a joint to use fluorescence techniques exclusively to fully characterize the fouling (Chen et al. 2018).

IR spectral analysis produces structural characterization about membrane semipermeable particles' functional groups allowing for their separation and differentiation (Huang et al. 2017). AT-FTIR and IR spectroscopy are particularly successful for recognizing bands of foulants bonds (Saha et al. 2007).

Such approaches have become useful tools for investigating membrane deposition processes, e.g., polysaccharide-promoted fouling in UF and membrane fouling in municipal wastewater treatment (Xiao et al. 2013). Further, this DATR-FTRIR spectroscopy combined with other techniques will give us geometrical and quantitative details on shock and so on.

Nevertheless, IR spectroscopy has drawbacks as far as the resolution level and penetration depth is concerned, and therefore one has to not only examine the IR spectra but also should couple the monitored technique with other additional methods to attain an advanced knowledge of the fouling (Saha et al. 2007).

Raman spectroscopy serves as an alternative vibrational structural information source to IR spectroscopy and is sensitive to vibrations which show a change of polarizability. It is the minimally invasive sort and conforms well to real-time monitoring of such membrane-fouling processes (Sistat et al. 2009).

Raman and SERS methods have a spontaneous approach to identify various littered foulants, including proteins and polysaccharides, on membrane surfaces. SERS boosts sensitivity but substrate selection may not always work and can adversely affect the membrane properties (Chen et al. 2015).

Besides its positive features, Raman spectroscopy can have the problems of low efficiency of quanta and sensitivity of the experiment. In addition, it is necessary to have a complex experimental setup and data analysis (Efeoglu and Culha 2013).

Combination of microscopy techniques including SEM and TEM offers a more elaborated view of film formation (Cui et al. 2011).

Visible light microscopy, IR microscopy and Raman microscopy are capable of schematic representation of foulants on membrane surfaces by means of monitoring fouling mechanisms inside (Bhargava 2012).

Microscopic techniques of fluorescence make it possible to have exquisite prerequisites of sensitivity and spatial resolution with the purpose of differentiating between diversities of the acid sulfate. The confocal and 3D microscopy accelerates the 3D reconstruction and real-time observation of biofouling (Hughes et al. 2006).

Fougelant and Raman microscopy enable the detection of spatial information and chemical composition of the foulants adhered onto the membranes without the use of labels. Both ATR-IR and SRS microscopies could be useful to obtain information about the evolution and interactions of foulants with the membrane

Nevertheless, microscopy techniques have their own limitations and they may not be able to provide adequate resolution and depth of imaging. There are many other techniques required for complementary analysis and advanced techniques of data analysis necessary to fully characterize the fouling digital content (Chen et al. 2018). The Table 9.2 shows the spectroscopic methods used for analysis of membrane fouling.

9.4 Membrane-Fouling Mitigation Strategies

A lot of work is made to tackle the fouling phenomenon in MBR and improve the performance of membranes. That can be found by improving the membrane performance using optimized process conditions and other membrane characteristics modifications. A detailed presentation of their role in fouling installations is discussed in several review papers (AlSawaftah et al. 2021; Guo et al. 2012; Rudolph et al. 2019). But it is important to note that the present paper is solely about reducing the misfunctioning of the membrane by finding out fouling mitigation methods. They involve pretreatment processes and membrane cleaning ones. This section will be dedicated to reducing fouling in MD process by two different methods that will be discussed in detail in the coming sections: (i) Possible pretreatment to be carried out to remove foulants from feed water so that it sloughs off formation of fouling and minimize its formation (ii) Undefined membrane processes: reverse osmosis, nanofiltration, ultrafiltration, etc., water softening: deionization, chemical coagulation, tri-valent ions, and acids (iii) Other treatment techniques commonly used to de-foul a membrane. Through such approaches, the foulants may be partially or completely erased. Undefined, flushing, backwashing, air bubbling) Hence, whatever fouling type will be available, those two mitigation techniques will work out if well used. However, the complexity of the thing comes in choosing the proper methods for fouling removal, especially from the solutions containing multiple foulants. For inorganic salts, examples of pretreatment methods are better than cleaning membrane fouling, because the membrane can become fouled with organic fouling and highly hard to clean. Conversely, cleaning of organic fouling is easy as it can be performed by mechanical cleaning while non-organic fouling needs chemical cleaners. The organic fouling removal becomes easier if the fouling concerns are only the organic fouling factors in the feed solution. In fact biogas film is frequently an aliased type of fouling where the two work in unison. The two types of fouling sometimes degrade to this problem such that pretreatment options become ineffectual. Microorganisms occurring in feed water, they use organic substance, grow EPS and thus promote biofilm growth. EPS has similar properties to the feed materials that are used by other microorganisms and maintains continuity of biofilm formation. Moreover, organic residues will constantly be adsorbed on the membrane surface to create a fertile condition to foster biofouling (Guo et al. 2012). The fact is that before-treatment and cleaning membrane-fouling techniques of high efficiency are still very limited.

Table 9.2 Indicates spectroscopic methods for membrane-fouling analysis

Spectroscopic method	Characteristics	References
UV–Vis spectroscopy	• Measures absorbance at various wavelengths • Identifies structural changes in foulants • Parameters like UV254, SUVA254, SUVA280, SUVA365, SUVA436 indicate aromatic contents and chromophoric groups in DOC • E2/E3 and E4/E6 ratios reflect humification degree and molecular weight distribution • Limited by overlapping spectra and inability to identify fluorescent fouler like polysaccharides	Chen et al. (2018),Wei et al. (2009), Uyguner and Bekbolet (2005), Yan et al. (2013), Wang et al. (2016)
Fluorescence spectroscopy	• Rapid and sensitive • Techniques like EEM-PARAFAC, excitation-emission spectra, and synchronous fluorescence identify sources and ingredients of biofouling and humic fouling • Used in applications like MBR and UF to determine fouling degree • Limited by non-fluorescent components remaining unidentified	Chen et al. (2018), Henderson et al. (2009), He and Hur (2015)
IR spectroscopy	• Structural characterization of functional groups • Techniques like AT-FTIR identify bands of foulant bonds • Used for investigating membrane deposition processes • Limited by resolution level and penetration depth, requiring complementary techniques	Huang et al. (2017), Xiao et al. (2013)
Raman spectroscopy	• Provides vibrational structural information • Minimally invasive, suitable for real-time monitoring • Techniques like SERS enhance sensitivity • Challenges include low quantum efficiency, sensitivity, complex setup, and data analysis	Sistat et al. (2009), Chen et al. (2015), Efeoglu and Culha (2013)
Microscopy techniques	• SEM and TEM offer detailed views of film formation • Visible light microscopy, IR microscopy, and Raman microscopy provide schematic representation of foulants • Confocal and 3D microscopy enable real-time observation of biofouling • Techniques like ATR-IR and SRS microscopy detect spatial information and chemical composition without labels • Limited by resolution and depth of imaging, requiring complementary techniques	Cui et al. (2011), Hughes et al. (2006)

9.4.1 Pretreatment Approaches

Pretreatment approaches are used to evade the fouling from occurring. Pretreatment operations mainly employ different strategies to break down the foulants components in the feed solution and thus minimize the possibility of fouling by improving the feed solution quality. The utilization of the pretreatment technology in the MD process has not that big meaning in the pressure-driven treatment processes. Nevertheless, what was confirmed (Field 2010) is that pretreatment is a way to control the processing of MD. This, thus, justifies the meaningful decision on pretreatment technology that can appreciably reduce membrane-fouling formation, increase the efficiency of permeate water, enhance MF performance, and increase membrane life (Prihasto et al. 2009). It is a critical step in pre-MD processing, especially if using actual feed water (ii). i.e., (industrial wastewaters, seawater, etc.) contains an array of foulants. Before the purification step, a specific method is used either to remove the ones that block the pores or stop big particles and other particles from reaching the membrane surface by accumulating there (Sim et al. 2018). Multiple techniques including improved pretreatment, flux optimization and membrane properties can be adopted for the reduction of fouling incidences. Such techniques as optimum flow rate conditions optimization and membrane surface modification are not under the purview of this paper. They are referred to as fouling pretreatment techniques. In this chapter only the feed pretreatment methods are covered to decrease fouling of the membrane. Pretreatment methods commonly active in water treatment include coagulation, flocculation, water softening andthe addition of antiscalants to diminish scaling and enhance treatment efficiency. (a) sedimentation/clarification, pH adjustment, water softening, and anti-scaling) (b) temperature. e. for example Ethering, nano-filtration, softening (thermal water), or, finally, a mixture of these methods. e. On a molecular level, there are also different types of bonds (covalent and non-covalent, mechanical and thermal). In the following segments, the common pre-treating techniques for MD systems would be described.

9.4.1.1 Mechanical Pretreatment Process

Ultrafiltration and membrane filtration using NF, UF, and MF preceded by mechanical pretreatment to filter colloid, suspension, microorganism, organic hustle and burden from the solution. MD process is less costly than membrane-based filtration concerning occurrence of membrane-fouling issues and damaging of the membranes. According to the authors of Karakulski and Gryta (2005), they evidenced that MD foulings of the organic nature were reduced by NF pretreatment. Carbonate is still present at a small amount in NF permeate stream; however, it was rapidly rejected by the acidization process. Thus, it was sufficient to commensurate the rate of HCI into the NF permeate flow with its pH equal to 5. Besides, the inorganic fouling can be prevented or terminated as the negative charge of the NF surface hinders the transportation of sulfate ions leading to $CaSO_4$ scale on the membrane. This can be used to suppress precipitate formation and reduce the concentration of less soluble ions such as divalent cations (Ca^{2+}, Mg^{2+}) on the membrane surface in the nanofiltration (NF) system. Besides the mechanical pretreatment process, these three primary processes include nanofiltration, ultrafiltration, and microfiltration. Along with the process of removing water's hardness, nanofiltration helps to get rid of organic substances too. Microfiltration is also put under the purview of MD systems as a pretreatment method that removes cells and solutes. Through microfiltration as a pretreatment step, almost nothing of inorganic scales such as calcium scale was remained in groundwater. While El-Abbassi et al. (2012) revealed that microfiltration works better than coagulation/flocculation for removing TOC and TS from raw olive mill wastewater, the former is carried out better (the latter being more effective). For another series of experiments which requires either saline water or brine, membrane filtration was also tested. Song et al. illustrated that DCMD can be operated by very brackish water on daily basis for longer than 3 months without fouling. Prior to introducing this water to the MDS, it received pretreatment in a membrane filtration unit where it was subjected to macroparticles and biological matters. In contrast, biofilm process only took place after 28 h of operation using raw seawater from Mediterranean Sea without filtration as feed (Son et al. 2018). The ultrafiltration (UF) technique, although not very common, can efficiently remove the suspended solids and colloids (Guo et al. 2012). MD process with ultrafiltration can be employed for other uses like the recovery of high quality products out of the fine water. Zhong et al. (2021) also had an experiment which involved doing a complete separation from traditional Chinese medicine (TCM) wastewater by the combined method of ultrafiltration and membrane distillation (MD). The search for more valid results leads them to add additional pretreatments before performing microfiltration. Jansen et al. (2013) and Gryta (2010) unit of microfiltration membrane (MFM, 10 μm pore size) was connected consecutively with vacuum pump to degas the solution and was followed by liquid gap membrane distillation (LGMD) unit. The seawater was utilized as the feedstock and the final distillate was generated with brilliant quality. Yet, the reason of the degassing impact was not uncovered; however, the positive results were registered. The same pretreatment technique, which is common to RO and which involves chemistry and membrane filtration, was recommended by other researchers for treating MD.

9.4.1.2 Chemical Pretreatment Process

Some of the chemical pretreatment methods include: coagulation/flocculation, pH correction, precipitation, softening, antiscalants, and disinfection. All chemical processes are directed toward one method shared with reducing membrane fouling by adjusting and nullifying the charge of the disparate material. It is recommended that these processes may be applied to the various categories of fouling such as inorganic fouling, colloidal fouling, biological fouling, and oxidizing process fouling (Laqbaqbi et al. 2017). For instance, the colloids and suspended solids, these require pH adjustment so that their foulants are away from their isoelectric point. The process will thus reduce formation of gel layer on the particle and allow the nanoparticles to aggregate into heavy and large nanoprobes that will not pass through the holes on the membrane.

9.4.1.2.1 Coagulation

Coagulation is a non-complicated and cost-effective process that is used as a pretreatment. The coagulation/flocculation idea prevails in the given circumstance since various particles existing in the feed water are going to be stabilized, meaning that colloidal particles, divalent ions together with bacteria are going to come together and cluster to form larger foulants so that there will be a reduced likelihood of the formation of scales on the surface and the pores of the membrane. Others can also be divalent ions coagulations. These ions can combine with other ions to form chained macromolecules that cling on to the surface of the membrane. This type of molecule is exchanged; however, the monovalent ions that prefer to be left out during the process of precipitation cannot be eliminated. Wang et al. (2008) has looked into the impact of the pretreatment coagulation technologies trying to tune the performance of the effluent of MD process where cooling water has been used as the feed solution after re-circulation. Most of the times, RCW will be presented in various forms of contaminants to remove corrosion, biofouling and scaling effects (Choudhury et al. 2012). Therefore, it is possible to note that several different pretreatment methods are needed to suppress the occurrence of fouling. Wang et al. explained various pretreatment methods such as coagulation, precipitation, filtration, acidification and degasification for the removal of the biofouling. PACl which is a coagulant was used on RCW and it was effective in removing almost entirely NOM, TP, TOC, antiscalant additives and SS from the feed solution. In the process, the foulants were detached through the mechanism of scaling them up to become larger ones of magnesium-calcite that settle on the membrane surface and subsequently do not get to the pores of the membrane. Such deposits have the effect of reducing the degree of membrane wettability, the lowering of the falling permeate flux and an improvement in the quality of permeate water. Microfiltration was employed for filtration of the feed solution to get it rid of large-sized foulants. In an attempt to cut down CO_2 content in RCW, acid treatment was used, further followed by degassing. Hence, it serves to reduce the content of CO_2 significantly. The data showed a flux permeate of 23% despite coagulation pretreatment prior to MD process. In another study, polyaluminum chloride (PACl) which was used as a coagulant for biologically treated coking wastewater (BTCW) during a MD process. The findings demonstrated that (PACl) proved to be a possible option for removal of microbial water without the risk of membrane contamination. In addition to PAM, BTCW was also controlled using coagulants and this process also lowered the contaminant level. Compared to the monomers (PAM), though, the polymer aggregates tend to form a dense cake film on the membrane's surface. The study conducted another research concluding that the MLC method was applied before the MF system to achieve FGDW wastewater. Materials that were also employed in this process include lime, magnetic seeds, and others that under the name of the others included in the table above, were employed for removal of Suspended Solids (SS). Water to be fed was subjected to PACl and PAM flocculants and hydrated lime was used as a chemical conditioner. Lime was the first coagulant used in FGDW that did not remove the ions from its feed solution. The magnetic seeds were added first and in turn the coagulants were added. In accordance with the results, it could be stated that the rejection of salts and perstabilization organic substance was rather high. On the other hand, after 150 h of operation water trenching and wetting were found to be the main type of fouling where the inorganic scale was present along with some biological and organic fouling. The inorganic filth is considered to be the "bricks," whereas colloids, organic matter, and microorganisms are considered to be the "concrete" that form a thick plaster of various foulants (Zheng et al. 2020).

9.4.1.2.2 Acidification (pH Adjustment)

The feed solution is often applied to pH correction which certainly fulfills the definition of one of the known fouling pretreatment methods. Almost in all situations, the pH adjustment means dipping the feed into an acidic solution through reducing the solution pH. It is thereby that by maintaining acidic conditions in the water, the alkaline salts like calcium carbonate ($CaCO_3$), the major component of inorganic fouling will be dissolving in the acidic solution instead of depositing on the membrane surface. Several pieces of research (He et al. 2009; Ahmad et al. 2012) show that the feed solution should be pretreated with HCl to set the pH level of the solution to 4 or 5, thus minimizing scale formation with a stable flow rate during the DCMD process. The reason that is placed for a stable flux is that hydrogen ions which originate from HCl solution and entered the feed will neutralize carbonate (CO_3^{2-}) and bicarbonate (HCO_3^-) ions using the following equation (Malaeb and Ayoub 2011): $H^+ + HCO_3^- \rightarrow H_2O + CO_2$ is another point in favor of acidification, however, Karakulski and Gryta (2005) considered that silica colloids fouling was independent of acidified feed solution. The research by Bush et al. (2018) also investigated silica deposition in the MD process. The intervention to reduce silica fouling includes pretreatment methods. The filter flux efficiency was affected to an insignificant extent when feed-water pH was lower than 5 or higher than 10; only the membrane fouling took place. NaOH solution at a pH greater than 11 has an effect of enhanced silica particles dissolution. Thus, both acidification and alkalization processes are advised to be applied to get rid of the scouring caused by silica fouling. The experiment proved that the reaction rates were the highest at neutral pH ranging between 6 and 8. During this plugging, the channels are blocked and the capacity for diffusion is decreased. Among other things, the silica gel layer will decrease evaporation rates at the boundary between the scale layer and membrane surface. It is due to the fact that silica gel has the characteristics of adsorption and pore structure. Moreover, the challenges of silica cleaning are not easy to handle and silica fouling happens more on a cleaned membrane with the new membrane.

This implies silica pretreatment and regular surveillance of feed water are the best measure of the successful long-term endeavor of MD.

9.4.1.2.3 Water Softening

Just like pH adjustment, water softening also makes use of lime, soda ash (Na_2CO_3) or caustic soda (NaOH) for $CaCO_3$ precipitating out of the treated water instead of dissolving it. It is most commonly performed prior to microfiltration with the intention of preventing abrasion of the membrane by the larger elements. This process, however, comes with one major disadvantage: long settlement period for the process. To ensure the high standard of this step, an accelerated precipitation softening (APS) technique was proposed by Qu et al. (2009). The APS pretreatment process begins with pH adjustment using caustic soda and ferric sulfate seeding, followed by membrane filtration. Qu et al. (2009) have treated RO concentrate with ASP and then reintroduced it to the DCMD process. The overall conclusion reached was that the APS pretreatment process achieved excellent results by removing 92% of the calcium present in the feed solution. Flowingly, membrane surface will receive much less $CaCO_3$ and $CaSO_4$ scales.

9.4.1.2.4 The Use of Antiscalant

Another pretreatment step to combating inorganic fouling in the MAB process is antiscalant. The primary function of sequestants is to fractionate these scale inhibitors, including carbonate as well as phosphate, sulfate, and fluoride scale. It is, however, employed to filter out a number of ions from colloids and metal oxides (Gryta 2008; He et al. 2009). The anti-scaling product is a chemical additive added in the feed solution to break the strong adhesion feature of the scaling. Different antiscalant methods that are used to suppress scale include changing CO_2 concentration, reducing precipitation rate, retarding crystal growth, and reshaping the crystal structure (He et al. 2009). It is shown that antiscalants application (AS) minimizes the scale development in many water treatment technologies, not only at low feed temperatures, but also at high feed ones. The MD process usually works at elevated feed temperatures and thus requires antiscalant addition. Additionally, it is a low-cost method whenever the feed chemical quantity used is less than 10 ppm. Inversely, the high concentration of antiscalant increases the surface tension of water as organic nature of antiscalant is composed of hydrophilic property similar to water. On this basis, the rate of wetting of the skin will increase. The study by Qu et al. (2020) instead studied three types of anti-scalants and the conclusion was that the temperature affects the performance of antiscalant and a higher feed temperature will adverse the inhibition process of the different scalants. DCMD technique used to observe the gypsum scaling reduction efficiency of MDC220, PTP-0100, and SHMP antiscalants by using the synthetic RO feed effluent solution. As for all three anti-scaling agents, there was no reduction in feed surface tension as well as severe interference with next stage of crude separation.

Additionally, the effective level of MDC220 antiscalant must be at least 5 mg/L which is used to prevent gypsum scaling. Antiscalants may relate to the type of feed solution they are used on. One of the earliest water treatment antiscalants was sodium hexametaphosphate (SHMP) which prevent scaling of $CaCO_3$ and sulfate scale inhibiting. SHMP is considered a very delicate antiscale in operation since dose quantity is the most important factor. If SHMP dosing is inaccurate, SHMP hydrolysis happens and the inhibition of the scaling is lost. In addition to the above, calcium phosphate scale development is enhanced. Gryta (2012) did this by replacing sodium polyphosphates as antiscalant and thus can prevent depositing calcium carbonate scales, but however the layer of deposits becomes less porous instead. Furthermore, a flux decline that occurs when the amount of antiscalant is increased is detected in the desalination process by DCMD. In the pretreatment of the softening of water usually carried out by acidification process, the same process is applicable for removal of low-porous layer. This fact in turn reflects that the impermeable layer is formed only on the outer surface of the membrane as the pore remains unaffected. Nowadays, the antiscalants in desalination procedures are mostly organophosphonates, polyelectrolytes, and condensed polyphosphate (Ketrane et al. 2009). In essence, anti-scaling mixture must contain a blend of HEDP and polyacrylate or phosphoric acid and PAA (Gryta (2012).

In the present study, the impact of different antiscalants on $CaCO_3$ and gypsum scales was the focus. The five forms of antiscaling showed favorable results and showed no evidence of membrane loss. The impressive antiscalant effect was demonstrated by the extension of the crystal formation induction period and hence slowing down the precipitation process even at the absence of large dosage 0. 6 mg/L. Multiple induction periods were derived for antiscalants with various antiscalants. In addition, sodium polyacrylate and polyacrylic acid-based compounds, i.e., K752 antiscalant, were found out to be greatly influencing the course of gypsum scale formation. For $CaCO_3$ scale, all antiscalants achieved similar results; no advancement of GHR antiscale (nitrogen-containing organophosphorus compound) over the others was clear. Along with the benefits which antiscale agents provided for the DCMD process, the drawbacks cannot be ignored as well. Moreover, as it was pointed out in this paragraph, antiscalants work by generating scale and then they become foulants that come with building on surfaces of the membrane. Certain antiscalants with amphiphilic attributes and which act as surfactants very often lower water surface tension and raise the incidence of pore flooding by the working membrane. Thus prolonging a membrane's lifetime (Al-Shammiri et al. 2000). Moreover, antiscalants also belong to this category. Biofilms can be created by cells which are very critical for microbes as they provide energy and nutrients which are most essential for them. Additionally, other antiscaling agents that include phosphate ions facilitate the growth of microbes and algae (Al-Shammiri et al. 2000). The width of antiscalant is also dependent on the concentration of feed solution. The interaction of some chemical residuals after the previous process of the antiscalants can form the additional foulant layer or reduce the effectiveness of the antiscalant. Such as illustrated through the example of feed solutions having iron molecules which get extra fouling (Tijing et al. 2015).

Furthermore, the application of antiscalant also is not easy to be assessed in processes like DCMD systems, in which temperature and feed are always changing. Hence, further investigations on this point should be done. In short, the efficiency of any of the chemical pretreatment processes depends on the various factors like the chemicals used, feed temperature, dosage point, membrane attributes, and the particulate properties.

9.4.1.3 Thermal Pretreatment Process

9.4.1.3.1 Boiling Process

Thermal pretreatment is a method that involves altering the water temperature in a particular way, these temperatures have to be raised to boiling point. In general, there are such thermal pretreatment steps which take away (HCO_3) from the reaction solution. This leads to a decrease in organic scaling on membrane surface due to reduced hydrolytical Fig. 4. Scaling of the membrane surface with time. Another solution was provided for a prerogative boiling process of the feed accompanied by filtration-based solution which was reported to decrease tendency of biofilm formation in some feed solutions with protein foulants (Gryta 2008). Studies were made by Gryta (2008) and according to their findings, the concentration of TOC in the 30-min boiled saline water was found to be 60% and the turbidity was 79% which passed the filter paper (Gryta 2008). Thus, the thermal pretreatment technique could eliminate the sudden technique shrivel. Thus, the membrane fouling is reduced by a filtration process after thermal pretreatment as has been supported by the research findings.

9.4.1.3.2 Thermal Water Softening

The process of thermal water softening is the process whereby bicarbonates (HCO_3) are removed from their solution by boiling. In feed preparations of water with high water hardness such as ground water softening and thermal water softening, hardness is useful in precipitating bicarbonate ions. It suggests that there is no HCO^-3 ions' scale and no CO_2 degassing on the surface of the membrane and no $CaCO_3$ deposition that is connected with alkaline formation. The experiment relating to the thermal softening pretreatment process with groundwater concentrations in the MD system was conducted with the use of the hollow fiber membrane by Gryta et al. and co-workers in 2010. The obtained results prove that the boiled feed solution should be filtered with a 0. The values of the parameter in the tissue of animals losing 2 μm decrease the concentrations of bicarbonate ions from 2 up to 3 times of the initial content. Sometimes thermal water softening pretreatment has been used in water treatment mostly. Nonetheless, it has some disadvantages also. This kind of scale was developed when the membrane surface was in contact with the solution that contained lesser amount of CO_3 ions and then exposed to heat. The residue buildup, in turn, reduces the permeate flux and adversely affects the efficacy of the process, respectively (Cho et al. 2018). On the other side, the corrosion because of pH value on the membrane can be avoided by a simple pretreatment of acidification where the pH of feed solution is made lower to 4. One disadvantage which is associated with the application of thermal water softening is that it is very costly as far as energy consumption is concerned; this is because the MD process strives to work at minimal feed temperature with a view of minimizing the energy consumed.

9.4.1.3.3 Cleaning Process

Membrane cleaning process consists of cleaning a foul membrane when the fouling pretreatment techniques are unsuccessful in minimizing the fouling. We can clean the membrane through several techniques like removing of particles completely or partially ensuring that the pores do not get clogged and particles do not accumulate on the membrane surface. The process of periodic membrane cleaning is undoubtedly the main measure for extending the membrane lifetime in all membrane water treatment processes (Yang et al. 2013). Some popular membrane cleaning methods are, g. through chemical, enzymatic, physical, and biological) techniques, the biofouling are cleaned off the membrane. On the other hand, the chemical method and physical method are the two most frequently used cleaning methods. Cleaning performance is quantifiable by means of resistance removal and permeates flux recovery (Sohrabi et al. 2011). The chemical cleaning approach includes employing chemical agents such as caustic soda, oxidants, acids, chelates, or proprietary surfactants, which act to weaken interactions forces between the foulants and the membrane surface. The primary methods are built on large quantities of chemicals that might damage the membrane and diminish its lifespan character. Further, after cleaning the membrane with chemical agents a waste stream is generated to carry these used chemical agents. The secondary treatment technology is the one that is required (Laine et al. 2003). Nevertheless the physical cleaning method involves utilization of hydraulic or mechanical power in order to ease the detachment of the foulants from the surface of the membrane Van (Oss 1995). Membrane cleaning method is not only the removal from the membrane surface of built-up particles but also for the purpose of rejuvenating the membrane as against the case with the membrane flux operation. The wetted membranes can be reliably reused after being easily cleaned because the wettability of membrane is the key factor of membrane fouling. Before it is used again, to make the wet membrane again have the form of the original, it should be fully dried. However, the main problem associated with drying of the membrane is that drying process is always a complicated one, because during the drying process the porous structure of the membrane may be easily affected. Besides, osmotic reentry will not restore cells to the pre-swelling state (Jiang et al. 2017). The primary difficulty of concentrations remaining after evaporating water while drying the membrane is that some of it lodges itself in the membrane pores. In this case, you can chemically wash the membrane, then dry it to be as good as new as well too.

9.4.1.3.4 The Chemical Cleaning Methods

In all of the chemical cleaning processes, some of the chemical agents are used in different proportions. They are not only include metal chelating agents (chelating agents). Besides the chelating agents, there are different sorts of surfactants, enzymes, oxidants, caustic soda, acids, and alkalis (Al-Amoudi and Lovitt 2007; Porcelli and Judd 2010). These chemicals are specifically designed to counter affect the driving forces that hold the foulants to the membrane surface for easy release. Thus, the foulants would be defouled from the membrane surface. It requires a large amount of chemical for cleaning methods that solely depend on chemical. Thus, the choice of a chemical agent that will work best in a specific situation turns into a critical issue in chemical methods. The following selection step involves other factors that are not restricted to chemical or economical but also has respect to foulants components and the fouling type.

9.4.1.3.5 Acid Cleaning

In this chemical treatment process, the strong as well as weak acids may be chosen based upon the kind of foulant. Acid cleaning is, probably, the most familiar method of chemical washing. In the case of chemical cleaning through acids, extra preventive measures must be taken to avoid damaging the membrane. Furthermore, foulants can also react with acids to form more deposit on the surfaces of tubes. Sulphuric, hydrochloric, and nitric acids are aimly used for the removal of the inorganic foul, particularly $CaCO_3$ (Gryta 2007) For instance, hydrogen chloride HCl is famous for acidizing of basic crystal salts such as $CaCO_3$ (Gryta 2008). As for the surface fouling, washing the membrane using acid however does it. Although restricting and increasing the salt particle in the membrane pore decreases initial permeate flow, it is often impossible to achieve the starting point again. The procedure of acid cleaning includes the use of distilled water as a feed solution for the rinsing of the membrane. Subsequently, the solution used to replace water also consists of an acid. Acids in the feed solution dissolve salt deposits off the membrane surface and thus excluded the salts from the membrane surface (Criscuoli et al. 2008). Discharge has to be done after performing the permeate after the acid cleaning. As for the fact that HCl as volatile constituent may wet with water vapor and condense to the side, this had been noted by Gryta (2012) and coauthor. For the experiments of this study, 3 wt% HCl cleaning solution has been used for dissolving the $CaCO_3$ scale at the same time getting the flux back to the initial value. Experiments that include use of a microscope on the clean membranes reveal features similar to those of unused membranes according to Gryta (2008). As it is comparable to Yang et al. (2012) the authors had extracted similar results with 5 wt% HCl solution. However, one another study it as stated that dilution and rinsing of the membrane with 2–5% HCl solution lead to a reducing in the permeate flux throughout. Acid rinsing treatments are repeated typically every 30–70 h of DCMD operation with PP hollow fiber membrane. This may be due to another research carried out by Gryta (2007) which used concentrated HCl solution for cleaning the iron oxide scale. Thus, removing the iron oxides was relatively difficult, and hence more concentrated (more than 5 wt%) acid cleaning was needed. This research utilizes 18 wt% and 36 wt% HCl solutions to clean PP hollow fiber membranes. However, highly concentrated HCl solution (36 wt%) was able to clean all the iron oxides off the membrane but at the same time reduce permeate flux significantly, damage the membrane seriously, and increase membrane wettability. In prior work, permeate flux was declined by 21% after the acid cleaning process was done. The author suggested to alter the process through air drying the membrane after washing it with 36 wt% HCl to allow the pores of the membrane not to be filled with the acid solution. This is because be particle filling pore with acid solution hence the increase of the area of the wetted membrane. Therefore, halving the amount of a solution that should be used to remove iron oxide could be deemed as more effective which was the case with the 18 wt % HCl solutions. As the feed solution contains lesser acid conc., the permeate flux again returns to the initial level as the membrane is slightly fouled.

9.4.1.3.6 Chelating Agents and Surfactants

These are chemicals that react with metal compounds such as the main pollutant iron oxide to form a stable compound that can be easily displaced by water. They are very sensitive to pH and can be damaged because of the feed temperature or flow rate (Ang et al. 2006). Of all the chelating agents, it is evidenced that EDTA (Ethylene diamine tetra acetic acid) is the most common chelating agent used in chemical cleaning approaches (Sohrabi et al. 2011). Apart from the chemical agents that dissolve biofilms such as: surfactants, acids, and enzymes, there are other chemical products that can be used to eliminate or treat environmental pollutants. There are surfactants located in the organic compounds which have hydrophobic and hydrophilic groups. This characteristic assists the organics to be trapped by the outer hydrophobic side of the bi-surfactant while the inner hydrophilic side persists moistened in water (Sohrabi et al. 2011). The process would make it possible to have an occurrence of solid–liquid interface which brings about the termination of the organic matter adsorption on the membrane surface. One of the chemical cleansers often used is a Sodium dodecyl sulfate or simply called SSD. Without a doubt, surfactants can be applied for the treatment of colloids as well as for dissolving organically fouled particles. SDS's primary use is for the elimination of objectionable nonvolatile colloidal matter on succeeded seawater; however, right cleaning parameters should be met (Garcia-Fayos et al. 2015).

9.4.1.3.7 Base Cleaning

Being alkaline chemical compounds, for instance, sodium hydroxide solution is perfect as means of mechanical and biofouls control (Rabie et al. 2001). On the one hand feed solutions that contain blend of organic and inorganic fouling, Two-step cleaning mechanism is employed that consists of acidic and base cleaning. Curcio et al. (2010) are adopted citric acid solution as the first step and NaOH as the second step to pretreatement $CaCO_3$ and HA fouling. The sanitization procedure posts full recovery in permeate flux and membrane hydrophobicity. The other studies (of the 26th and 193rd kinds, respectively) that have been carried out reached the same conclusions (Srisurichan et al. 2005). For disassembly of naturally-growing biofilm whose constituents comprise organic matter and bacteria Krivorot et al. (2011) employed sodium hydroxide only. The extra step is to apply 70% ethanol solution to the system as a disinfectant. This would ensure the removal of any living organism from the system. At this point, the solution is washed down with distilled water. The conclusion showed that the tangent flow was restored to its initial flux as a result. But, Ang et al. (2011) knew that only NaOH single solution does not work when organic foulants are present with inorganic ones. Nevertheless, apart from using EDTA or SDS to clean feed solution, a fast echo clean at the basic feed solution is also an effective method for removing organic fouling. Several scientists focused on the further discovery of the natural cleaning agents, which presented the unique longer-lasting effect without damaging the environment. Al-Guillen-Burrieza et al. (2014) worked with six different solvents to clean PTFE flat sheet membranes which are widely used for treating brine or saline water. To ensure accuracy of results, the study employed five chemical solvents with different levels of acidity and was conducted using the distilled water as well. This study showed that the best solvents have been found to be pH level equals 2 with 0% content. a mixture with 8wt% citric acid could be used for instance. 1% of oxalic acid. That is to mention, the majority of foulants (the expression "the main part" is more understandable in this case – from the point of view of an ordinary reader). e. A SEM analysis on the surface of PVDF membranes (A, Al, Fe, NaCl and Mg) showed that the removal of these metals becomes more significant when highly acidic solvents are used. Besides that, the chemical cleaning method caused a nearby membrane to be defective. Therefore, membrane wetting was shown to be applicable. Additionally membrane dry out rates are increasingly relative to the membrane wetting. The other researchers, it was found out, tested with other acids, a different method to remove membrane fouling. Many tests involved cleaning of the membrane with nitrous acid that would result in lowering biofouling and $CaCO_3$ precipitation. It was also understood from other studies that the porous membrane can be effectively cleaned and brought back to its function just with the process of water rinsing. By way of examples, various studies have been conducted as to save the effectiveness of chemical cleaning procedures (Piasecka et al. 2015). It varies with the nature of the situation fouling type becoming the major factor ranked the cleaning efficiency. In addition to this, some other factors as for instance the resistance of the membrane to chemical cleaning and the deposit location that may reduce the cleaning process efficiency are of much concern in the case studies (Guillen-Burrieza et al. 2014). Generally, the application of chemical cleaning methods on commercial MD membranes is not a cost-effective process due to the various limitations and side effects. Modified membranes which permit chemical cleaning are more advantageous for their prevention of fouling as well as ease of maintenance. Membrane foul can be worsened by chemical cleaning and its overall result can be membrane wettability. Yet matters like chemical flux and feed pressure would be helpful in the improvement of the chemical cleaning process. If a food diet is going to be fed at higher temperature, their solubilities in the feed liquid may get enhanced instead of accumulating them on the membrane surface. The smaller the level of pressure employed, the lower will be foulant flux toward the membrane surface and on which they tend to form deposits (Al-Amoudi and Lovitt 2007). Besides, synergies of chemical agents in terms of their efficiency might be also modified by their combination with other chemicals. There is another situation with regard to chemical agents that they are specific for a certain type of foulants either and simultaneous application of two or more chemical agent in the same solution of feed will lead to the improper work of the whole cleaning process (Ang et al. 2011). In conclusion, acid cleaning solution is a very effective solvent in case of carbolic acid type water scale. Even the inorganic foulants including iron oxide requests stronger acid solutions to be rinsed off the membrane surface (Gryta 2007). Thus, few studies have raised the point that the acid cleaning is not very efficient on the case of wastewater feed solutions. Such (matter) could be caused by the occurrence of organic substance and microorganisms. Organic material from natural sources often acts as a binding agent for inorganic particulate and helps in the establishment of a biofilm (Gryta et al. 2001). Moreover, chemical techniques can be applied also to deal with colloidal fouling. Si particles could be separated by NaOH solvent; however, this is believed to be a tough method. As a result, a preferred chemical method for application chemistry on colloids is the use of surfactants.

9.4.1.4 Physical Cleaning Methods

The physical methods of cleaning involve the facilitation of the mechanical removal, developing too much pressure in the membrane, or making too much flow across its surface. Hydraulic and mechanical forces are where physical cleaning comes in as it helps to form voids that are necessary for releasing the fouling layer (Shi et al. 2014). This is achieved through the discharge of water from the hose-pipe and later the use of brush or sponge to scrape the membrane (Ahmad et al. 2012). The physical method of cleaning involves flushing the membrane with back-flowing gas, bubbling with sparging water, inconsistent flow promotion, reversing still water, using water to rinse the membrane, and pulsing. At times, two more methods, electrical field and ultrasonic cleaning, can be added and regarded as physical cleaning methods (Ahmad et al. 2012). Some of the main physical cleaning methods are explained in the sections below.

9.4.1.4.1 Backwashing

Backwash is a procedure in which water flows from back to front. This occurs from the permeate side, a state where it feeds the feed stream on the other side of the membrane. It is generally accepted that a filtration procedure includes cleaning the membrane surface layer by removing the deposit that has accumulated. Sealing of the membrane by foulants is governed only through the removal of foulants from their surface (Ahmad et al. 2012). In manifest instances, a backwash might not be so useful when the foulants accumulate deep into the membrane pores. To obtain cleaner soils for longer periods, the backwash pressure should be greater than the operating pressure. As the force of pressing is not implemented in MD process, only the tiny amount of force is needed to deform the structure of membrane, which is always subjected by the pressure on membrane surface. The choice between the backwashing schemes, i.e., continuous or pulsed is plus of flotation process. It is also called pagina baccula or blanstrike (Ahmad et al. 2012). Short duration pulsation (i.e., about 0). The regeneration process using CBW instead of single-minute CBW is also more effective. Still, even continuous or short-termed backfiltration has proved to be efficient in sediment removing. Backwashing as a method for small colloids removal was, however, mentioned for the first time. 3D printing was able to fabricate pore size (45 μm) and separately proteins where it has effective results with colloidal fouling (93). Nevertheless, a good number of studies including one by Sagiv and Semiat (2005) to establish the coefficients of backwashing have been only proved to succor in tubular and hollow fiber membranes due to their high-pressure endurance requirements. A desilting device that is constantly backwashing completely removes biological foulants as well. Meindersma et al. reported that reversing the direction of the feed solution manifests in the accumulation of the pollutant on the membrane nitro cells and a decline in the flux. Water out from the pond was chosen as the sole feeding in the AGMD hollow fiber membrane system that was the subject of the research. After 2200 h of operation the directional flow of energy was reversed. The assessment indicated that implementation of such an easy method could erase the membrane biofouling and return the flow. Sometimes air is the direction backwashing is used to wash off scales and crystal layers attached to the membrane (Warsinger et al. 2017). These emulsions are mostly used to give back the wet breathable skin. Consequently, this approach must be utilized when the wetted membrane is composed of wet soil grains packed in the pores of water membranes. The effectiveness of this separation will be increased by pressing out the liquid from pores of membrane in such fashion that the other particles will not fall or precipitate. In addition to air-warming, atmosphere also assists to empower plasma molecule cleaning. Firstly, researching the best optimal temperature for air backwashing to run with could be between 60 and 70%. Secondly, the air backwashing component of brine solution might be experimented in a submerged vacuum membrane distillation and crystallization (VMDC) system. Among all the other air stream temperatures tested, e.g., 70 °C and 80 °C, only 60 °C brought about optimum temperature for crystal formation because only 60 °C is unable to activate the complexity of temperature to get involved in the solubility of specific solutes. Thus, the additional cleaning process beyond air backwashing was recommended for the dissolved scale and inorganic fouling control which were the major sources of the contaminants.

9.4.1.4.2 Gas Bubbling

Thus, the technique of gas bubbling/gas sparing represents one of the methods of air treatment that is employed by this technology. Blow-down or bubbling. This process was utilized by a few researchers as an off-loading method to desludge biofilms. The same kind of gas bubbling in some of the research is gas sparging that is quite outstanding for this process. The principles of this process are to improve the variation of shear rate and convective power toward the membrane surface and consequently alleviate the issues of concentration polarization and temperature polarization. Consequently, it reduces fouling formation. When proceeding, it is added into feed solution which is composed of two phases and is provided below (e.g., liquid–gas flow). Based on our findings, we can conclude that air bubbles influence the ion and molecular transport across the membrane that can reduce concentration polarization and temperature polarization. They also raise the external shear rate and the fluidity of the environment at the membrane surface. In that way, the removal of organics in the membrane surface may be cut down. According to Amed et al. (2012), the gas injection factor does not affect mesh flux extent. By reducing the amount of kinematics, the redness will be minimized. Compared with this, the non-depositional fouling will be diminished. Moreover, the nature of irreversible fouling will not be changed, and the amount of scum accumulation will be increased. The size of the bubbles in a cleaning procedure tells how efficient a washing method is. Bubbles with a fairly consistent and fine size would generally lead to better effects than the ones that were coarser. The aforementioned bubbles can bring uniform flow circulation, intensive mixing, and increase the shear rate on the membrane surface, therefore increasing the chance of larger wastewater molecules going through the pores. It was revealed that lower-sized micromeries can help with future flow and prevent precipitation of salts, mostly when bubbles exist at high rate. Another research from Ding et al. the gas flow, the bubbling duration, and the duration of methanol dosing were investigated on an MDF process to determine its efficacy in reducing fouling. Using DCMD with PTFE flat sheet membrane in this experiment, the traditional Chinese medicine (TCM) extract sample has been passed on whereas a fan has been used to connect the feed with the injection of air bubbles. For the initial phase of this research, a gas bubbling act implemented, a permeate flux drop was recorded. The evidence here is likely the ability of gas molecules to attach themselves to the surface of the membrane. Correspondingly, it lowers the contact place or the place where the membrane surface and the vapor molecules interact with each other. Furthermore, membranes can entrap the bubbles that block pores and prevent the vapors from crossing over to the permeate side. Data revealed that the growing of gas bubbling riddance rate and duration is suggested to reduce fouling. The same, based on what said before, applies to the gas bubbling from external fouling only. These results can be considered within the frame of the periodical work by Kim et al. Therefore, gas bubbling conveys many benefits, but gas–liquid circulation is still far from being uncomplicated. It will need broader inquiry due to its complexity. Particles and solids in the medium were used to create feedstock. The research ended by saying that backwashing is more effective than introducing air bubbles into the system in cleaning the fouled membranes and then enhancing their performance. However, it was suggested that a method which is more effective than the use of air bubbles would be to combine backwashing together with air bubbling in removing the accumulated film. The air bubbling and backwashing triplet exhibited the most efficient membrane cleaning method and zone fouling removal rate after comparison of the five-testing cleaning methods. The high efficiency may be accredited to the type of contribution solution used. When we are dealing with suspended solids or colloids, backwashing is being recommended while bubbling of gas is a better solution for inorganic fouling problems.

9.4.1.4.3 Membrane Elbowing

Washing the membrane is an easy fouling drains technique. This consists of forward–backward flushing that is used to prevent accumulation of the particles on the membrane surface. The same process of getting rid of unwanted material through a backward flushing is also involved here as backwashing had been explained in the previous section. While carrying out forward membrane flushing, the pumping action of the membrane is performed either by feed solution or permeate flow. These are materials that flow through with the system's rinsing solution more rapidly than they do during the process itself, which makes these materials flow all over the membrane surfaces thus creating turbulence all over the system. Heavy hydraulic pressure levels are needed to promote such significant turbulence because the pressure gradient should be high. Following which, this deposit on the membrane surface is discontinued and the solution is released in the feed solution. On the contrary, particles can desorb from the membrane surface by washing process procedure but not from the pores themselves. To get a more environmentally friendly process reaction, Ramon et al. proposed to use chemical and physical methods as a combination of them. Using a hydraulic cleaning technique like chemical cleaning can be beneficial by loosening up grid-fouling layer and getting intrusion debris out of membrane pores, while hydrodynamic cleaning which depends mainly on the sheer force of the fluid pushing the particles at a distance will always be important. Different authorities had discussed the cleaning and flushing of a contaminated membrane surface as membrane flushing. Nghiem et al. (2011) has tested azide-free tubing as a means of upflow cleaning for PTFE flat sheet membrane in DCMD system. Flocculation of the inorganic fouling was subjected to the synthetic saline water that served as the feed solution. On the membrane surface there were scales of $CaCO_3$ and silicate and freshwater hungry $CaSO_4$. The most severe of the three layers, $CaSO_4$. Hence, the idea of $CaSO_4$ fouling type was studied extensively. $CaSO_4$ particle have been noticed to build up with enhanced feed temperature according to the definition which suggests that there is an induction period delay. When membrane induction time and cleansing the surface of nucleation spots can be systematically regulated, then inorganic-fouling mitigation becomes possible. Hence, a certain ROUTINE// FLUSHING of membrane would be the best. i.e., over 20 h of work day (1 month), the tank was filled with three cycles of cleaning water every day. The detaching of the membrane was executed with deionized water (DI). After regular membrane flushing; the results showed a constant permeate flux for the whole MD process.

9.5 Conclusions

Using membrane technology is common and predominant in the present era of water treatment and wastewater reclamation and in the food processing industry. Nevertheless, the problems associated with membrane fouling remain to be solved which prevents the practical application of this technique. In this chapter, membrane fouling is defined, explained in further detail, and the effects of membrane fouling and possible methods of prevention are outlined.

This is due to a variety of processes that bring about membrane fouling such as physical depositing, adsorbing, or bio-growth. To prevent fouling, it is essential to comprehend these mechanisms to provide the proper fouling mitigation techniques. Membrane fouling is a major issue in many membrane systems as it impairs the operation, results in lower fluxes, higher energy requirements and more frequent cleaning or replacement, and higher costs in total.

A number of techniques and preventive measures have been attempted to address the membrane-fouling issue. The paper also examines the preventive and corrective measures discussed to include surface modifications, chemical cleaning processes, backwashing, and improved monitoring and control methods. Furthermore, advanced solutions like application of nanomaterials in coatings and cocktail concepts inspired by barnacles can prevent fouling.

In conclusion, one can strongly advocate recognition of membrane fouling as a critical factor, which influences the concentration of membrane technologies in different fields in terms of performance and cost-effectiveness. From the evaluation of the foulants and foulings that occurred during this investigation, it can be concluded that the fouling mechanisms have a controlling effect on the performance and service life of membrane systems if well managed through operational strategies. It will also make practicable the application of the membrane technology in water and food processing hence making a contribution to the advancement of sustainable technologies.

References

Ahmad AL, Che Lah NF, Ismail S, Ooi BS (2012) Membrane antifouling methods and alternatives: ultrasound approach. Sep Purif Rev 41(4):318–346

Al-Amoudi A, Lovitt RW (2007) Fouling strategies and the cleaning system of NF membranes and factors affecting cleaning efficiency. J Membr Sci 303(1–2):4–28

Al-Juboori RA, Yusaf T (2012) Biofouling in RO system: mechanisms, monitoring and controlling. Desalination 302:1–23

AlSawaftah N, Abuwatfa W, Darwish N, Husseini G (2021) A comprehensive review on membrane fouling: mathematical modelling, prediction, diagnosis, and mitigation. Water 13(9):1327
Al-Shammiri M, Safar M, Al-Dawas M (2000) Evaluation of two different antiscalants in real operation at the Doha research plant. Desalination 128(1):1–16
Ang WS, Lee S, Elimelech M (2006) Chemical and physical aspects of cleaning of organic-fouled reverse osmosis membranes. J Membr Sci 272(1–2):198–210
Ang WS, Tiraferri A, Chen KL, Elimelech M (2011) Fouling and cleaning of RO membranes fouled by mixtures of organic foulants simulating wastewater effluent. J Membr Sci 376(1–2):196–206
Bush JA, Vanneste J, Gustafson EM, Waechter CA, Jassby D, Turchi CS, Cath TY (2018) Prevention and management of silica scaling in membrane distillation using pH adjustment. J Membr Sci 554:366–377
Chen P, Cui L, Zhang K (2015) Surface-enhanced Raman spectroscopy monitoring the development of dual-species biofouling on membrane surfaces. J Membr Sci 473:36–44
Chen W, Qian C, Zhou KG, Yu HQ (2018) Molecular spectroscopic characterization of membrane fouling: a critical review. Chem 4(7):1492–1509
Cho H, Choi Y, Lee S (2018) Effect of pretreatment and operating conditions on the performance of membrane distillation for the treatment of shale gas wastewater. Desalination 437:195–209
Choudhury MR, Hsieh MK, Vidic RD, Dzombak DA (2012) Corrosion management in power plant cooling systems using tertiary-treated municipal wastewater as makeup water. Corros Sci 61:231–241
Criscuoli A, Zhong J, Figoli A, Carnevale MC, Huang R, Drioli E (2008) Treatment of dye solutions by vacuum membrane distillation. Water Res 42(20):5031–5037
Cui L, Yao M, Ren B, Zhang KS (2011) Sensitive and versatile detection of the fouling process and fouling propensity of proteins on polyvinylidene fluoride membranes via surface-enhanced Raman spectroscopy. Anal Chem 83(5):1709–1716
Curcio E, Ji X, Di Profio G, Fontananova E, Drioli E (2010) Membrane distillation operated at high seawater concentration factors: Role of the membrane on $CaCO_3$ scaling in presence of humic acid. J Membr Sci 346(2):263–269
Donlan RM (2002) Biofilms: microbial life on surfaces. Emerg Infect Dis 8(9):881
Efeoglu E, Culha M (2013) In situ-monitoring of biofilm formation by using surface-enhanced Raman scattering. Appl Spectrosc 67(5):498–505
El-Abbassi A, Kiai H, Hafidi A (2012) Phenolic profile and antioxidant activities of olive mill wastewater. Food Chem 132(1):406–412
Field R (2010) Fundamentals of fouling. Membr Water Treat 4:1–23
Garcia-Fayos B, Arnal JM, Gimenez A, Alvarez-Blanco S, Sancho M (2015) Static cleaning tests as the first step to optimize RO membranes cleaning procedure. Desalin Water Treat 55(12):3380–3390
Gryta M (2007) Influence of polypropylene membrane surface porosity on the performance of membrane distillation process. J Membr Sci 287(1):67–78
Gryta M (2008) Alkaline scaling in the membrane distillation process. Desalination 228(1–3):128–134
Gryta M (2010) Desalination of thermally softened water by membrane distillation process. Desalination 257(1–3):30–35
Gryta M (2012) Polyphosphates used for membrane scaling inhibition during water desalination by membrane distillation. Desalination 285:170–176
Gryta M, Tomaszewska M, Grzechulska J, Morawski AW (2001) Membrane distillation of NaCl solution containing natural organic matter. J Membr Sci 181(2):279–287
Guillen-Burrieza E, Ruiz-Aguirre A, Zaragoza G, Arafat HA (2014) Membrane fouling and cleaning in long term plant-scale membrane distillation operations. J Membr Sci 468:360–372
Guo W, Ngo HH, Li J (2012) A mini-review on membrane fouling. Biores Technol 122:27–34
He W, Hur J (2015) Conservative behavior of fluorescence EEM-PARAFAC components in resin fractionation processes and its applicability for characterizing dissolved organic matter. Water Res 83:217–226

He F, Sirkar KK, Gilron J (2009) Studies on scaling of membranes in desalination by direct contact membrane distillation: $CaCO_3$ and mixed $CaCO_3/CaSO_4$ systems. Chem Eng Sci 64(8):1844–1859

Henderson RK, Baker A, Murphy KR, Hambly A, Stuetz RM, Khan SJ (2009) Fluorescence as a potential monitoring tool for recycled water systems: a review. Water Res 43(4):863–881

Henthorne L, Boysen B (2015) State-of-the-art of reverse osmosis desalination pretreatment. Desalination 356:129–139

Huang BC, Guan YF, Chen W, Yu HQ (2017) Membrane fouling characteristics and mitigation in a coagulation-assisted microfiltration process for municipal wastewater pretreatment. Water Res 123:216–223

Hughes D, Tirlapur UK, Field R, Cui Z (2006) In situ 3D characterization of membrane fouling by yeast suspensions using two-photon femtosecond near infrared non-linear optical imaging. J Membr Sci 280(1–2):124–133

Jansen AE, Assink JW, Hanemaaijer JH, Van Medevoort J, Van Sonsbeek E (2013) Development and pilot testing of full-scale membrane distillation modules for deployment of waste heat. Desalination 323:55–65

Jiang S, Li Y, Ladewig BP (2017) A review of reverse osmosis membrane fouling and control strategies. Sci Total Environ 595:567–583

Kalafatakis S, Zarebska A, Lange L, Hélix-Nielsen C, Skiadas IV, Gavala HN (2020) Biofouling mitigation approaches during water recovery from fermented broth via forward osmosis. Membranes 10(11):307

Karakulski K, Gryta M (2005) Water demineralisation by NF/MD integrated processes. Desalination 177(1–3):109–119

Karakulski K, Kozlowski AAMAW, Morawski AW (1995) Purification of oily wastewater by ultrafiltration. Sep Technol 5(4):197–205

Ketrane R, Saidani B, Gil O, Leleyter L, Baraud F (2009) Efficiency of five scale inhibitors on calcium carbonate precipitation from hard water: effect of temperature and concentration. Desalination 249(3):1397–1404

Krivorot M, Kushmaro A, Oren Y, Gilron J (2011) Factors affecting biofilm formation and biofouling in membrane distillation of seawater. J Membr Sci 376(1–2):15–24

Ladewig B, Al-Shaeli MNZ (2017) Fouling in membrane bioreactors. Fundam Membr Bioreact Mater Syst Membr Fouling 39–85

Laine JM, Campos C, Baudin I, Janex ML (2003) Understanding membrane fouling: a review of over a decade of research. Water Sci Technol Water Supply 3(5–6):155–164

Laqbaqbi M, Sanmartino JA, Khayet M, García-Payo C, Chaouch M (2017) Fouling in membrane distillation, osmotic distillation and osmotic membrane distillation. Appl Sci 7(4):334

Maddah H, Chogle A (2017) Biofouling in reverse osmosis: phenomena, monitoring, controlling and remediation. Appl Water Sci 7:2637–2651

Malaeb L, Ayoub GM (2011) Reverse osmosis technology for water treatment: state of the art review. Desalination 267(1):1–8

Piasecka A, Bernstein R, Ollevier F, Meersman F, Souffreau C, Bilad RM, Vankelecom I (2015) Study of biofilms on PVDF membranes after chemical cleaning by sodium hypochlorite. Sep Purif Technol 141:314–321

Porcelli N, Judd S (2010) Chemical cleaning of potable water membranes: a review. Sep Purif Technol 71(2):137–143

Prihasto N, Liu QF, Kim SH (2009) Pre-treatment strategies for seawater desalination by reverse osmosis system. Desalination 249(1):308–316

Qu D, Wang J, Wang L, Hou D, Luan Z, Wang B (2009) Integration of accelerated precipitation softening with membrane distillation for high-recovery desalination of primary reverse osmosis concentrate. Sep Purif Technol 67(1):21–25

Qu F, Yan Z, Yu H, Fan G, Pang H, Rong H, He J (2020) Effect of residual commercial antiscalants on gypsum scaling and membrane wetting during direct contact membrane distillation. Desalination 486:114493

Rabie HR, Côté P, Adams N (2001) A method for assessing membrane fouling in pilot-and full-scale systems. Desalination 141(3):237–243

Rudolph G, Virtanen T, Ferrando M, Güell C, Lipnizki F, Kallioinen M (2019) A review of in situ real-time monitoring techniques for membrane fouling in the biotechnology, biorefinery and food sectors. J Membr Sci 588:117221

Sagiv A, Semiat R (2005) Backwash of RO spiral wound membranes. Desalination 179(1–3):1–9

Shi X, Tal G, Hankins NP, Gitis V (2014) Fouling and cleaning of ultrafiltration membranes: a review. J Water Process Eng 1:121–138

Sim LN, Chong TH, Taheri AH, Sim STV, Lai L, Krantz WB, Fane AG (2018) A review of fouling indices and monitoring techniques for reverse osmosis. Desalination 434:169–188

Sistat P, Huguet P, Deabate S (2009) Confocal Raman microscopy for membrane content visualization. Monit Vis Membrane-Based Process 127–149

Sohrabi MR, Madaeni SS, Khosravi M, Ghaedi AM (2011) Chemical cleaning of reverse osmosis and nanofiltration membranes fouled by licorice aqueous solutions. Desalination 267(1):93–100

Son M, Yang W, Bucs SS, Nava-Ocampo MF, Vrouwenvelder JS, Logan BE (2018) Polyelectrolyte-based sacrificial protective layer for fouling control in reverse osmosis desalination. Environ Sci Technol Lett 5(9):584–590

Srisurichan S, Jiraratananon R, Fane AG (2005) Humic acid fouling in the membrane distillation process. Desalination 174(1):63–72

Tang CY, Chong TH, Fane AG (2011) Colloidal interactions and fouling of NF and RO membranes: a review. Adv Coll Interface Sci 164(1–2):126–143

Tijing LD, Woo YC, Choi JS, Lee S, Kim SH, Shon HK (2015) Fouling and its control in membrane distillation—a review. J Membr Sci 475:215–244

Uyguner CS, Bekbolet M (2005) Evaluation of humic acid photocatalytic degradation by UV–vis and fluorescence spectroscopy. Catal Today 101(3–4):267–274

Van Oss CJ (1995) Hydrophobicity of biosurfaces—origin, quantitative determination and interaction energies. Colloids Surf, B 5(3–4):91–110

Virtanen T, Reinikainen SP, Kögler M, Mänttäri M, Viitala T, Kallioinen M (2017) Real-time fouling monitoring with Raman spectroscopy. J Membr Sci 525:312–319

Wang J, Qu D, Tie M, Ren H, Peng X, Luan Z (2008) Effect of coagulation pretreatment on membrane distillation process for desalination of recirculating cooling water. Sep Purif Technol 64(1):108–115

Wang T, Qu G, Ren J, Yan Q, Sun Q, Liang D, Hu S (2016) Evaluation of the potentials of humic acid removal in water by gas phase surface discharge plasma. Water Res 89:28–38

Warsinger DM, Tow EW, Swaminathan J (2017) Theoretical framework for predicting inorganic fouling in membrane distillation and experimental validation with calcium sulfate. J Membr Sci 528:381–390

Wei J, Gao B, Yue Q, Wang Y, Li W, Zhu X (2009) Comparison of coagulation behavior and floc structure characteristic of different polyferric-cationic polymer dual-coagulants in humic acid solution. Water Res 43(3):724–732

Xiao F, Xiao P, Zhang WJ, Wang DS (2013) Identification of key factors affecting the organic fouling on low-pressure ultrafiltration membranes. J Membr Sci 447:144–152

Yan M, Wang D, Korshin GV, Benedetti MF (2013) Quantifying metal ions binding onto dissolved organic matter using log-transformed absorbance spectra. Water Res 47(7):2603–2611

Yang XH, Tian R, Ma SJ, Lv HL (2012) Study on membrane fouling experiment of stacked AGMD module in low temperature. Adv Mater Res 396:458–462

Yang JY, Li YS, Huang B (2013) Research on refurbishing of the used RO membrane through chemical cleaning and repairing with a new system. Desalination 320:49–55

Yiantsios SG, Sioutopoulos D, Karabelas AJ (2005) Colloidal fouling of RO membranes: an overview of key issues and efforts to develop improved prediction techniques. Desalination 183(1–3):257–272

Zator M, Ferrando M, López F, Güell C (2007) Membrane fouling characterization by confocal microscopy during filtration of BSA/dextran mixtures. J Membr Sci 301(1–2):57–66

Zheng L, Jiao Y, Zhong H, Zhang C, Wang J, Wei Y (2020) Insight into the magnetic lime coagulation-membrane distillation process for desulfurization wastewater treatment: from pollutant removal feature to membrane fouling. J Hazard Mater 391:122202
Zhong W, Zhao Y, Chen S, Zhong J, Guo L, Zheng D, Li S (2021) Resources recycle of traditional Chinese medicine (TCM) wastewater 1: effectiveness of the UF-MD hybrid system and MD process optimization. Desalination 504:114953

Concluding Remarks

Membrane technology has become a key component in modern water treatment and environmental management due to its efficiency, reliability, and ability to produce high-quality treated water. The chapters in this book highlight the wide range of applications of membrane processes, including wastewater treatment, drinking water purification, desalination, and industrial separation processes.

Despite significant advancements, challenges such as membrane fouling, operational costs, and energy consumption continue to influence system performance. Addressing these limitations requires continued research in advanced membrane materials, system optimization, and sustainable process design.

Future developments in nanomaterial-based membranes, hybrid treatment systems, and improved monitoring technologies are expected to enhance the efficiency and sustainability of membrane processes. By integrating technological innovation with sustainable practices, membrane technology will continue to play a vital role in achieving global water security and environmental protection.

A. Mandpe et al. (eds.), *Membrane Filtration and Membrane Bioreactors for Wastewater Treatment*, Advances in Wastewater Research,
https://doi.org/10.1007/978-981-95-6706-5

Papier | Fördert gute Waldnutzung
FSC® C083411

Kolibri 360 GmbH
Ferdinand-Jühlke-Straße 7
99095 Erfurt, Deutschland
produktsicherheit@kolibri360.de